Fachwörterbuch Feuerwehr und Brandschutz

Torsten Schmiermund

Fachwörterbuch Feuerwehr und Brandschutz

Deutsch-Englisch/Englisch-Deutsch

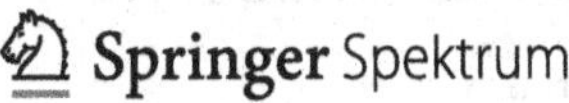

Torsten Schmiermund
Frankfurt am Main, Deutschland

ISBN 978-3-662-64119-4 ISBN 978-3-662-64120-0 (eBook)
https://doi.org/10.1007/978-3-662-64120-0

Die Deutsche Nationalbibliothek verzeichnet diese Publikation in der DeutschenNationalbibliografie; detaillierte bibliografische Daten sind im Internet über http://dnb.d-nb.de abrufbar.

Planung/Lektorat: Désirée Claus

Springer Spektrum ist ein Imprint der eingetragenen Gesellschaft Springer-Verlag GmbH, DE und ist ein Teil von Springer Nature.
Die Anschrift der Gesellschaft ist: Heidelberger Platz 3, 14197 Berlin, Germany

Vorwort

Dieses Fachwörterbuch entstand im Zuge der Übersetzungsvorbereitungen für das Buch „Das Chemiewissen für die Feuerwehr" (Springer Verlag, 2019) und stellt das z. Zt. umfangreichste Fachwörterbuch für den Bereich Feuerwehr und Brandschutz dar.

Die mehr als 14 000 Einträge resultieren aus über 7 000 deutschen bzw. englischen Begriffen.

Bei dem vorliegenden Werk handelt es sich um ein Wörterbuch – nicht um ein Lexikon, d. h. die Fachterminologie wird als bekannt vorausgesetzt. Erklärungen zu den einzelnen Begriffen sind nicht enthalten.

Einige Begriffe lassen sich naturgemäß nicht direkt übersetzen, so z. B. aus dem Bereich der Feuerwehrfahrzeuge oder der Organisation der Feuerwehr. Hier wurden die nahekommenden Entsprechungen verwendet.

Wenn Sie Korrekturen und Anmerkungen haben oder Ergänzungen wünschen, dann können Sie sich per E-Mail direkt an den Autor wenden: fw_chemie@aol.com.

Torsten Schmiermund
August 2021

Preface

This technical dictionary was created in the course of translation preparations for the book „Das Chemiewissen für die Feuerwehr" (Springer Verlag, 2019) and is currently the most comprehensive technical dictionary in the field of firefighting and fire protection.

The more than 14 000 entries result from over 7 000 German and English terms.

This work is a dictionary – not an encyclopaedia, i.e. the technical terminology is assumed to be known, explanations on the individual terms are not included.

Some terms are naturally not directly translatable, e.g. from the field of fire engines or the organisation of the fire service. The closest equivalents have been used here.

If you have corrections and comments or wish to make additions, then you can contact the author directly by e-mail: fw_chemie@aol.com.

Torsten Schmiermund
August 2021

Inhaltsverzeichnis

I Deutsch-Englisch

II English-German

Serviceteil

Deutsch-Englisch

Inhaltsverzeichnis

A

Abbau, biologischer – biodegradation

Abbaubarkeit, biologische – biodegradability

Abbrand – burning combustion

Abbrandeigenschaft – combustion property

Abbrandfaktor – burning rate factor

Abbrandfestigkeit – burning stability

Abbrandgeschwindigkeit – burning rate, burning velocity

Abbrandgeschwindigkeit, flächenbezogene – area burning rate

Abbrandgeschwindigkeit, lineare – linear burning rate

Abbrandgeschwindigkeit, mittlere – mean burning rate

Abbrandgeschwindigkeit, reale – real burning rate

Abbrandgeschwindigkeit, relative – relative burning rate

Abbrandrate, flächenbezogene – area burning rate

Abbrandrate, lineare – linear burning rate

Abbrandrate, massenbezogene – mass burning rate

Abbrandrate, mittlere – mean burning rate

Abbrandrate, reale – real burning rate

Abbrandrate, relative – relative burning rate

Abbrandverhalten – burning behaviour, burn down behaviour

Abbrandverlust – burn down loss

abbrandverzögernd – fire-retardant

abbrennen – to burn down, to burn out, to burn off

Abbrennen (der Oberfläche) – surface combustion

Abbrennen von Feuerwerkskörpern – burn off pyrotechnics

Abbruch (Gebäude, Anlagen) – demolition

Abbruch der Verbrennung – stop of combustion

ABC (atomar, biologisch, chemisch) – ABC (atomical, biological, chemical) = NBC (nuclear, biological, chemical) = CBRN (chemical, biological, radioactice, nuclear)

ABC-Ausbildung – NBC training

ABC-Ausrüstung – NBC equipment

ABCDE-Merkschema: *A*chtung, *A*real, *A*nsatz – *B*edenkzeit, *B*arriere (Selbstschutz), *B*arriere (Fremdschutz) – Zählen, Kommunizieren – *d*urchdachte Grundstrukturen, *D*irigieren, *D*elegieren – *E*valuieren, *E*rstkontakt Notaufnahm – ABCDE-mnemonic: *a*ttention, area, *a*pproach – *b*reak, *b*arrier (self-protection), *b*arrier (external protection) – *c*ount, *c*ommunicate – *d*evelop, *d*irect, *d*elegate – *e*valuate, *e*mergency departments

ABC-Einsatz – NBC mission, NBC operation

ABC-Einsatzkräfte – NBC forces, NBC task forces, NBC emergency forces

ABC-Gefahren – NBC hazards

ABC-Gefahrenlage – NBC hazard situation

T. Schmiermund, *Fachwörterbuch Feuerwehr und Brandschutz*, https://doi.org/10.1007/978-3-662-64120-0_1

ABC-Gefahrstoffe – hazardous materials type „NBC“ (nuclear, biological, chemical)

ABC-Löschpulver – extinguishing powder type ABC

ABC-Pulver – extinguishing powder type ABC, ‚ABC powder‘

ABC-Schutzanzug – NBC protection suit

ABC-Zug – chemicals incident unit (CIU)

abdecken – cover

Abdeckplane – protection cover

Abdeckung – cover, coverage

abdichten – make tight, seal

Abdichtmasse – sealing compound, sealant

Abdichtung – seal, sealing

Abfackeln – flaring

Abfall – waste, rubbish

Abfall, biogefährdender – biohazardous waste

Abfall, gefährlicher – hazardous waste

Abfall, hochradioaktiver – high active waste (HAW)

Abfall, infektiöser – infectious waste

Abfall, klinischer – clinical waste

Abfall, radioaktiver – active waste, radioactive waste

Abfall, schwach radioaktiver – low active waste (LAW)

Abfallbehälter – waste container

Abfallbeseitigung – waste disposal

Abfalleimer – waste bin

Abfallsack – waste bag, rubbish bag

Abfallverbtrennung – waste incineration

Abflammen (der Oberfläche) – surface combustion

Abführen von Rauch und Wärme – smoke venting

Abfüllstation – filling station

Abgase – exhaust gases

Abgasschlauch – exhaust gases hose

Abgastemperatur – off-gas temperature

abgereichert (Chemie) – down-blended

abgereichert (Kernphysik) – depleted

abgeschirmte Strahlenquelle – shielded radiation source

Abgrenzung – demarcation

abhängige Löschwasserversorgung – dependent water supply

Abholpunkt – pick-up point

Abhörgefahr – risk of unlawful interception

Abhörsicherheit – security against interception

Abklingbecken (Kerntechnik) – holding basin, fuel cooling station

abklingender Brand – decaying fire

Abklingphase (eines Brands) – fire decay

Abklingzeit (eines Brands) – cooling down period

Abklingzeit (Kerntechnik) – decay time

abkoppeln – uncouple

abkühlen – cool, cool down

Abkühlen – cooling

abkühlen lassen – let cool, allow to cool

Abkühlung – cooling

ablassen (Druck reduzieren) – relieve, vent

ablassen (jemanden herunterlassen) – lowering

ablassen (Menge reduzieren) – drain, discharge

Ablasshahn – discharge cock

Ablauf – process, development, order of events

Ablauf (für eine Flüssigkeit) – drain

Ablaufplan – operating plan

Ableitung (von elektr. Strom) – electrical leakage, leakage current

Ablesung (Gerät, Messwert) – reading, read out

ablösende Kräfte – relief-forces

Ablösezeitpunkt – time of relief

Ablösung – relay, relief

Ablösung von Einsatzkräften – redemption of forces

Abluft – waste air, used air, exhaust air

Abluftöffnung – exhaust air opening

Abluftschacht – exhaust duct

Abmagern (Löscheffekt) – reducing

Abmarsch – alarm assignment, attendance, departure

Abmarsch, erster – first alarm

Abmarsch, zweiter – second alarm

Abmarschfolge – predetermined attendance

Abmarschzeit – time of departure

abmelden – deregister, sign-out

Abmeldung – notice of departure

Abmessungen (Höhe, Breite, Tiefe) – dimensions (height, width, depth)

Abnahme (Minderung) – decrease, reduction

Abnahmebescheinigung – certificate of approval

Abnahmeprüfung – acceptance test

Abplatzung (z. B. bei Beton) – bursting

AB-Pulver – AB powder

abräumen – clear up

Abrieb – abrasion

Abriegeln – bolt, encircle

abriegeln – blocking

Abrollbehälter (AB) – abroll container, demountable pod, roller container

Abrollbehälter Einsatzleitung (AB-ELW) – command and control unit, command pod

Abrollbehälter Schlauch (AB-Schlauch) – hose laying unit, hose pod

Abrollbehälter Wasser/Schaum (AB-Schaum) – foam distribution unit, foam pod

Abrollbehälter-Transportsystem – roller container transport system (ACTS – abroll container transport system)

Absauganlage – exhaust equipment

Absaughaube – exhaust hood, suction hood

Absaugung – suction, exhauster

Abscheider – separator

Abschied (aus dem Dienst) – discharge

Abschied, ehrenhafter – honourable discharge

Abschirmung – shielding, screening

Abschirmung (gg. Strahlung) – radiation shielding

Abschleppseil – towing rope

Abschluss, flammendichter – fire resistant closure, fire resistant damper

Abschluss, rauchdichter – smoke-proof closure

abschmelzen – to melt off

Abschnitt – sector

Abschnitt (der Brandbekämpfung) – firefighting sector

Abschnittsleiter – fire sector commander

abschöpfen – skim off

Abschottung – firestop seal

abschwächen – buffer

Abseilachter – figure eight descender

Abseilen – rope descend, rope down, abseiling, rappelling

Abseilgeräte – rope-down equipment

Abseilpunkt – lower-off point

Absicherung – safeguarding

Absicherung der Einsatzstelle – safeguarding of scene of operation

Absicht – intent, intention

absieben – strain

absolute Atommasse – atomic mass, mass of atom

absoluter Nullpunkt (Temperatur) – absolute zero

Absorberstab (Kerntechnik) – absorbing rod

Absorption – absorption

Absorptionsgranulat – absorbance granules

Absorptionsmittel – absorbent

Absorptionsmittel – absorbance agent

Absorptionsquerschnitt – absorption cross-section

Absorptionsvermögen – absorption capacity, absorbing capacity

Abspaltung – separation

Absperrarmatur – shut-off device, cut-off device

Absperrband – barricade tape

Absperrgrenze – barrier border, barrier limit

Absperrklappe – shut-off flap, cut-off flap

Absperrmaßnahmen – measures to fence off

Absperrmaterial – barrier material

Absperrschieber – shut-off gate, cut-off gate

Absperrung – barrier, barricade

Absperrventil – shut-off valve, cut-off valve

Abstand – distance, space

Abstand halten! – Keep clear!

Abstandsgesetz – distance law, spacing law, inverse-square law

Abstandshalter – spacer

abstoßend – repellent

Abstrahlung – emitted radiation

Abstrahlung – emitted radiation

Absturzgefahr – fall danger

Absturzsicherung – fall arrester, safety catcher

Absturzsicherung – fall protection

Absturzsicherungsgeschirr – fall arrest harness, fall protection harness, full-body fall arrest harness

Abstützung – prop

Abteilung, aktive (Einsatzabteilung) – active division

Abteilungsleiter – section chief

Abtransport – transporting away

Abtrennung (räumlich) – partition

Abtrennung (Substanzen) – separation, isolation (from a mixture)

Abtrennung (zwischen Etagen) – fire break

Abtrennung eines Elektrons – electron detachment

abtropfen – to drip

Abtropfen – dripping

Abtropfen, brennend – burning dripping

Abwägung – weighing, consideration

Abwärme – waste heat

abwartende Behandlung – expectant treatment

Abwasser – waste water, sewage, effluent water (domestic origin)

Abwasserableitung – waste-water effluent, waste-water discharge

Abwasseraufbereitungsanlage – wastewater treatment plant (WWTP)

Abwasserpumpe – waste-water pump

Abwasserreinigungsanlage (ARA) – sewage treatment plant

abwehrender Brandschutz – fire-fighting, fire defence

Abweichung – deviation

Abzeichen – badge, insignia

Abzieher (Bodenwischer) – squeegee

Abzug – exhaust hood, fume hood, hood

Abzugshaube – fume hood, hood

Abzweigung – branching-off (pipelines), branch

Acetylcholinesterase-Hemmer – acetylcholinesterase-inhibitor (AChEI)

Acetylen – acetylene, ethine

Acetylenruß – acetylene black

Achslast – axial load

Achslast, zulässige – permissible axle load

Achterknoten (Schlaufe) – figure-eight knot loop

Achterknoten (Stopperknoten) – figure-eight knot

Acidität – acidity

Adapter – adapter

adiabatische Kompression – adiabatic compression

administrativ-organisatorische Maßnahmen – administrative-organisational-measures

Adrenalin – adrenaline, epinephrine, adrenine

Adsorption – adsorption

Adsorptionsvermögen – adsorption capacity, adsorbing capacity

A-Einsatz – mission type A (atomic), operation with dangers due to radioactive substances

Aerosol – aerosol

Aerosollöschverfahren – method of extinguishing by aerosol

Aerosolnebel – aerosol mist

Aerosolspray – aerosol spray

Aerosolteichen – aerosol particle

Affengriff – monkey grip

Affenkette – chain sinnet

Aggregate – aggregates

Aggregate der Feuerwehr – aggregates of the fire brigade

Aggregatzustand – state of aggregation, physical state, state of matter

Aggregatzustandsänderung – change of state

Airbag – airbag

Airbag-Abdeckung – airbag cover

Akkubohrer – cordless drill

Akkumulator – accumulator

Akkumulatorsäure – accumulator acid, storage battery acid

Akkuschrauber – cordless screwdriver

aktive Abteilung (Einsatzabteilung) – active division

Aktivierungsenergie – energy of activation

Aktivierungswärme – heat of activation

Aktivierungszeit – activation time

Aktivität – activity

Aktivität, spezifische – specific activity

Aktivitätskonzentration – activity concentration

Aktivitätsverlust – activity loss, loss in activity

Aktivitätsverlust, vollständiger – loss of activity

Aktivkohle – activated carbon, absorbent carbon

Aktivkohlefilter – activated-carbon filter, carbon filter

akustisch – audible

akustische Signale – acoustic signals

akute Erkrankung – acute disease, acute illness

akute Giftwirkung – acute toxic effect

akute respiratorische Erkrankung (ARE) – acute respiratory disease (ARD)

akute Strahlenerkrankung – acute radiation syndrome (ARS)

akute Toxizität – toxicity, acute

ALARA-Prinzip: so niedrig wie vernünftigerweise erreichbar – ALARA-principle: as low as reasonably achievable

Alarm – alarm, alert

Alarm- und Ausrückeordnung (AAO) – alarm and response regulations

Alarm, blinder – false alarm with good intent

Alarm, böswilliger – malicious false alarm

Alarm, sofortiger – immediate alarm

Alarm, unechter – false alarm

Alarm, verzögerter – delayed alarm

Alarmanlage – alarm system

Alarmdosimeter – alert dosimeter

Alarmempfänger – alarm receiver

Alarmfahrt – alarm response, responding

Alarmglocke – alarm bell

Alarmhorn – alarm horn

alarmieren – alarm, alert, call

Alarmierung – alarming

Alarmierungseinrichtung – alarm equipment

Alarmierungsstichwort – alarm key word

Alarmierungszeit – alarm time

Alarmordnung – alarm regulations

Alarmorganisation – alarm organisation

Alarmplan – intervention planning

Alarmschwelle – alarm threshold

Alarmsignal – alarm signal

Alarmsirene – alarm siren

Alarmspitze – national contact point

Alarmstufen – alarm grades

Alarmübermittlung, -übertragung – alarm transmission

Alarmübertragungseinrichtung – alarm transmission equipment

Alarmübung – fire drill, alarm drill

Alarmventil – alarm valve

Alarmzeichen – alarm signal

Alarmzustand – alarm condition

alkalibeständig – alkaliproof, lye-proof

Alkalipatrone – alkali cartridge

alkalisch – alkaline

Alkane – alkanes

Alkene – alkenes

Alkine – alkines

Alkohol für Desinfektionszwecke – rubbing alcohol

alkoholbeständig – alcohol-resistant

alkoholbeständiges Schaummittel – alcohol-resistant foam compound

allergischer Schock – allergic shock

Alleskleber – general purpose adhesive

allgemeine Begriffe – general terms

allgemeine Gaskonstante – general gas constant

Allradantrieb – four-wheel drive

Alpha-Einfang – alpha capture

Alpha-Strahler – alpha emitter

Alpha-Strahlung – alpha radiation

Alpha-Teilchen – alpha particle

Alpha-Zerfall – alpha decay, alpha disintegration

Altbau – old building

Altersabteilung – retirement division

Altersheim – nursing home

Alterung – ageing, aging

alterungsbeständig – resistant to ageing, non-aging

Altöl – used oil, waste oil

Altpapier – waste paper

Altweiberknoten – thief knot

Aluminiumkeil – aluminum wedge

Amalgame – amalgams

ambulant – outpatient

Ambulanzliege – wheeled stretcher

Ammoniak (NH_3) – ammonia

ammoniakfest – ammonia-resistant

Ammoniaklösung – ammonia solution

Amplitudenmodulation (AM) – amplitude modulation (AM)

Amtsarzt – public medical officer, public health officer

Amtshilfe – administrative assistance

Amtsleiter (AL) (der Feuerwehr) – fire chief, chief of the fire department

anaphylaktischer Schock – anaphylactic shock

Anbau – annex, extension

andauernder Brand – ongoing fire

andauerndes Brennen – sustained combustion

Änderungen – changes, modifications

Anemometer – wind gauge, anemometer

Anfangsphase eines Brandes – initial phase of fire

anfeuchten – humidify, prewet

angeregter Zustand – exited state

angereichert (Kernphysik) – enriched

angereichertes Uran – enriched uranium

Anglerknoten – fisherman's knot

angrenzender Raum – adjoining room

Angriff – attack

Angriffsgeräte – attack equipment

Angriffsrichtung – direction of attack

Angriffstrupp (A-Tr) – attack crew, attack team, attack squadron

Angriffstruppführer, -in (A-TrFü, A-TrFü'in) – attack crew leader, attack team leader, attack squadron leader

Angriffstruppmann, -frau (A-TrM, A-TrFr) – attack crew member, attack team member, attack squadron member

Angriffsweg – firefighting way, attack route, attack path

Angstreaktion – fear reaction, fright reaction

anhaltende Flammenbildung – sustained flame

Anhängeleiter (AL) – trailer ladder

Anhängemasse – trailer load

Anhängerpumpe – trailer pump

Anhebepunkt – lift point

Anion – anion

anionisch – anionic

Ankerstich, doppelter – cow hitch, lark's head, lark's foot, girth hitch

Ankertauknoten – anchor bend, fisherman's bend

ankohlen – to char

Ankohlung – charring

Anlage – plant, unit, system

Anlage zur Löschmittelförderung – installation for delivery of extinguishant

Anlage, bauliche – built environment

Anlage, ortsfeste – fixed installation

Anlagensicherheit – plant safety

Anlagenteil – system part, system component

Anleiterwinkel – proper angle for a ladder

Anmarschzeit (der Feuerwehr) – fire brigade attendance time

Annahme – assumption

Annihilation – annihilation

Anomalie des Wassers – anomaly of water

Anordnung (Befehl) – order

Anordnung (räumlich) – arrangement

anorganisch – inorganic

anorganische Chemie – inorganic chemistry

Anreicherungsanlage – enrichment plant, enrichment facility

Anrufer nicht auffindbar – Gone On Arrival (GOA)

ansaugen – to aspirate, to suck

Ansaugen – suction, aspiration

Ansaugen (Vorgang) – suction

Ansaugleistung – suction capacity

Ansaugleitung – suction line

Ansaugrohr – suction pipe, suction tube

Ansaugschlauch – suction tube

Ansaugstutzen – intake socket, suction support

Anschlagmittel – lifting means

Anschlagpunkt – anchor point

Anschluss – connection

Anschluss (Gas, Strom, Wasser) – connection line

Anschlüsse (Armaturen, Hähne) – fixtures

Ansprechdruck – response pressure

Ansprechempfindlichkeit – sensivity of response

Ansprechtemperatur – response temperature

Ansprechverhalten – response behaviour

Ansprechzeit – response time

Anstauen – dam up

Anstauen – water rising

ansteckend – infectious

Ansteckungsgefahr – risk of infection

ansteckungsgefährlicher Stoff – infectious substance

Anstellleiter – standard ladder, single pole ladder

Anstiegsgeschwindigkeit – rate-of-rise

Anstrengung – effort

Anstrich – coat, coating

Anstrich, der eine Dämmschicht bildet – intumescent coating

Antenne – aerial, antenna

antibakteriell – bactericidal, antibacterial

Antikatalysator – inhibitor, anticatalyst

antikatalytischer (Lösch)Effekt – inhibiting effect

Antreteordnung – line-up order

Anweisung – instruction

Anweisungen erteilen – issue orders

Anwendung – application

Anwendungsbereich – field of application, scope

Anwendungsgrenzen – application limits

Anwesenheitsliste – attendance list

Anzeigeeinrichtung – indicating device

Anziehung – attraction

Anziehung (der Elektronen durch den Atomkern) – electron-nucleus attraction

Anziehung von Ionen – ionic attraction

Anziehung, elektrostatische – electrostatic attraction

Anziehung, zwischenmolekulare – intermolecular attraction, molecular attraction

Anziehungskraft – attractive force

anzünden – to set fire to, to ignite

Anzündholz – kindling

Aorta – aorta

Äquivalentdosis – equivalent dose

Äquivalentdosisleistung – equivalent dose rate

Arbeit, körperliche – physical work

Arbeiten in Höhen – working at heights

Arbeiter-Samariter-Bund (ASB) – Worker's Samaritan Foundation

Arbeitsbereich – scope of work

Arbeitsbühne – manlift

Arbeitsdruck – operating pressure, service pressure

Arbeitsgerät – working tool

Arbeitshandschuhe – working gloves, work gloves

Arbeitshygiene – work hygiene

Arbeitskorb (einer Drehleiter) – fire-fighting cage

Arbeitsleine – working line

Arbeitsleine (Verwendung a. d. Saugleitung) – drop line

Arbeitsmediziner – occupational physician

arbeitsmedizinische Untersuchung – occupational health check

Arbeitsmethode – procedure, operating method

Arbeitsmittel – means of production

Arbeitsplatz – workplace, place of work

Arbeitsplatzkonzentration, maximale (MAK) – maximum permissible workplace concentration

Arbeitsraum – workroom

Arbeitsschutz – occupational safety and health

Arbeitssicherheit – work safety, operational safety

Arbeitsstellenscheinwerfer – working light

Arbeitsstoffe – working substances

Arbeitsstoffe, biologische – biological working substances

Arbeitsunfall – occupational accident

Armaturen – fittings

Armaturen, wasserführende – water carrying fittings

Ärmelabzeichen – sleeve badge

Armierung (Beton) – concrete reinforcement

Art und Weise – manner

Art und Weise, geeignete – adequate manner

Arterie – artery

Arterienklemme – artery forceps, artery clamp

Arzneimittel – drugs, medicaments

Arzneimittelvergiftung – drug poisoning

Asbest – asbestos

Asche – ash, ashes

Aschenbecher – ash tray

Astsäge – branch saw

A-Teil einer Steckleiter – base module (9-runged) of a scaling ladder

A-Teil-Einschub einer Steckleiter – attachment module (2-runged) of a scaling ladder

Atem – breath

Atem- und Herzstillstand – cardiopulmonary arrest, cardiorespiratory arrest

Atemanschluss – connecting piece for breathing

Atemanschluss – face piece

atembar – inhalable

Atembeutel – breathing bag

Atemfrequenz (AF) – respiratory rate

Atemgerät – breathing apparatus

Atemgift – respiratory poison, breathing poison

Atemgrenzwert – maximum voluntary ventilation (MVV)

Atemluft – breathing air

Atemluftflasche – breathing air cylinder

Atemluftreserve – breathing air reserve

Atemluftvorrat – breathing air supply

Atemminutenvolumen (AMV) – minute respiratory volume (MRV)

Atemnot – laboured breathing

Atemorgane – respiratory system

Atemschlauch – breathing tube

Atemschutz – respiratory protection, breathing protection

Atemschutzausbilder – breathing apparatus instructor

Atemschutzausrüstung – breathing protection equipment

Atemschutzfilter – breathing air filter

Atemschutzfiltergeräte – filter appliances

Atemschutzgerät (ASG) – self-contained breathing apparatus (SCBA), respiratory protective device

Atemschutzgerät mit Filter – filter respirator

Atemschutzgerät mit Überdruck – positive pressure breathing apparatus

Atemschutzgerät, umluftunabhängiges – self-contained breathing apparatus (SCBA)

Atemschutzgeräteträger (AGT) – wearer of breathing apparatus set, smoke diver, draegerman {US}

Atemschutzgerätewagen (GW-A) – breathing apparatus tender (BA-tender) (BAT), breathing apparatus appliance (BA-appliance)

Atemschutzgerätewart (AS-Gw) – breathing apparatus maintainer

Atemschutzgerätewerkstatt – breathing equipment workshop

Atemschutzmaske – respirator, breathing mask

Atemschutztauglichkeit – breathing apparatus fitness

Atemschutzüberwachung (ASÜ) – breathing protection control, respiratory protection monitoring

Atemschutzüberwachungssystem – respiratory protection monitoring system

Atemschutzübungsanlage – breathing apparatus training facility, breathing exercise facility, respirator training facility

Atemschutzvollmaske – full facepiece respirator

Atemspende – breathing transmission

Atemstillstand – respiratory arrest, breathing arrest, apnea

Atemwege – airway

Atemwegsverätzung – respiratory tract burn

Atemwiderstand – respiratory resistance

Atemzeitvolumen – maximum voluntary ventilation (MVV)

Atemzentrum – respiratory center

Atemzug – breath

Atemzugvolumen (AZV) – tidal volume (VT)

atmen – breath, respire

Atmosphäre – atmosphere

Atmosphäre, explosionsfähige – explosible atmosphere

atmosphärische Entladung – atmospheric discharge

atmosphärischer Luftdruck – atmospheric pressure

Atmung – breathing, respiration

Atmungsorgane – respiratory system

Atom – atom

Atom im Grundzustand – normal atom, ground-state atom

Atom, angeregtes – excited atom, activated atom

Atom, endständiges – terminal atom

Atom, ionisiertes – ionized atom

atomar – atomic, atomical

Atomart – atomic species

Atomausstieg – nuclear phase-out, nuclear power phase-out

Atombau – atomic structure

Atombindung – atomic bond, homopolar bond, covalent bond

Atombindungszahl – covalency, covalence

Atombombe – atomic bomb, nuclear bomb, A-bomb, fission bomb

Atombunker – fallout shelter

Atomenergie – nuclear energy, atomic energy

atomgetrieben – nuclear-powered

Atomgewicht – → Atommasse

Atomgröße – atomic size

Atomhülle – atomic shell, extranuclear region, electronic region

Atomkatastrophe – nuclear disaster

Atomkern – nucleus, atomic nucleus, core of an atom

Atomkerne – atomic nuclei

Atomkraft – nuclear power

Atomkraftwerk (AKW) – nuclear power plant (NPP)

Atommasse – atomic mass

Atommasse, absolute – atomic mass, mass of atom

Atommasse, relative – relative atomic mass (RAM)

Atommasseneinheit (amu) – atomic mass unit (amu)

Atommasseneinheit, vereinheitlichte (*u*) – unified atomic mass unit (*u*)

Atommeiler – nuclear reactor, nuclear plant

Atommodell – atomic model

Atommüll – radioactive waste, nuclear waste

Atommülltransport – transport of nuclear waste, transport of radioactive waste

Atomphysik – atomic physics

Atomreaktor – atomic reactor, nuclear reactor

Atomspaltung – nuclear fission

Atomunfall – nuclear accident

Atomzerfall – atomic disintegration, radioactive disintegration, nuclear disintegration

ätzend – caustic, corrosive

ätzende Dämpfe – corrosive vapours

ätzende Gase – caustic gases

ätzende Stoffe – corrosive substances

ätzfest – etch-proof

Ätzwirkung – caustic effect

Aufbauzeit – construction time

Aufbereitung – conditioning

Aufbewahrungskiste – locker

aufbinden (Knoten) – untying

Aufenthaltsraum – staff room, recreation room

Auffahrbohlen – ramp planks

Auffahrunfall – shunt

Auffang- und Haltegurt – work positioning and restraint belt

Auffangbehälter (für Chemikalien) – dunk tank

Auffangwanne – retention pond, retention sump, catch pot, catcher

Aufflammen – flashing, flaring up

Aufflammung – deflagration

Aufgaben – tasks, missions

Aufgabenverteilung – designation of responsibilities

aufklappbar – openable

Aufladelöscher – chargeable fire extinguisher

Aufladung, elektrostatische – electrostatic charging, static electrification

Auflaufbremse – overrun brake

Aufnahme (in den Körper) – intake, ingestion

Aufräumarbeiten – clearing-up operations

aufräumen (einer Brandstelle) – mop up

aufrichten einer Leiter – heel a ladder

Aufrichtwinkel – erecting angle

Aufsaugmittel – absorbent

Aufschlämmung – slurry

Aufstellfläche – set-up area

Aufstellung (Fw-Fahrzeuge) – setting up of the vehicles

Aufstiegsgeschwindigkeit (Tauchen) – ascent speed

Auftrag – mission

auftragsbezogene Zusammenarbeit – mission-concentrated cooperation

Auftragserledigung – task fulfilment

Auftragsinhalte – objective of the mission

Auftragstaktik – mission-tactics

Auftrieb (in Luft) – lift

Auftrieb (in Wasser) – buoyancy

Auftriebskraft – buoyancy force

aufwallen – to boil up, to bubble up

Aufzug – elevator, lift

Aufzugskabine – cage, lift cage

Aufzugsschacht – lift shaft

Auge (Seil) – loop, turn

Augendusche – eye-wash station

augenreizend – lachrymatory, irritating the eye

Augenreizstoff – eye irritant, lachrymator

Augenreizung – eye irritation

Augenschutz – eye protection

Augenspülflasche – bottle of eyewash solution, eye dropper bottle

Augenspülflüssigkeit – eye rinsing liquid

Augenwaschflasche – bottle of eyewash solution, eye dropper bottle

Aus- und Weiterbildung – training and supplementary education

Ausatemluft – exhaled air

Ausatemventil – exhalation valve

Ausatemventilscheibe – exhalation valve disk

Ausatemwiderstand – exhalation resistance

ausatmen – breathe out, exhale

Ausatmung – expiration, exhalation

Ausbeute – yield

Ausbeute, stöchiometrische – stoichiometric yield

Ausbilder – instructor

Ausbilder im Atemschutz – breathing apparatus instructor

Ausbildung – exercise, instruction, training

Ausbildungsnormen – training standards

ausbreiten – to spread out, to propagate, to diffuse

Ausbreitung – spreading, diffusion, propagation

Ausbreitung von Flammen auf der Oberfläche – surface spread of flame

Ausbreitung von Gaswolken – spread of gas clouds

Ausbreitungsgeschwindigkeit – propagation velocity

Ausbreitungsgeschwindigkeit elektromagnetischer Wellen – propagation velocity of electromagnetic waves

Ausbreitungsgrenze – propagation limit

ausbrennen – to burn out

Ausdehnung (Erweiterung) – expansion

Ausdehnung (Verlängerung) – extension

Ausdehnungskoeffizient – coefficient of expansion

Ausdehnungsverhalten – expansion behaviour

Ausdehnungsvermögen – dilatability

Ausfall der Funkverbindung – radio failure

Ausfallquote – failure rate

Ausfallsicherheit – meantime between failures (MTBF)

Ausfluss (aus einem Strahlrohr) – nozzle discharge, discharge out jet pipe

ausführbar – executable, practicable

Ausgang (Tür ins Freie) – final exit

Ausgangsdruck – outlet pressure, output pressure

ausgasen – degas

Ausgehuniform – dress uniform

ausgelaufene Stoffe – spilled substances

ausgetretene Stoffe eindämmen – contain spillages

Ausladung (z. B. beim Kran) – radius

Auslandseinsatz – operation abroad

Auslassöffnung (z. B. für Löschmittel) – agent outlet

Ausleger (bei einem Kran) – jib

Auslöseeinrichtung – tripping device

Auslöseschwelle (med.) – trigger threshold

Ausnahme – exception

Ausnahmegenehmigung – exceptional permission

Auspuffbremse – exhaust brake

Ausrückebereich – fire district, turn-out reach

Ausrückefolge – turn-out sequence

Ausrückeordnung – pre-determined attendance, turn-out order

Ausrückezeit – response time, reaction time

Ausrüstung – equipment

Ausrüstung, persönliche – personal equipment

Ausrüstung, zusätzliche – supplementary equipment

Ausschluss – exclusion

Ausschreibung (Beschaffung) – request for proposal (RFP)

Außenangriff – outside (fire) attack

Außendruck – external pressure

Außenelektron – valence electron

Außengewinde – external thread, male thread

Außentemperatur – outside temperature, exterior temperature

Außentreppe – outside stair

Außentreppenhaus – external staircase

Außenwand – exterior wall, outer wall

außer Betrieb – inoperative

außer Dienst – off duty

äußere Bestrahlung – external irradiation

Ausstieg – escape hatch

ausstoßen – emit

ausstrahlen – emit

ausströmen (Gas) – effusion

ausströmendes Gas – escaping gas

Austauchstufe (Tauchen) – decompression stop depth

Austausch – exchange

austauschbar – exchangeable

Austritt – spillage

Austrocknung – desiccation

ausweglose Situation – hopeless situation

Auswerfen brennbarer Flüssigkeiten – ejection of combustible liquids

ausziehen (verlängern) – extend

Auszugslänge (einer Leiter) – extension length of ladder

Authentifizierung – authentication

Autobahn – motorway, freeway

Autobahnauffahrt – slip road, on-ramp

Autobatterie – → Bleiakkumulator

Autogenschneiden – autogenous (gas) cutting, oxygen-acetylene cutting

Autogenschweißen – autogenous (gas) welding, oxygen-acetylene welding

Autokatalysator – autocatalytic agent

Autokatalyse – autocatalysis, self-catalysis

automatische Brandschutzanlage – automatic fire protection installation

automatische Feuerlöschanlage – automatic fire-fighting installation

automatischer Brandalarm – automatic fire signal

automatischer Brandmelder – automatic fire detector

automatischer Feuermelder – automatic fire alarm (AFA)

Autooxidation – autooxidation, spontaneous oxidation

Autoradiographie – radioautographic, autoradiography

Autorität aufgrund der Funktion – authority of post

Avogadro-Gesetz – Avogadro's law

Avogadro-Konstante – Avogadro's constant, Avogadro's number

Axt – axe

B

Bach – stream, creek, brook

Bagatellschaden – minor damage, minimal damage

Bahnüberführung – railway overbridge

Bahnübergang – level crossing, grade crossing

bakterielle Erkrankung – bacterial disease, bacteriosis

Bakterium, Bakterien – bacterium, bacteria

Balken – beam, girder

Balkenschuh – joist hanger

Bandfalldämpfer – webbing brake, strap fall attenuator, belt-type fall impact absorber

Bandmaß – tailor's tape

Bandschlinge – sling, runner

Bandschlingenknoten – European death knot (EDK), offset overhand bend (OOB), thumb knot, open-hand knot, water knot

Bandschnalle (für Ehrenzeichen) – service ribbon

Bannerträger – bannerman, flag-bearer

Baracke – hut

Barotrauma – barotrauma

Bartaxt – bearded axe

Barterlass – beard decree

Base (Lauge) – base

Basis – base

basisch (alkalisch) – basic

Basiseinheit – base unit

Batterie – battery

Batterietester – battery tester

Bauantrag – planning application

Bauart – type of construction

Bauartzulassung – type approval

Bauaufsichtsbehörde – building control department

Baubehörde – building control agency

Baudenkmal – historical building, historical monument

Baugenehmigung – house-building permission

Baugesetz – building law

Bauholz – timber

Baukastensystem – modular system

Baukunde – building science

bauliche Anlage – built environment

bauliche Maßnahme – structural measure

bauliche Sicherungsmaßnahme – physical safeguarding measure

bauliche Voraussetzungen – building requirements

baulicher Brandschutz – structural fire protection

Bauordnung – building regulations

Bauprodukte – building products

Baustelle – construction site

Baustoff – building material

Baustoffklasse – building material class

bautechnischer Explosionsschutz – structural explosion protection

Bauteil – component

Bauten, fliegende – fairground rides

Bauwesen – civil engineering

T. Schmiermund, *Fachwörterbuch Feuerwehr und Brandschutz*, https://doi.org/10.1007/978-3-662-64120-0_2

BC-Löschpulver – extinguishing powder type BC

BC-Pulver – BC-powder

B-Druckschlauch – fire hose type B

Be- und Entlüftungseinrichtungen – ventilation devices

Beanspruchung, thermische – heat stress

Beatmung – respiration, ventilation, breathing

Beatmung mit Beatmungsbeutel – bag valve mask ventilation, BVM ventilation, bagging, Ambu™ bag ventilation

Beatmung, künstliche – artificial respiration

Beatmungsbeutel – bag valve mask (BVM)

Beatmungsgerät – medical ventilator, respiration apparatus, respirator

Beatmungshilfe – respiratory aid

Beatmungsmaske – respiration mask, artificial respiration mask

Becquerel – becquerel

Bedachung – roofing, roof covering

Bedachung, harte – non-inflammable roof-covering

Bedieneinrichtung (BE) – control device

Bedientableau – control panel

Bedienung – operation

Bedienungsanleitung – operation instructions, manual, instruction manual, operating manual

Befehl – order

Befehl ausführen – execute a command

Befehl befolgen – obey an order

Befehl entgegennehmen – take an order

Befehl, auf höheren – on orders from above

Befehl, besonderer – special order

Befehl, dienstlicher – official order

Befehl, mündlicher – oral order

Befehl, schriftlicher – written order

Befehls- und Meldesystem – order and report system

Befehlsgebung – issue of orders

Befehlsgliederung – structure of an order

Befehlskette – chain of command

Befehlsstelle – command post

Befehlsstelle, bewegliche – mobile command post

Befehlsstelle, ortsfeste – fixed command post

Befehlssystem – order system

Befehlsweg – chain of command

Befestigung – fixture

Befestigungsgurt – fastening strap

Befestigungspunkt – anchorage point, mounting point, fastening point

Beflammungsdauer – flame application time

befördern, auf einer Trage – carry on a stretcher

Beförderung (im Dienstgrad) – advancement

Beförderungseinheit – transport unit

Beförderungspapiere – transport documents

Befreiung – extrication

Befugnisse – competences

Begehbarkeit – walk-on stability

Begleitschaden – collateral damage

begrenzender Faktor – limiting factor

Begriffe – terms

Begriffe, allgemeine – general terms

Behälter – containers, tanks

Behälter, zerbrechlicher – breakable container

Behälterexplosion – container explosion

Behältergerät (BG) – container apparatus

Behandlung – treatment

Behandlung, abwartende – expectant treatment

Behandlung, minimale – minimal treatment (MT)

Behandlung, sofortige – immediate treatment (IT)

Behandlung, verzögerte – delay treatment (DT)

Behandlungsplatz (BHP) – treatment station

Behandlungsprozess – treatment process

Behelfsbrücke – emergency bridge, temporary bridge

Behelfsunterkunft – makeshift shelter

Behinderungen (Hindernisse) – obstructions

Behörde – agency, authority

Behörden und Organisationen mit Sicherheitsaufgaben (BOS) – institutions and organisations in the field of safety and security, authorities and organisations with security tasks

Beifahrerairbag – front passenger airbag

Beil – hatchet

Beiltasche – hatchet bag

B-Einsatz – mission type B (= biological) → biological assignment, operation with dangers due to biological substances

Beißkeil – bite protection mouth wedge

Beißzange – pinchers

bekämpfen – tackle, combat

belastbar – loadable

Belastbarkeit – loading capacity

Belastung (Beanspruchung) – loading, strain

Belastung (Traglast) – weight

Belastung (Ursache) – strain

Belastung (Zustand) – stress

Belastungsanzeige (Drehleiter) – safe load indicator

Belastungszustand – stress

Beleuchtung – illumination

Beleuchtungsanlage – lightning installation

Beleuchtungsgerät – lightning device, lightning apparatus

Beleuchtungsgerät, tragbares – portable lightning apparatus

Belüftung – aeration, ventilation

Belüftungsanlage – ventilation system

Belüftungsöffnung – ventilation hole

Bemessungsgrundlage – rating basis

Benetzbarkeit – wettability

benetzen – wet, moisten

Benetzungsmittel – wetting agent, wetter

benötigte Information – needed information

Benzin – gasoline

Benzinkanister – gasoline canister

Benzinmotorsäge – gas chain saw, petrol-engined saw

Beobachtungen – observations

Beobachtungen, eigene – personal observations

Berechnung – calculation

Berechnungsdruck – calculation pressure

berechtigt – empowered

Bereich – area, zone, section, sector

Bereich, explosionsgefährdeter – explosion-hazardous area

Bereich, geschützter – protected area

Bereich, strahlungsgefährdeter – high radiation area

Bereich, ungeschützter – unprotected area

Bereich, zwischenmenschlicher – interpersonal area

Bereiche, tief liegende – low areas

Bereitmeldung – ready-message

Bereitschaft (Dienst) – attendance

Bereitschaftsraum (Zimmer) – on-call room

Bereitschaftszeit – time of readiness

Bereitstellung – preparation

Bereitstellungsfläche – staging area

Bereitstellungsraum (BSR) – assembly area

Bergebehälter – salvage receptacle

Bergefass – salvage drum, salvage cask

bergen – salvage, rescue, recover, save

Bergnot – distress in mountains

Bergrettung – mountain rescue

Bergsteigerhelm – climber's helmet

Bergung – salvage

Bergungsaktion – salvage operation, salvage mission

Bergungsarbeiten – salvage operations, salvage work

Bergungseinheit – salvage unit

Bergungseinsatz – salvage operation, salvage effort

Bergungsgeräte – salvage devices

Bergungskosten – salvage charges, salvage costs

Bergungskran – recovery crane

Bergungsmannschaft – salvage crew

Bergungsschiff – salvage ship

Bergungsseilwinde – salvage winch

Bergungsspezialisten – salvage experts

Bergungstrupp – salvage team

Berieselungsanlage – sprinkler system

Berstdruck – bursting pressure

Berstdruck für Druckschläuche – bursting pressure for fire hoses

Berstscheibe – bursting disk

Beruf – profession, job

Berufsfeuerwehr (BF) – professional fire service

Berufsfeuerwehr (BF) – paid fire brigade, whole-time fire brigade, professional fire brigade

Berufsfeuerwehrmann/-frau – whole-time firefighter, full-time firefighter

Berufskrankheit – occupational disease

Berufslebensdosis (Radioaktivität) – lifetime (cumulative) occupational radiation exposure

berühren – touch, contact

Berühren, Gefahr beim – contact hazard

Berührung mit der Haut – skin contact

Berührungsschutz – touch protection, protection against contact

Besatzung – crew

Beschädigung – damage

Beschichtung – coating

Beschleunigungsstreifen – acceleration lane

Besen – broom

besondere Einsatzsituationen – special operational situations

besonderer Befehl – special order

beständig – persistent

Beständigkeitsliste – resistance list

bestehende Gefahren – existing dangers

Bestimmung (einer Substanz oder Größe) – determination

Bestimmung (Regelung) – ruling

Bestimmung, gesetzliche – legal requirement

Bestimmungsgrenze – determination limit, limit of detection

Bestimmungsverfahren – determination procedure

Bestrahlung – irradiation

Bestrahlung, äußere – external irradiation

Bestrahlung, innere – internal irradiation

Bestrahlungsdichte – radiant-flux density

Bestrahlungsdosis – irradiation dose

Bestrahlungsintensität – irradiance, fluence rate, irradiation intensity

betankt – refuelled

Beta-Strahler – beta emitter

Beta-Strahlung – beta radiation

Beta-Teilchen – beta particle

betäubend wirkende Substanz – narcotic

Beta-Zerfall – beta decay, beta disintegration

Beteiligte – participants

Beton – concrete

Betreiber – operator, user

Betreuer – supervisor

betriebliche Voraussetzungen – production requirements, business requirements

betrieblicher Brandschutz – operational fire protection

Betriebsart – operation mode

betriebsbereit – enabled, ready for use, operable

Betriebsdruck – operating pressure, service pressure (pump)

Betriebsfeuerwehr (BtF) – factory fire brigade, company's fire brigade, on-site fire brigade

Betriebsmittel, elektrische – electric equipment

Betriebsmittel, explosionsgeschützte elektrische – explosion-proof electric equipment

Betriebssicherheit – operational safety

Betriebsspannung – operating voltage

Betriebsstoffe – consumables

Betriebsstörung – break down

Betriebszeit – operating time

betroffene Einheit – concerned unit

Beulenpest – bubonic plague

Beurteilung – assessment, evaluation

Beurteilung der Brandgefährdung – assessment of fire hazard

Beurteilung der Lage – assessment of the situation

Beurteilungswerte – assessment values, rating values

Bevölkerung – population (pop.)

Bevölkerungsschutz – civil protection

Bewältigung der Schadenslage – mastering of the disaster

bewegliche Befehlsstelle – mobile command post

Bewegungsenergie – kinetic energy

Bewegungsfläche – movement area

Bewehrung (Beton) – concrete reinforcement

Bewertung – evaluation

Bewertung der Lage – evaluation of the situation

Bewertungsfaktor – quality factor (QF)

Bewertungssystem (z. B. bei Prüfungen) – grading system

bewusstlose Person – unconscious person

Bewusstsein – consciousness, conscious mind

Bezeichnung – designation

Bezeichnung, taktische – tactical designation

Bezettelung – badging

Bezirksschornsteinfegermeister – district master chimney sweeper

Biegefestigkeit – flexural strength

Bildung (Entwicklung) – formation, generation

Bildung einer Holzkohleschicht – formation of a charcoal layer

Bildungsenergie – heat of formation

Bildungswärme – heat of formation

Bildzeichen – icon, pictorial symbol

Billiarde – quadrillion (1 000 000 000 000 000 = 10^{15})

Billion – trillion (1 000 000 000 000 = 10^{12})

Binärwaffen – binary weapons

Bindemittel – binder, binding agent

binden (Knoten) – tying

Bindestrick – binding cord

Bindung, chemische – chemical bond

biogefährdender Abfall – biohazardous waste

biologische Abbaubarkeit – biodegradability

biologische Arbeitsstoffe – biological working substances

biologische Kampfmittel – biological warfare agents (BWA)

biologische Waffe – biological weapon

biologischer Abbau – biodegradation

biologischer Kampfstoff – biological warfare agent

Biosicherheitslabor – biohazard containment laboratory (L1-L4)

Biosicherheitsstufen (S1-S4) – biosafety level, biohazard containment classes (S1-S4)

Biostoffe – biomaterials, biological substances

Biostoffverordnung – Biological Agents Ordinance

Biowaffe – biological weapon

Biowaffenkonvention – biological weapons convention (BWC)

Blasen auf der Haut – skin blisters

Blaulicht – flashing blue light, emergency light

Blaulicht (als RKL) – blue strobe light, blue rotating light

Blaulicht, mit eingeschaltetem … – with blue lights switched on

Blausäure – hydrogen cyanide, prussic acid

Blei – lead

Bleiabschirmung – lead shielding

Bleiakkumulator („Autobatterie") – lead storage battery, lead accumulator, lead-acid battery

Bleiblock (Radioaktivität) – pig, container of lead for radioactive materials

Bleigurt – weight belt (diving)

Bleischürze – lead apron

Bleistift – pencil

Bleistiftspitzer – pencil sharpener

Bleiwolle – lead wool

blinder Alarm – false alarm with good intent

Blindgänger – blind shell, unexploded bomb, unexploded ordnance (UXO)

Blindkupplung – cap coupling, blank cap

Blindstutzen – blank cap

blinken – to flash

Blinkleuchte – warning lamp

Blitz – lightning

Blitzableiter – lightning arrester (LA), lightning protector

Blitzentladung – lightning discharge

Blitzintubation – rapid sequence intubation (RSI)

Blitzleuchte – flashing light

Blitzlicht – strobe light

Blitzschlag – lightning bolt, stroke of lightning

Blitzschutz – lightning protection

Blitzschutzmaßnahmen – lightning protection measures

Blitzunfall – lightning strike

Blutdruck – blood pressure

Blutdruckmessgerät – sphygmomanometer

Blutdruckmessung – measuring blood pressure

Bluterguss – haematoma, hematoma

Bluthochdruck – hypertension

Blutkampfstoffe – blood warfare agents

Blutkreislauf – circulatory system

Blutübersäuerung – acidosis

Blutvergiftung – sepsis

BM-Strahlrohr – multi-purpose jet pipe type BM

Bockleiter – double ladder

Boden – floor, ground

Bodenfeuer – surface fire

Bogen – arch

Bohle – plank

Bohrfutterschlüssel – drill chuck key

Bohrhammer – drill hammer

Bohrmaschine – drill, drilling machine

Bohrsches Atommodell – Bohr atom model

Bolzenschneider – bolt clipper

Bombe, schmutzige – dirty bomb

Bombenanschlag – bomb attack

Bordausrüstung – on-board equipment

böswilliger Alarm – malicious false alarm

Brand – fire (blaze)

Brand unter Kontrolle (BuK) – stop message: fire under control

Brand, abklingender – decaying fire

Brand, andauernder – ongoing fire

Brand, offener – open fire, outside fire

Brandabschnitt – fire compartment, fire cut

Brandabschnittsbildung – formation of fire compartments

Brandabschnittstür – fire compartment door

Brandalarm – fire signal

Brandalarm, automatischer – automatic fire signal

Brandanalyse – fire analysis

Brandanschlag – fire attack, arson attack

Brandart – fire manner

Brandatmosphäre – fire atmosphere

Brandausbreitung – fire spread, fire propagation, fire spreading

Brandausbreitungsgefahr – risk of fire spread

Brandausbreitungsgeschwindigkeit – rate of fire spread

Brandausbruch – outbreak of fire

Brandausbruchsstelle (BA-Stelle) – place of fire origin

Brandauswirkung – fire impact

Brandaxt – fire axe

Brandbekämpfer – fire fighters

Brandbekämpfung – fire fighting

Brandbekämpfungsabschnitt – fire-fighting sector

Brandbekämpfungsausrüstung – fire-fighting equipment

Brandbekämpfungstaktik – fire-fighting tactics

Brandbelastung – fire load density

Brandbeobachter – fire lookout (a person)

Brandbericht – fire report

Brandbeschleuniger – fire accelerant, combustive agent

brandbeständig – fire-resistant, fire-proof

Branddauer – fire duration, burning length

Branddirektor – fire director

Brände in geschlossenen Räumen – compartment fires

Brandeffekte – fire effects

Brandeinwirkung – fire exposure

Brandeinwirkung, Verhalten bei (Bauteile) – fire performance

Brandentdeckung – fire detection

Brandentdeckungszeit – detection time

Brandentstehung – fire origin, fire emergence

Brandentstehungsraum – room of fire origin

Brandentstehungsstelle – point of origin

Brandentwicklung – fire development, fire growth

Brandentwicklungsdauer – time of fire growth

Brandereignis – fire event, fire incident

Branderkennung – fire detection

Brandermittler – arson investigator, fire investigator

Brandermittlung – fire investigation

Brandfläche – fire area

Brandfluchthaube – evacuation hood, fire escape hood, escape smoke hood

brandfördernd – supporting combustion

brandfördernd (Substanz) – oxidizing (substance)

Brandforschung – fire research

Brandfrüherkennung – early fire detection

Brandfrüherkennungssystem – early fire detection system

Brandgase – fire gases, combustion gases

Brandgasvolumen – fire gas volume

Brandgefahr – risk of fire

Brandgefährdung – fire hazard

Brandgefahrenklasse – fire hazard class, class of fire danger

brandgefährlich – presenting a fire hazard

Brandgefährlichkeit – fire hazard

Brandgeruch – smell of fire, odour of fire

Brandgeschehen – quantity of fires and fire looses

Brandhäufigkeit – fire frequency

Brandhaus – fire drill house, fire house

brandhemmend – fire-resistant

Brandherd – seat of fire, fire source

Brandintensität – fire load density

Brandkanal (z. B. in Heu) – fire duct

Brandkatastrophe – fire disaster

Brandkenngrößen – fire characteristics

Brandklasse – fire class

Brandklasse A: feste, glutbildende brennbare Stoffe – fire class A: solid combustible materials

Brandklasse B: brennbare flüssige Stoffe – fire class B: flammable liquids

Brandklasse C: brennbare Gase – fire class C: flammable gases; {US}: class B

Brandklasse D: brennbare Metalle – fire class D: combustible metals

Brandklasse E: Elektrobrände (zurückgezogen) – fire class E: electrical fires (retracted); {US}: class C

Brandklasse F: Speiseöl/-fett – fire class F: cooking oils/fats; {US}: class K (kitchen)

Brandlast – fire load

Brandlast, flächenbezogene – fire load density

Brandlastberechnung – calculation of fire load

Brandlastdichte – fire load density

Brandlehre – fire theory

Brandleistung – heat release rate (HRR), rate of heat release

Brandmauer – fire-resisting wall, fire wall

Brandmeldeanlage – fire detection and alarm system, fire alarm system

Brandmelder – fire detector

Brandmelder, automatischer – automatic fire detector

Brandmeldetelefon – fire emergency telephone

Brandmeldezentrale (BMZ) – fire alarm control panel (FACP), fire alarm centre

Brandmeldung – fire detection

Brandmeldung (telefonisch) – fire call

Brandmittel – incendiary agents

Brandmodell – fire model

Brandmodell, physikalisches – physical fire model

Brandnest – fire pocket

Brandobjekt – fire object

Brandopfer (Person) – fire victim

Brandort – fire place

Brandphase – fire phase, burning phase

Brandprüfung – fire test

Brandquelle – fire source

Brandrauch – fire smoke

Brandrauchdurchzündung – backdraft

Brandraum – fire room, fire compartment

Brandraumtemperatur – temperature of fire room

Brandreste – fire scene residue

Brandrisiko – fire risk

Brandrückstände – debris

Brandsachverständiger – fire expert

Brandsatz – incendiary composition, incendiary agent, incendiary material

Brandschaden – fire loss, damage caused by fire

Brandschau – fire inspection, fire audit

Brandschicht – burnt layer

Brandschlacke – fire clinker

Brandschneise – fire break, fire lane

Brandschutt – fire debris

Brandschutz – fire protection, fire safety, fire prevention

Brandschutz (BS) – fire control and protection

Brandschutz, abwehrender – fire-fighting, fire defence

Brandschutz, baulicher – structural fire protection

Brandschutz, betrieblicher – operational fire protection

Brandschutz, vorbeugender – fire prevention (FP), fire protection, fire safety

Brandschutzabschluss – fire-resisting closure

Brandschutzanlage – fire protection installation

Brandschutzanlage, automatische – automatic fire protection installation

Brandschutzanstrich – fire retardant paint

Brandschutzaufklärung – fire protection enlightenment

Brandschutzausbildung – fire protection training

Brandschutzbeauftragter – fire protection officer

Brandschutzbescheinigung – fire certificate

Brandschutzbeschichtung – intumescent coating

Brandschutzbestimmungen – fire protection regulations, fire regulations

Brandschutzeinrichtungen – fire-protection appliance

Brandschutzerziehung – fire protection education

Brandschutzfassade – fire-protection facade

Brandschutzgeschichte – fire prevention history

Brandschutzgesetz – fire protection law, fire law, fire protection act

Brandschutzhelfer – fire protection helper, fire protection assistant

Brandschutzingenieur (BSIng) – fire safety engineer, fire prevention engineer

Brandschutzingenieurwesen – fire-safety engineering

Brandschutzklappe – fire damper, fire flap

Brandschutzklassen – fire protection classes

Brandschutzleiter (Beruf) – chief fire prevention officer

Brandschutzmaßnahme (konkret) – fire protection measure

Brandschutzmaßnahmen (allgemein) – fire protection method

Brandschutznormen – fire standards

Brandschutzordnung (BSO) – fire safety regulations

Brandschutzposten – fire piquet, fire picket

Brandschutz-Produkte – fire protection products

Brandschutzrecht – fire protection law

Brandschutzschiebetür – fire protection sliding door

Brandschutzschott – firestop seal

Brandschutztechnik – technics of fire protection

brandschutztechnische Kennzahlen – fire protection indexes

Brandschutztor – fire gate

Brandschutztür – fire door, emergency fire door

Brandschutzübung – fire drill

Brandschutzventil – fire protection valve

Brandschutzverglasung – fire resistant glazing

Brandschutzverkleidung – fire protection cladding

Brandschutzverordnung – fire regulations

Brandschutzvorhang – fire protection curtain

Brandschutzvorschriften – fire safety regulations, fire protection regulations

Brandschutzwand – fire barrier, fire wall

Brandschutzzeichen – fire safety signs

Brandserie – fire series, fire run

Brandserie – series of fires

Brandsicherheit – fire safety

Brandsicherheitswache – fire picket {GB}, fire watch {US}

Brandsimulation – fire simulation

Brandsimulationsanlage – hot fire training system

Brandsperre – fire barrier

Brandspuren – burn marks, fire traces

Brandstabilität – fire stability

Brandstatistik – fire statistics

Brandstelle (BSt) – scene of fire, fire ground, site of fire

Brandstifter – incendiary, arsonist

Brandstiftung – arson

Brandstiftung, fahrlässige – negligent arson, careless arson

Brandstiftung, vorsätzliche – deliberate arson, nonnegligent arson

Brandstreifen – backfire

Brandszenario – fire scenario

Brandtemperatur – temperature of fire

Brandtote – fire fatalities, fire casualties

Brandtrichter – fire funnel

Brandumfang – fire size, fire area

Brandunterdrückung – fire suppression

Brandunterdrückungsanlage (BUA) – fire suppression system (FSS)

Branduntersuchung – fire investigation

Brandursache – cause of a fire, fire cause, cause of conflagration

Brandursachenermittler – fire cause investigator

Brandursachenermittlung – fire investigation, inquiry of fire cause

Brandursachenforschung – studies of fire causes

Brandverhalten – burning behaviour, fire properties, reaction of fire

Brandverhalten (Material) – fire performance

Brandverhalten, extremes – extreme fire behaviour

Brandverhütung – fire prevention, preventive fire protection

Brandverlauf – fire course

Brandverlaufskurve – fire progression curve, fire progress curve

Brandverletzte – fire casualties

Brandverletzung – burn, burn wound

Brandversicherung – fire insurance

Brandversicherungspolice – fire policy

Brandverursacher – person who caused a fire

Brandwache – fire picket {GB}, fire watch {US}

Brandwand – fire wall, fire-resisting wall

Brandwärme – heat of fire

Brandwolke – cloud of/from fire

Brandzone – fire zone

Brauchwasser – process water, industrial water

Brechbeutel – vomit bag

brechen (bei Übelkeit) – vomit

brechen (Zerbrechen) – break

Brechmittel – emetic

Brechstange – jimmy, pry bar, crow bar

Brechwerkzeug – breaking tool

Bremsanlage – brake system, braking system

Bremsblock (Kerntechnik) – moderating block

Bremse – brake

Bremsflüssigkeit – brake fluid

Bremsknoten (HMS) – crossing hitch, Italian hitch, Munter hitch, HMS

Bremsstrahlung (Kerntechnik) – bremsstrahlung

Bremssubstanz (Kerntechnik) – moderator

brennbar – combustible

brennbar, leicht brennbar – easily combustible

brennbar, normal brennbar – normally combustible

brennbar, schwer brennbar – hardly combustible

brennbare Dämpfe – flammable vapours

brennbare feste Stoffe – combustible solid substances

brennbare flüssige Stoffe – combustible liquid substances

brennbare Flüssigkeiten – combustible liquids

brennbare Gase – combustible gases

brennbare gasförmige Stoffe – combustible gaseous substances

brennbare Gemische – combustible mixtures

brennbare Lösemittel – flammable solvents

brennbare Masse – combustible load

brennbare Stoffe – combustible substances

brennbarer Staub – combustible dust

Brennbarkeit – combustibility

Brennbarkeitsprüfung – flammability test

Brenndauer – burning time

Brennelement – fuel assembly

brennen – burn

Brennen – burning

Brennen der Oberfläche – surface burn

Brennen, andauerndes – sustained combustion

brennend – alight

brennendes Abtropfen – burning dripping

Brennpunkt – inflammation point, (firing point)

Brennschneiden (Acetylen-Sauerstoff) – autogenous gas cutting, autogenous cutting, oxygen-acetylene cutting, flame cutting

Brennstäbe (Kerntechnik) – fuel rods

Brennstoff – fuel, combustible material

brennstoffarme Verbrennung – fuel-lean combustion

brennstoffreiche Verbrennung – fuel-rich combustion

Brennstoffzelle – fuel cell

Brennverhalten – burning behaviour

Brennwert – gross heat of combustion, gross calorific value, caloric value

Brennwert, spezifischer – specific caloric value

Brille – glasses, spectacles

Brillenträger – spectacle wearer

Brisanz – brisance, shattering capability

britische Pferdestärke – British horsepower (BHP)

Bruch (Material, Knochen) – fracture

Brunnen – well

Brustbund (Knoten) – bowline

Brustgurt – chest harness

Brüstung – breastwork, breast rail

Brüter – breeder reactor

B-Teil einer Steckleiter – attachment module (7-runged) of a scaling ladder

Büchsenöffner – can opener, tin opener

Buchstabieralphabet – phonetic alphabet

Buchstabiertafel – phonetic alphabet

Bucht (Seil) – bight

Bügelsäge – bow saw

Bühne (Theater) – stage

Bulin – → Pfahlstich

Bunde (Knoten) – bends

Bündelfunk – trunked radio, multi-user band radio

Bündelfunksystem – trunked radio system

Bundesamt für Bevölkerungshilfe und Katastrophenschutz (BBK) – Federal Office for Civil Protection and Disaster Assistance

Bundesamt für Strahlenschutz (BfS) – Federal Office for Radiation Protection (Germany)

Bunsenbrenner – Bunsen burner

Buntmetall – non-ferrous metal

Bürgerliches Gesetzbuch (BGB) – German Civil Code

Bürgermeister – mayor

Bürolocher – puncher, office puncher

Buschbrände – bush fires

C

Carbonsäure – carboxylic acid

Castor-Behälter – castor container

CBRN (chemisch, biologisch, radioaktiv, nuklear) – CBRN (chemical, biological, radioactive, nuclear)

CBRNE (chemisch, biologisch, radioaktiv, nuklear, explosiv) – CBRNE (chemical, biological, radioactive, nuclear, explosive)

C-Druckschlauch – fire hose type C

C-Einsatz – mission type C (chemical), chemical assignment, operation with dangers due to chemical substances

Chemie, anorganische – inorganic chemistry

Chemie, organische – organic chemistry

Chemieunfall – chemical accident

Chemiewaffe – chemical weapon

Chemiewaffeneinsatz – deployment of chemical weapons

Chemiewaffenentsorgung – disposal of chemical weapons

Chemiewaffenkonvention – chemical weapons convention

Chemiewaffenübereinkommen (CWÜ) – chemical weapons convention

Chemiewehr – chemicals incident unit (CIU)

Chemikalie(n) – chemical(s)

Chemikalienbindemittel – chemical absorbent agent

Chemikalienbinder – chemical absorbent

Chemikalienschrank – chemical cabinet, chemical safety cabinet

Chemikalienschutz – chemical protection

Chemikalienschutzanzug (CSA) – hazmat suit, chemical protective suit, chemical protection suit, self-containment suit

Chemikalienschutzanzug (CSA), gasdichter, Typ 1a-ET – type 1a gas-tight suit for emergency teams

Chemikalienschutzanzug (CSA), gasdichter, Typ 1b-ET – type 1b gas-tight suit for emergency teams

chemische Bindung – chemical bond

chemische Gleichung – chemical equation

chemische Kampfmittel – chemical warfare agents (CWA)

chemische Kampfstoffe – chemical warfare agents (CWA)

chemische Reaktion – chemical reaction

chemische Waffe – chemical weapon

chemisches Gesetz – chemical law

chemisches Gleichgewicht – chemical equilibrium

Chlor – chlorine

Chlorgasvergiftung – chlorine poisoning

Chlorkohlenwasserstoffe (CKW) – chlorocarbons

Chlorwasserstoff – hydrogen chloride

T. Schmiermund, *Fachwörterbuch Feuerwehr und Brandschutz*, https://doi.org/10.1007/978-3-662-64120-0_3

chronische Erkrankung – chronic disease, chronic ailment

chronische Giftwirkung – chronic toxic effect

CM-Strahlrohr – multi-purpose jet pipe type CM

Compton-Effekt – Compton effect

Compton-Streuung – Compton scattering

Coronavirus-Erkrankung 2019 – corona virus disease 2019 (COVID-19)

D

Dach – roof

Dachboden – loft

Dachdeckung – roof covering

Dachentlüftung – roof vent

Dachfenster – dormer window, skylight

Dachfläche – roof surface

Dachgaube – dormer

Dachgeschoss – attic

Dachkennzeichnung (Fahrzeuge) – roof identification

Dachkonstruktion – roof construction

Dachlatte – roof batten

Dachleiter – roof ladder

Dachmonitor – deck gun

Dachpappe – roofing felt

Dachräumung (z. B. von Schnee) – clearing of roofs

Dachrinne – gutter

Dachstuhl – roof truss

Dachstuhlbrand – roof fire

Dachziegel – roof tile

Dalton'sche (Gas)Gesetze – Dalton's (gas) laws

Dalton-Gesetz – Dalton's law

Dämmschicht – insulation layer, insulating layer, thermal insulating layer

Dämmschichtbildner – intumescent coatings

Dämmstoffe – insulating materials

Dämmung – insulation

Dampf, Dämpfe – vapour, vapours

Dampfblasenkoeffizient – void coefficient

Dampfdichte – vapour density

Dampfdichteverhältnis – ratio of vapour densities

Dampfdruck – vapour pressure

Dämpfe, ätzende – corrosive vapours

Dämpfe, brennbare – flammable vapours

Dämpfe, giftige – toxic vapours

dämpfen (abschwächen) – buffer

Dampfexplosion – vapour explosion, steam explosion

Dampfleitung – steam pipeline

Dampf-Löschanlage – steam fire-extinguishing installation

Dampflöschverfahren – steam extinguishing method

Dampf-Luft-Gemische – vapour-air-mixtures

Dampfphase – vapor phase, vapour phase

Dampfsperre – vapour barrier

Darmmilzbrand – gastrointestinal anthrax

Das ist eine Übung! – This is an exercise.

Das ist *keine* Übung! – This is *not* a drill! ; This is *not* an exercise!

Datenblatt – data sheet

Datenträger – data carrier

Datenübertragung – data transfer

Dauer bis zum Eintreffen – attendance time

Dauerdruck-Feuerlöscher – stored-pressure fire extinguisher, permanent-pressure fire extinguisher

Dauerdruck-Wasserlöscher – pressurized water fire-extinguisher

Dauerton – continuous tone

T. Schmiermund, *Fachwörterbuch Feuerwehr und Brandschutz*, https://doi.org/10.1007/978-3-662-64120-0_4

D-Druckschlauch – fire hose type D

Decke (im Gebäude) – ceiling

Decke (zum Wärmen) – blanket, quilt

Deckeffekt (ein Löscheffekt) – cover effect

Deckung – cover

Deckvermögen (Löschmittel Schaum) – covering capacity, blanketing ability

Deflagration – blast, deflagration

Deflagrationsdruck – deflagration pressure

Deflagrationsgrenzdruck – deflagration limit pressure

Dehnfuge – dilatation joint

Deichverteidigung – dike defence

Dekompressionskammer – decompression chamber

Dekon G (G = Gerät) – decontamination of equipment

Dekon P (P = Personen) – decontamination of persons/people

Dekon-Platz – decontamination station (decon station), decontamination area (decon area)

Dekontamination – decontamination

Dekontamination, erweiterte – enhanced decontamination

Dekontamination, Not- – emergency decontamination

Dekontamination, Standard- – standard decontamination

Dekontaminationsdusche – decontamination shower

Dekontaminationsplatz – decontamination station, decontamination area

demontierbar – demountable, removable

Denk- und Handlungsablauf – process of thinking and acting

Deponie – landfill

Deponiegas – landfill gas (LFG)

desensibilisiert – desensitized

Desensibilisierungsmittel – desensitizer

Desinfektion – disinfection

Desinfektionsmittel – disinfectants

Desorption – desorption

detaillierte Information – detailed information

Detergenzien – detergents

deterministische Strahlenwirkung – deterministic radiation effect

Detonation – detonation

Detonationsbereich – detonation range

Detonationsdruck – detonation pressure

Detonationsgeschwindigkeit – detonation velocity

Detonationsgrenzen – detonation limits

Detonationswelle – detonation wave

Deuterium (^{2}H, D) – deuterium, heavy hydrogen

Deuteriumoxid (D_2O) – deuterium oxide

Deutsches Rotes Kreuz (DRK) – German Red Cross

dezentrale Wasserversorgung – decentralized water supply

dicht (fest verschlossen) – tight, sealed tight

Dichte – density

Dichteanomalie – density anomaly

Dichtemessgerät – densimeter

Dichteverhältnis – ratio of densities

Dichtewert – density value

Dichtheitsprüfung – leakage test

Dichtkeil – sealing wedge

Dichtkissen – sealing pad, sealing cushion

Dichtmanschette – gasket

Dichtmasse – sealing compound, sealant

Dichtstopfen – sealing plug

Dienst, außer – off duty

Dienst, im – on duty

Dienst, operativer – operational duty

Dienstanweisung – standing order

Dienstgrad – rank, service grade, grade

Dienstgradabzeichen – rank badges, rank insignias, rank marking

Dienstkleidung – duty-dress

dienstlicher Befehl – official order

Dienstreise, auf einer sein – away on duty (A.O.D.)

Dienststelle – duty station

Dienststellung – position

Dienstvergehen (Feuerwehr) – offence against regulations of the fire service

Dienstvorschrift – regulation, duty regulation, duty instruction

Dienstzeit – term of service

Differentialmelder – rate-of-rise detector

Differentialsperre (Kfz) – slip differential

Differenzdruck – differential pressure

Diffusion – diffusion, scattering

diffusionsdicht – diffusion-resistant

Diffusionsflamme – diffusion flame

Diffusionsgeschwindigkeit – diffusion rate, diffusion velocity

diffusionshemmend – diffusion-impending

Diffusionskoeffizient – diffusion coefficient

Diffusionsvermögen – diffusibility, diffusivity

digitaler Meldeempfänger (DME) – digital paging device

Digitalfunk (Begriff) – digital radio

Digitalfunk (digitaler Bündelfunk) – terrestrial trunked radio (TETRA)

Digitalfunkgerät – digital radio transmitter

Digitalfunknetz – digital radio network

Diktiergerät – dictaphone

dimensionslos – non-dimensional

DIN (Deutsches Institut für Normung e. V.) – German Institute for Standardisation

DIN-Norm – German Standard

Dioxine – dioxins

direkte Kühlung – direct cooling

direkter Löschangriff – direct attack, direct fire attack

Direktmodus (Digitalfunk) – direct mode operation (DMO)

Dispergiermittel – dispersing agent

Dispersion – dispersion

Dispersionsgrad – degree of dispersion

Disponent (Leistellenmitarbeiter) – dispatcher

Disproportionierung – disproportionation, dismutation

Dissousgas – dissous gas

Dissoziation – dissociation

Dissoziation, thermische – thermal dissociation

Disziplinarstrafe – disciplinary action, non-judicial punishment

D-Löschpulver – extinguishing powder type D, metal fire extinguishing powder

Docht – wick

Dochtwirkung – wick effect

Donator – donor

Doppelklebeband – double-sided adhesive tape

Doppelknoten – double knot

Doppelschlauchanschluss – Siamese connection

doppelter Ankerstich – cow hitch, lark's head, lark's foot, girth hitch

doppelter Mastwurf – double clove hitch

doppelter Pfahlstich – bowline on a bight

doppelter Schotenstich – double sheet bend

Dosenöffner – can opener, tin opener

Dosierbereich – dosing range

Dosierventil – metering valve

Dosiervorrichtung – dosing device

Dosimeter – dosimeter

Dosimeterplakette – dosimeter badge

Dosimetrie – dosimetry

Dosis – dose

Dosis, effektive (ED) – effective Dose (ED)

Dosis, letale (LD) – lethal dose (LD)

Dosis, mittlere letale (LD_{50}) – median lethal dose (LD_{50})

Dosis, tödliche – lethal dose

dosisäquivalent – dose equivalent

Dosiskontrolle – dose control

Dosisleistung – dose rate

Dosisleistungsänderung – dose rate change

Dosisleistungskonstante – dose rate constant

Dosisleistungsmesser (DLM) – dose rate meter, dosimeter

Dosisleistungsmessgerät (DLM) – dose rate meter, dosimeter

Dosisleistungswarner (DLW) – dose rate warning device

Dosisleistungswarngerät (DLW) – dose rate warning device

Dosismessung – dosimetry

Dosisminderung – dose reduction

Dosisverringerung – dose reduction

Dosiswarngerät (DWG) – dose warning device

dotiert – doped

D-Pulver – extinguishing powder type D, metal fire extinguishing powder

Drahtbürste – wire brush

Drahtglas – wired glass

Drahtschere – cutting pliers, wire cutters

Drahtschlinge – wire loop

Drahtseil – cable

Drehimpuls – angular momentum, moment of momentum

Drehleiter (DL) – turntable ladder (TL), aerial ladder

Drehleiter mit Korb (DLK) – turntable ladder with rescue cage, aerial ladder platform (ALP)

Drehleiterkorb – aerial ladder platform

Drehmoment – torque, rotational force

Drehmomentschlüssel – torque handle, torque wrench

Drehzahl (Upm – Umdrehungen pro Minute) – number of revolutions (rpm – revolutions per minute)

Dreibeinstativ – tripod

Dreibock – tri jack

Dreieckstuch (Erste Hilfe) – triangular cloth, Esmarch mitella

Dreifach-Verteiler – three-way distributor

Dreikantschlüssel – triangle wrench

Driftweite – drift distance

dringlich – urgent

Dringlichkeit – urgency

Drohne – drone

Drohneneinsatz – deployment of drones

Drosselventil – throttle valve

Druck – pressure

Druck, dynamischer – dynamic pressure

Druck, hydrostatischer – hydrostatic pressure

Druck, kritischer – critical pressure

Druck, statischer – static pressure

Druck-/Unterdruck-Messgerät – pressure-vacuum gauge

Druckabfall – pressure drop

druckabhängig – pressure-dependent

Druckanstieg – pressure rise

Druckanstieg, maximaler – maximum pressure-rise

Druckanstieg, zeitlicher – rate of pressure, pressure rise at to time

Druckanstieg, zeitlicher, maximaler – maximum rate of pressure

Druckbegrenzungsventil – pressure limiting valve, pressure relief valve

Druckbehälter – pressure tank, pressure vessel

Druckbehälterzerknall – pressure vessel burst, pressure vessel rupture

Druckbelüftung – forced ventilation

Druckbelüftungsanlage – forced ventilation system

druckbeständig – pressure-resistant

Druckbilanz – pressure balance

druckdicht – pressure-tight

Druckentlastung – pressure relief

Druckentlastungsklappe – pressure relief flap

Druckentlastungsöffnung – pressure relief vent

Druckentlastungsventil – pressure relief valve

Druckentspannung (Druck ablassen) – pressure release

Druckerhöhung – pressure increase

Druckerhöhungspumpe – booster pump

druckfest – pressure resistant

Druckfestigkeit – pressure resistance

Druckgas – compressed gas

Druckgasbehälter – compressed gas tank

Druckgasflasche – compressed gas cylinder

Druckgasflasche mit Atemluft – air cylinder

Druckgasflaschenbatterie – compressed gas cylinder set

Druckgasflaschenventil – cylinder valve

Druckgaspatrone (im Feuerlöscher) – gas cartridge

Druckgefälle – pressure gradient

Druckhöhe – pressure head

Druckhöhe, manometrische – manometric pressure height

Druckknopfmelder – press button fire alarm box, push button alarm, manual alarm point

Druckkupplung – delivery hose coupling, pressure coupling

Druckleitung – delivery hose

drucklos – depressurized, pressure-less

Druckluft – compressed air

Druckluftbremsanlage – air brake system

Druckluftbremse – air brake, pneumatic brake

Druckluft-Fluchtgerät – emergency escape breathing device (EEBD)

Drucklufthammer – pneumatic hammer

Druckluftkompressor – air compressor

Druckluftleitung – compressed air line, compressed air pipe

Druckluftmeißel – pneumatic chisel

Druckluftschaum (DLS) – compressed-air foam (CAF)

Druckluftschaumanlage – compressed-air foam system (CAFS)

Druckluftschlauchgeräte – compressed-air breathing apparatus

Drucklufttauchgerät (DTG) – scuba set, self-contained underwater breathing apparatus (SCUBA)

Druckluftversorgung – compressed air supply

Druckmesseinrichtung – pressure gauge device

Druckmesser – manometer

Druckmessgerät – manometer, pressure gauge, pressure meter, pressure measurement unit

Druckmessung – pressure measurement

Druckminderer – pressure regulator, pressure reducer

Druckprüfung – pressure test

Druckreduzierventil – reduction valve, pressure reducing valve

Druckregelung – pressure control

Druckregelventil – pressure control valve

Druckregler – pressure regulator

Druckschlauch – delivery hose, fire hose, pressure hose

Druckschlauch, gummiert – rubber-lined deliver hose

Druckschlauch, ungummiert – unlined delivery hose

Druckschlauchlänge – pressure hose length

Druckschlauchlänge, zulässige – permissible pressure hose length

Druckschwankung – pressure variation, pressure fluctuation

Druckstoß – pressure surge, pressure shock

Druckstutzen (Pumpe) – pressure socket piece

Drucküberwachung – pressure monitoring

Druckverlust – pressure loss

Druckwasserreaktor (DWR) – pressurized water reactor (PWR)

Druckwelle – blast wave, pressure wave, shock wave

Dual-Use-Güter – dual-use goods

Dübel – dowel

Duft/Geruch – smell, odour, scent

Duft, angenehmer – fragrance, scent, pleasant smell

Duft, unangenehmer – unpleasant smell

Dung (Mist) – manure

Düngemittel – fertilizer

Dunghaken – manure hoe

Dunst – fume

Durchbruch – penetration

Durchbruch, vollständiger – full penetration

Durchdringung – penetration

Durchdringungszeit – penetration time

Durchfall (Erkrankung) – diarrhoea, enteritis

Durchfluss – flow, flow rate

Durchflussgeschwindigkeit – flow velocity

Durchflussmenge – flow rate

Durchflussmessung – flow measurement

durchführen – carry out

Durchführung – realization

durchlässig – permeable

Durchlässigkeit – permeability

Durchmesser – diameter

Durchmesser, kritischer – critical diameter

Durchnässung – soddenness

Durchschnitt (Werte) – mean, average, arithmetic mean

Durchzündung – blow up

Duroplaste – duroplasts, thermosetting plastics

Dusche – shower

Düse – nozzle

Düsenweite – nozzle diameter

Dynamik-Kernmantelseil – dynamic kernmantle rope

dynamischer Druck – dynamic pressure

E

Ebola-Fieber (EF) – Ebola virus disease (EVD)

Ebola-Virus – Ebola virus

eCall (Kfz) – eCall (automated emergency call system for motor vehicles)

Echtzeitbetrieb – real-time mode

Edelgas – noble gas

Edelmetall – noble metal

effektive Dosis (ED) – effective dose (ED)

effektive Verbrennungswärme – actual calorific value

Ehrenabteilung – honorary division

Ehrenamt – honorary post

ehrenamtlich – honorary, unsalaried

Ehrenbeil – hatchet of honour

ehrenhafter Abschied – honourable discharge

Ehrenmitglied – honorary member

Ehrenwache – guard of honour

Eichung – gauging, official calibration, standardization

eigene Wahrnehmungen – personal perceptions

Eigengefährdung – self-endangerment

Eigenkontrolle – internal checking

Eigenschaften – characteristics, properties

Eigenschaften, thermische – thermal properties

eigensicher – natural safe, fail-safe

Eigensicherheit – intrinsic safety (IS)

Eigenstrahlung – intrinsic radiation

Eigentum – proprietary, ownership

Eigentumssicherung – protection of property

Eignung – suitability, fitness

Eimer – bucket

Eimer aus Segeltuch – canvas bucket

Einatemluft – inhaled air

Einatemventil – inhalation valve

Einatemwiderstand – inhalation resistance

einatmen – breathe in, inhale

Einbau – installation, fitting

Einbaupumpe – mounted pump

Einberufung – call-up, induction

Einberufungsbescheid – call-up paper, induction paper

Einbindestutzen – binding socket piece

eindämmen – contain

Eindämmung – containment

eindeutig – unambiguous

Eindringtiefe (von z. B. Löschwasser) – depth of penetration

Einfangsquerschnitt – capture cross-section

einfetten – lubricate, oil, grease

Einfluss – influence

Einfüllstutzen – filling neck

Eingang – inlet

Eingangsdruck (Pumpe) – pressure at the input-side, inlet pressure, input pressure

Eingangsschlauch – inlet hose

eingeklemmte Person – jammed person

eingerastet – latched, locked-in

eingeschlossene Person – enclosed person

T. Schmiermund, *Fachwörterbuch Feuerwehr und Brandschutz*, https://doi.org/10.1007/978-3-662-64120-0_5

eingesetzte Einsatzkräfte – involved forces

eingesetzte Kräfte – involved forces

Eingreiftruppe – response group

Eingreifzeit – response time, intervention time

Einheit (Gruppe) – unit

Einheit (Messgröße) – entity

Einheit, betroffene – concerned unit

Einheit, taktische – tactical unit

Einheitensystem, internationales (SI) – international unit system (SI)

einheitlich – uniform

Einheitsführer – unit leader

Einheits-Temperatur-Kurve – standard time-temperature curve

Ein-Helfer-Methode (Erste Hilfe, HLW) – one-helper method

Einhorn – unicorn

Einkaufszentrum – shopping centre

Einkleidung – clothing procedure

Einmalgebrauch – single-use, disposable use

Einmalhandschuhe – disposable gloves, one-way gloves, single-use gloves

Einmalschutzanzug – single-use protective suit

Einnahme (z. B. Tabletten) – ingestion

einnehmen (z. B. Tabletten) – ingest

Einphasensystem – single phase system

einrasten – engage, to snap in

Einreißhaken – pull down hook, ceiling hook, fireman's hook, preventer

Einrichtungen des Brandschutzes – fire-protection appliance

Einrichtungen, sanitäre – sanitary facilities, sanitary installations

Einsatz – emergency operation, operation, action, mission, intervention

Einsatzablauf – operational procedure

Einsatzabschlussbesprechung – debriefing session

Einsatzabschnitt (EA) – operation sector

Einsatzabschnittsleiter (EALtr) – sector-commander

Einsatzabteilung – active division

Einsatzanzug – call-out kit, turn-out coat

Einsatzaufgaben – operational tasks

Einsatzauftrag – mission order

Einsatzauftragsziele – objectives of the mission

Einsatzausbildung – training for action

Einsatzbedingungen – operational conditions

Einsatzbefehl – operational order, action order

Einsatzbeginn – begin of the mission

Einsatzbereitschaft – availability, readiness for action

Einsatzbericht – mission report, call report

Einsatzbreite – range of use

Einsatzdauer – duration of the mission, duration of operation

Einsätze, tägliche – day-to-day operations

Einsatzerfolg – operational success

Einsatzfähigkeit – intervention preparedness of forces, operational ability

Einsatzfahrt – action run

Einsatzfahrzeug – emergency vehicle, fire appliance, fire engine

Einsatzfahrzeuge frei! (Schild) – emergency vehicles only! (sign)

Einsatzform – form of employment, form of operations

Einsatzfotograf – fire photographer

Einsatzgrundsätze – operating principles

Einsatzgruppe – task force

Einsatzkleidung – fire suit, fire-fighting tunic

Einsatzkräfte – operational forces, operational units

Einsatzkräfte, ablösende – relief-forces

Einsatzkräfte, eingesetzte – involved forces

Einsatzkräfte, unterstellte – units under command, subordinated forces

Einsatzkurzprüfung (Atemschutz) – short operational check, quick check

Einsatzleiter (ELtr) – person in command, incident commander, commanding officer, officer in charge, head of operations

Einsatzleiter Feuerwehr – person in command, incident commander fire service

Einsatzleiter Rettungsdienst – chief ambulance officer

Einsatzleiterhandbuch – commander's manuals

Einsatzleitung (Begriff) – incident command organisation

Einsatzleitung (EL) (Funktion) – incident command, operational command, mission control

Einsatzleitung (Team) – operation staff

Einsatzleitungssystem – command system

Einsatzleitwagen (ELW) – command support unit, command vehicle

Einsatzmaßnahmen – operational measures

Einsatzmittel – operational equipment, operational means

Einsatzpersonal – operational personnel

Einsatzplan – incident planning, plan of action

Einsatzregeln – rules of engagement (RoE)

Einsatzschwellenwerte – operational thresholds, operational threshold values

Einsatzschwerpunkt – focal point of the mission, focus of mission

Einsatzsicherung – protection of the operation

Einsatzsituationen – operational situations, mission situations

Einsatzsituationen, besondere – special operational situations

Einsatzstab – command staff

Einsatzstelle (ESt) – place of operations, scene of operations, area of disaster, fireground

Einsatzstellenfunk – radio communication of scene of operation

Einsatzstufe – level of intervention

Einsatzszenarien – operational scenarios

Einsatztagebuch – mission diary

Einsatztaktik – operational tactics

Einsatztauchtrupp – emergency diving squad

Einsatztoleranzwert (ETW) – operational tolerable value

Einsatztrupp – operational crew

Einsatzübung – action drill

Einsatzvorbereitung (im Vorfeld) – mission preparation

Einsatzvorbereitungen (aktuelle) – preparation of action

Einsatzwert – operational value

Einsatzzeit – duration of action, time of action

Einsatzzentrale – incident room, control centre

Einsatzziele – objectives in missions

Einschnappen – snapping

Einschub – slide-in module, shelf

Ein-Seil-Technik – single rope technique (SRT)

einspannen – clamp, fix

Einspeisepunkt – feeding point

Einspeisung – supply, feeding

Einspeisung für die Feuerwehr – fire department supply connection

einstellen – justify

Einstufung – classification

Einsturz – collapse

Einsturzgefahr – collapse danger

Eintreffdauer – attendance time

Eintreffmeldung – arrival note

Eintrittswahrscheinlichkeit – probability of occurrence

Einwegspritze – disposable syringe

Einweiser – signaller

Einwirkzeit – exposure time

Einzelbefehl – individual order

Eisensäge – metal saw

Eiserner Vorhang (Bühne, Theater) – fire curtain, protective curtain

Eisessig – glacial acetic acid

Eislast – ice load

Eisschlitten – ice rescue sled, ice sledge

Eisunfall – ice accident

EKG-Elektroden – paddles

Elastizität – elasticity, springiness

Elastomere – elastomers

Elektriker – electrician

elektrische Betriebsmittel – electric equipment

elektrische Entladung – electrical discharge

elektrische Feldstärke – electric field strength

elektrische Funken – electric sparks

elektrische Ladung – electrical charge

elektrische Leitfähigkeit – conductivity, electric conductivity

elektrischer Strom – electric current

elektrischer Widerstand – electric resistance

Elektrizität – electricity

Elektrobrand – electrical fire

Elektroenzephalogramm (EEG) – electroencephalogram (EEG)

Elektrofachkraft (EFK) – electrically qualified person

Elektrogeräte – electrical appliances, electrical devices

Elektrohandschuhe – dielectric gloves, protective insulating gloves

Elektrokardiogramm (EKG) – electrocardiogram (ECG)

elektromagnetische Strahlung – electromagnetic radiation

elektromagnetische Umweltverträglichkeit (EMUV) – electromagnetic environmental compatibility (EMEC)

elektromagnetische Verträglichkeit (EMV) – electromagnetic compliance (EMC)

elektromagnetisches Spektrum – electromagnetic spectrum

Elektron – electron

Elektronegativität – electronegativity

Elektronenfluss – electron flux

Elektronengas – electron gas

Elektronenpaar – electron pair

Elektronenschale – electron shell

Elektronenstrahl (ES) – electronic beam (EB), electron stream

Elektronenstrahlung – electron radiation, electron rays

elektronisches Personendosimeter (EPD) – electronic personal dosimeter (EPD)

Elektroschock – electrical shock

Elektroschutzhandschuhe – protective insulating gloves, dielectric gloves

Elektroschweißen – electric welding

elektrostatische Anziehung – electrostatic attraction

elektrostatische Aufladung – electrostatic charge

elektrostatische Entladung (ESD, ESE) – electrostatic discharge (ESD)

elektrostatische Ladung – electrostatic charge

Elektrounfall – electric accident, accident caused by electric current

Elektroverteiler – distributor of electricity

Elektrowerkzeug – electric tools

Elektrowerkzeugkasten – electric tools box

Element (chem.) – element

Elementarschaden (Versicherungen) – natural hazard

Elementarteilchen – fundamental particle, elemental particle

Embolie – embolism

Emission – emission

Empfänger (Gerät) – receiver

Empfänger (Person) – recipient

Empfindlichkeit – sensitivity

Emulgator – emulsifier, emulsifying agent

Emulsion – emulsion

Enddruck – discharge pressure

Ende-zu-Ende-Verschlüsselung – end-to-end encryption (E2EE)

Endlager – permanent disposal site, final disposal site

Endlagerung – final deposition

endotherm – endothermic

endotherme Reaktion – endothermic reaction

Endotrachealtubus – endotracheal tube

Endtemperatur – final temperature

Endverbleibserklärung (EVE) – end-user certificate (EUC)

Energie – energy, power

Energie, kinetische – kinetic energy

Energie, mechanische – mechanical energy

Energie, potentielle – potential energy

Energiedosis – absorbed dose

Energiedosisleistung – absorbed dose rate

Energieerhaltungssatz – energy theorem, law of conservation of energy

Energiefreisetzung – energy release

Energiefreisetzungsrate – energy release rate

Energieniveau – energy level

Energiequelle – power source, energy source

Energieversorgung – power supply

Energieversorgungsunternehmen (EVU) – power supply company (PSC)

engmaschig – close-meshed

Entdeckungszeit – detection time

entfetten – degrease

entflammbar – flammable, inflammable

entflammbar, normal – normally inflammable

entflammbares Gas – flammable gas

Entflammbarkeit – flammability, inflammability

Entflammbarkeitsgrenzen – inflammability limits, flammability limits

Entflammung – inflammation

Entflammungsdauer – duration of inflammation

Entflammungstemperatur – inflammation temperature

entgiften – detoxify

Entgiftung – detoxification, decontamination

entgraten – deburring

Enthalpie – enthalpy

Entkeimung – sterilization

Entladung – discharge

Entladung von statischer Elektrizität – electrostatic discharge

Entladung, atmosphärische – atmospheric discharge

Entladung, elektrische – electrical discharge

Entladung, elektrostatische (ESD; ESE) – electrostatic discharge (ESD)

Entladungsfunke – discharge spark, electrical discharge spark

Entladungsvorgang (Elektrizität) – discharge event

Entlassungsurkunde – certificate of dismissal

Entlüftung (Vorgang) – deaeration

Entlüftungseinrichtung – venting device, venting system

Entlüftungspumpe – venting pump, primal pump

Entrauchung – smoke venting, smoke clearance

Entrauchungsanlage – smoke exhaust ventilation system

Entrauchungsklappe (ERK) – smoke exhaust damper, smoke control damper

Entscheidung – decision

Entscheidungskraft – decisiveness

Entschluss – decision, resolve

Entschlussfassung – take a resolution

Entseuchung – disinfection

Entsorgung – disposal

Entstehung – origin, emergence, creation

Entstehungsbrand – initial fire, incipient fire

Entstehungsphase – initial phase

Entwarnung – all-clear

Entwarnungssignal – all-clear signal

entwässern – drain

Entwässerung – water removal, desiccation

Entwässerungsvorrichtung – installation for water removal

Entwicklung – development

Entwicklungsphase – development stage, development phase

Entwicklungszeit – development time

Entziehung der Fahrerlaubnis – driver's license revocation

entzündbar – ignitable, flammable

entzündbarer Staub – flammable dust

Entzündbarkeit – ignitability, flammability

entzünden – ignite, light

entzündet – ignited

entzündlich – inflammable, ignitable

entzündlich, hoch- – extremely flammable

entzündlich, leicht- – highly flammable

entzündlich, schwer- – hardly flammable

Entzündlichkeit – ignitability, ease of ignition

Entzündung (medizinisch) – inflammation

Entzündung (zum Brennen bringen) – ignition

Entzündung von Rauchgasen – flue gas ignition

Entzündungstemperatur – ignition temperature

Entzündungswahrscheinlichkeit – ignition probability

erbgutverändernd – mutagen

Erbrechen – vomiting, vomition

erbrechen – vomit

Erbrochenes – vomitus

Erdanker – ground anchor

Erdanziehungskraft – force of gravity, gravitational force

Erddruck – earth pressure

Erdgas – natural gas

Erdgas, verflüssigt – liquefied natural gas (LNG)

Erdgeschoss – ground floor

Erdöl – crude petroleum

Erdung (Elektrizität) – grounding, earthing

Erdungskabel – earth wire, ground wire

Erdungsstange – ground rod, earth rod

Erdungszubehör – grounding accessories, earthing accessories

Ereignis – incident

Ereignisfall – incident

Ereignisort – place of incident

Erfahrung – experience

erfahrungsgemäß – as matter of experience

Erfolg – success

Erfrierung – frostbite

Erfrierung dritten Grades – third-degree frostbite

Erfrierung ersten Grades – first-degree frostbite

Erfrierung zweiten Grades – second-degree frostbite

ergänzende Maßnahmen – supplementary actions

Ergebnis der Maßnahmen – outcome of measures

Erhaltungssatz – conservation law

Erhebung – survey

erhöhte Radioaktivität – increased radioactivity

Erholung – recovery

Erholungszeit – recovery time

Erkennen der Lage (Erkundung) – ascertain the situation

Erker – oriel

erkranken – sicken, falling ill

erkrankte Person – ill person, sick person

Erkrankung – disease (Dz.), sickness, affliction

Erkrankung, akute – acute disease, acute illness

Erkrankung, akute respiratorische (ARE) – acute respiratory disease (ARD)

Erkrankung, bakterielle – bacterial disease, bacteriosis

Erkrankung, chronische – chronic disease, chronic ailment

Erkrankung, ernste – serious illness

Erkrankung, obstruktive – obstructive disease, obstructive disorder

Erkundung – reconnaissance, ascertaining the situation

Erkundungsflug – reconnaissance flight

Erkundungsflugzeug – reconnaissance plane

Erkundungshubschrauber – reconnaissance helicopter

Erkundungstrupp – reconnaissance troop

Erkundungszeit – exploring time, investigation time

ermächtigt – empowered

Ermessensspielraum – latitude

Ermittlungsbeamter (bei Brandstiftung) – Fire Marshall

Ermittlungsverfahren (EV) – preliminary investigation (PI)

Ermüdung (Person) – tiredness, fatigue

Ermüdungsbruch – fatigue failure

ernste Erkrankung – serious illness

Ernte – harvest

Ersatz – substitute, replacement

Ersatzmaßnahmen – compensating measures

Ersatzstromversorgung – replacement power supply

Ersatzstromversorgungsanlage – replacement power supply system

Ersatzvornahme – executive fiat

Erscheinungsbild (z. B. Gefahrstoff) – appearance

Erschlaffung – relaxation

erschöpfliche Löschwasserversorgung – exhaustible water supply for fire-fighting

Erstangriff – initial attack

Erste Hilfe (EH) – first aid

Erste Hilfe leisten – administer first aid, give first aid

Erste-Hilfe-Kasten – first-aid case, first aid box

Erste-Hilfe-Set – first aid kit (FAK)

Erste-Hilfe-Stelle – first-aid station

Erste-Hilfe-Training (EHT, EH-Training) – first aid training (FAT)

erster Abmarsch – first alarm

erster Rettungsweg – first escape route, first fire escape

Ersthelfer (Laie) – first-aider (FA)

Ersthelfer (professionell) – first responder, emergency responder

ersticken – suffocate, smother

Ersticken (Löscheffekt) – smothering effect, suffocating effect

ersticken, Flammen – smother the flames

erstickend wirkende Substanz – asphyxiant

Erstickung – suffocation

Erstickung (bei Lebewesen) – asphyxiation

Erstickungsgefahr – risk of asphyxiation

Erstmaßnahmen – initial actions

Erstversorgung – on-scene care, primary survey

Erträglichkeitsgrenze – tolerability limit

Erträglichkeitsschwelle – tolerability threshold

Erwägungen – considerations

erwärmte Stoffe – elevated temperature substances

Erwärmung – heating, warming

Erweichungspunkt – softening point

erweiterte Dekontamination – enhanced decontamination

Erzeugnisse, pyrotechnische – pyrotechnics

Eschenholz – ash wood

Esmarch-Handgriff – jaw thrust manoeuvre

Essigsäure – acetic acid

Etage – storey, floor

Ethin (Acetylen) – ethine (acetylene)

Europäische Atomgemeinschaft (EURATOM) – European Atomic Energy Community

Europäische Norm (EN) – European Standard, EU standard

Evakuierte (Personen) – the evacuated

evakuierte Personen – evacuated persons

Evakuierung – evacuation

Evakuierung, kontrollierte – phased evacuation, controlled evacuation

Evakuierungsbefehl – evacuation order

Evakuierungsstrategie – evacuation strategy

Evakuierungstunnel – evacuation tunnel

Evakuierungsübung – evacuation practice, evacuation test

Evakuierungszeit – time of evacuation

exotherm – exothermic

exotherme Reaktion – exothermic reaction

exotherme Zersetzung – exothermic decomposition

Expansionsstück – expansion adapter

Experteninformation – expert's information

Expertenwissen – expertise, expert knowledge

explodieren – explode

explosibel – explosible

explosibles Gemisch – explosible mixture

Explosion – explosion

Explosion, physikalische – physical explosion

Explosion, primäre – primary explosion

explosionsartige Verbrennung – explosive combustion

Explosionsausbreitung – explosion spread

Explosionsbereich – explosion range

Explosionsdruck – explosion pressure

Explosionsdruck, maximaler – maximum explosion pressure

explosionsfähig – explosible

explosionsfähige Atmosphäre – explosible atmosphere

explosionsfähige Stoffe – → Explosivstoffe

explosionsfähiges Gemisch – explosible mixture

Explosionsflamme – flame of explosion

Explosionsfortpflanzung – explosion propagation

Explosionsgefahr – risk of explosion

explosionsgefährdeter Bereich – explosion-hazardous area

explosionsgefährdeter Raum – explosion hazardous room

Explosionsgefährdung – explosion hazard

explosionsgefährliche Konzentration – explosion hazard concentration

explosionsgeschützt – explosion protected

explosionsgeschützte elektrische Betriebsmittel – explosion-proof electric equipment

Explosionsgeschwindigkeit – rate of explosion

Explosionsgrenze, obere (OEG) – upper explosion limit (UEL)

Explosionsgrenze, untere (UEG) – lower explosion limit (LEL)

Explosionsgrenzen – explosion limits

Explosionsgrenzenwarngerät – explosion-limits warning device

Explosionsintensitätsgrößen – → Explosionskennzahlen

Explosionskennzahlen – explosion indexes, explosion hazard indexes

Explosionspunkt, oberer – upper explosion point

Explosionspunkt, unterer – lower explosion point

Explosionspunkte – explosion points

Explosionsschutz – explosion protection

Explosionsschutz, bautechnischer – structural explosion protection

Explosionsschutz, vorbeugender – preventive explosion protection

Explosionsstärke – explosion strength

Explosionsunterdrückung – explosion suppression

Explosionsunterdrückungsanlage – explosion suppression system

Explosionsursache – cause of explosion

Explosionsvolumen – explosion volume

Explosionswärme – explosion heat

Explosionswelle – explosion wave

Explosionswirkung – explosion action, explosive effect

Explosionszone – explosion zone

explosiv – explosive

Explosivstoffe – explosives

Expositionsdosis, letale – lethal exposure dose

Expositionsfaktor (EF) – exposure factor (EF)

Expositionszeit – time of exposure

Expositionszeit begrenzen – limit time of exposure

EX-Schutz – EX-protection

extremes Brandverhalten – extreme fire behaviour

Exzenter – flywheel

Exzenterschneckenpumpe – eccentric screw pump

F

Fachaufsatz – scientific paper

Fachaufsicht – functional supervision

Fachbehörde – specialized authority

Fachdienst – special service

Fachfirma, anerkannte – recognized contractor

Fachkommission – committee of experts

Fachkunde – technical knowledge

fachkundige Person – qualified person, expert

Fachliteratur – specialized literature, technical literature

Fachmann – professional, practitioner

Fachwerk – framework, truss, timbering

Fackel – torch

Fahndung – manhunt, tracing

fahrbare Schlauchhaspel – mobile hose reel

fahrbarer Feuerlöscher – trolley fire extinguisher

fahrbares Krankenbett – gurney

Fahrbefehl – run command

Fahrer eines Krankenwagens – ambulance driver

Fahrerairbag – driver airbag

Fahrerlaubnis – driving licence

Fahrgeschwindigkeitsanzeiger – speedometer

Fahrgestell (allgemein) – running gear

Fahrgestell (Kfz) – chassis

fahrlässig – negligent, careless

fahrlässige Brandstiftung – negligent arson, careless arson

Fahrlässigkeit – negligence

Fahrtrage – wheeled stretcher

Fährunglück – ferry crash

Fahrverbot – driving ban

Fahrzeug – vehicle

Fahrzeugbesatzung – vehicle crew

Fahrzeugbrand – vehicle fire

Fahrzeugbreite – vehicle width

Fahrzeugdach – vehicle roof

Fahrzeugdach, begehbares – roof usable as platform

Fahrzeugführer (Fahrer) – driver, vehicle driver, car driver

Fahrzeugführer (Führungskraft) – unit leader

Fahrzeuggewicht – vehicle weight

Fahrzeughalle – appliance room

Fahrzeughöhe – vehicle height

Fahrzeugkollision – traffic collision

Fahrzeuglänge – vehicle length

Fahrzeugmasse – vehicle mass

Fahrzeugzubehör – vehicle equipment

Faktor, begrenzender – limiting factor

Fall Gelb – case yellow

Fall Rot – case red

Falldämpfer – shock absorber, fall-impact absorber, energy absorber

Fallhaken – ladder paws

Fällheber – peavy

Fallhöhe – height of fall

Fällkeil – felling wedge

Fällkerbe – undercut, birds' mouth

Fallschacht – drop shaft

Fällschnitt – back cut

T. Schmiermund, *Fachwörterbuch Feuerwehr und Brandschutz*, https://doi.org/10.1007/978-3-662-64120-0_6

Falschalarm – false alarm

falsche Planung – faulty planning

falscher Notruf – bogus emergency call

falsch-negativ – false-negative

falsch-positiv – false-positive

Faltbehälter – collapsible container

Falttrage – flexible stretcher

Fanatiker – dead-ender

Fangleine – rescue rope, fireman's rope

Fangleinenbeutel – rescue-rope bag

Fangöse – catch eye

Fangseil – safety cable

Fangstoß (Höhenrettung) – impact force

Fangvorrichtung – retainer

Faraday'sches Gesetz – Faraday's law

Fassade – facade

Fasspumpe – barrel pump, drum pump

Fassschlüssel – drum wrench, bung wrench

Fassstek (Knoten) – barrel hitch

Fäustel – club hammer, lump hammer

Faustformel – rough guide

Faustregel – rule of thumb

Feder (Drahtspirale) – spring

Feder (Nut-Feder-Verbindung) – tongue

Federkörner – spring center punch

Fehlalarm – false alarm

Fehleinsatz – faulty mission

Fehlerstrom – fault current

Fehlerstromschutzschalter (FI) – residual current operated circuit breaker (RCCB), residual current protective device (RCD), ground fault interrupter (GFI)

Fehlfunktion – malfunction

Fehlverhalten – misconduct, misbehaviour

Fehlverhalten, menschliches – human error

Feinstaub – fine dust, fine particulate matter

Feinstaubmaske – filtering facepiece (FFP), mist mask, mist respirator mask

Feldküche – field kitchen, mobile kitchen

Feldstärke – field strength

Feldstärke, elektrische – electric field strength

Feldstärke, magnetische – magnetic field strength

Feldweg – unfenced path, unfenced track

Fenster – window

Fensterbrüstung – breast wall

Fenstermitte – window center

Fenstersims – window sill, window ledge

Fenstersturz (Baukunde) – window lintel

Fernbedienung (Gerät) – remote control

Fernbedienung (Tätigkeit) – remote operation

Fernglas – binoculars, field glass

Fernmeldedienst – signal service, communications service

Fernmeldenetz – communication network

Fernmelder – signaller

Fernmeldesystem – communication system

Fernmeldeverbindung – signal communication

Fernschreiber – teletype

Fernübertragung – remote transmission

Fernverkehrsstraße – trunk road

Ferse – heel

Fesselknoten – handcuff knot

fest angeschlossen – permanently connected

fest eingebaut – permanently installed

feste entzündbare Stoffe – flammable solids

feste explosive Stoffe – solid explosives

feste Löschmittel – solid fire-extinguishing agents

fester Zustand – solid state

Festigkeit – strength, solidity, steadiness

Festigkeit, mechanische – mechanical strength

Festkupplung – fixed coupling

Festlegungen – particularities

Festnahme – apprehension

Festpunkt – anchorage point

Feststellanlage – locking mechanism

Feststellbremse – locking brake

Feststoffe – solids, solid substances

Festverbindung (Standleitung) – dedicated line

Festziehen – tightening

Fettbrand – grease fire, fat fire, burning fat

Fettbrandlöscher – fat fire extinguisher

Fettexplosion – FCI-steam explosion between oil and water

Fettpresse – grease gun

feuchtigkeitsempfindlich – sensitive to moisture, moisture-sensitive

feuchtigkeitsunempfindlich – insensitive to moisture

Feuer – fire

Feuer fangen – conflagrate, catch alight

Feuer- und Rettungswache (FRW) – fire and rescue station

Feueralarm – fire alarm, fire alert

Feueralarm (Übung) – fire drill

Feuerausbreitung – fire spreading

Feuerbeschau – fire inspection, fire audit

feuerbeständig – fire-resistant

Feuerbeständigkeit – fire resistance

Feuerbrücke – fire bridge

Feuerdreieck – fire triangle

Feuererscheinung – fire apparition

feuerfest – fire-proof

Feuerfünfeck – fire pentacle

Feuergefahr – fire hazard

feuergefährlich – flammable

feuerhemmend – fire-retardant

feuerhemmende Tür – fire-retardant door

Feuerhose – firenado, fire tornado

Feuerleiter – fire escape

Feuerlöschanlage – fire-extinguishing system

Feuerlöschanlage, automatische – automatic fire-fighting installation

Feuerlöscharmatur – hose fitting

Feuerlöschboot – fire boat

Feuerlöschdecke – fire blanket

Feuerlöscheimer – fire bucket

Feuerlöscheinrichtung – fire extinguishing unit

Feuerlöscher – fire extinguisher, fire drencher

Feuerlöscher, fahrbarer – trolley fire extinguisher

Feuerlöscher, tragbare – portable fire extinguishers

Feuerlöscherkasten – fire-extinguisher cabinet

Feuerlöschgerät – fire-fighting appliance

Feuerlöschgerät (Ausrüstung) – fire-extinguishing equipment, fire-fighting equipment

Feuerlöschgeräte, tragbare – portable fire-extinguishers

Feuerlöschhubschrauber – fire-fighting helicopter

Feuerlöschkreiselpumpe (FP) – fire pump, fire-extinguishing pump, fire-fighting pump, centrifugal fire pump

Feuerlöschkreiselpumpe, Hochdruck-Ausführung – fire pump for high pressure (FPH)

Feuerlöschkreiselpumpe, Normaldruck-Ausführung – fire pump for normal pressure (FPN)

Feuerlöschkreiselpumpe, tragbar (TS) – portable fire pump normal pressure (PFPN)

Feuerlöschmittel – fire-extinguishing agent, fire-extinguishing medium, fire-extinguishing substance

Feuerlöschpulver – fire-extinguishing powder

Feuerlöschschaum – fire-extinguishing foam

Feuerlöschstrategie – fire extinguishing strategy

Feuerlöschsysteme – fire-extinguishing systems

Feuerlöschteich – fire pond, fire water pond

Feuerlöschübung – fire drill

Feuermelder – firebox, fire alarm

Feuermelder, automatischer – automatic fire alarm (AFA)

Feuerpatsche – fire swatter, fire beater

Feuerschneise – fire road

Feuerschutz – → Brandschutz

Feuerschutzabschluss – fire seal, fire barrier

Feuerschutzanzug – fireproof suit

Feuerschutzdamm – fire dam

Feuerschutzklappe – fire damper

Feuerschutzkleidung (FSK) – fire-fighting clothes, flameproof clothes

Feuerschutzmauer – fire wall

Feuerschutzmittel – fire retardant, fireproofing agent

feuersicher – fire-proof

Feuersprung – fire flash

Feuersturm – fire storm

Feuertornado – firenado (short for fire tornado)

Feuertreppe – fire stairs, emergency staircase

Feuerüberschlag – fire spark over, fire spreading

Feuerübersprung – flash-over

Feuerungsanlage – furnace

Feuerversicherung – fire insurance

Feuerversicherungspolice – fire insurance police

Feuerviereck – fire quad, fire quadrangle

Feuerwache (FW) – firehouse, fire station, station house

Feuerwehr – fire brigade {GB}, fire department {US}

Feuerwehr – firemanship

Feuerwehr (Oberbegriff) – fire service, fire department

Feuerwehr- und Rettungs-Trainingscenter (FRTC) – fire and rescue training center (FRTC)

Feuerwehr, nichtöffentliche – non-public fire brigade

Feuerwehr, öffentliche – public fire brigade

Feuerwehralarmübung – fire alarm drill

Feuerwehrangehörige (FwA) – fire fighters

Feuerwehranhänger (FwA) – trailer for fire-fighting equipment

Feuerwehranzeigetableau (FAT) – fire alarm display

Feuerwehraufzug – fire-fighting lift

Feuerwehraxt – fire-fighting axe, fireman's axe

Feuerwehrbedienfeld (FBF) – fire operator panel

Feuerwehrbeil – fireman's hatchet

Feuerwehrboot – fire boat

Feuerwehrchef – fire chief, chief of the fire department

Feuerwehrdienst – fire brigade service, fire duty, fire brigade duty

Feuerwehr-Dienstanzug – fire brigade dress uniform

Feuerwehrdienstvorschrift (FwDV) – fire service regulation

Feuerwehreinsatz – fire run, fire brigade operation

Feuerwehrfahrzeug – fire appliance, fire engine, fire tender, (fire truck)

Feuerwehrfrau – firewoman

Feuerwehrfrau, freiwillige – volunteer fire-fighter, volunteer firewoman

Feuerwehrführungskraft – fire officer

Feuerwehrfunk – fire brigade radio

Feuerwehrgerätehaus (FGH) – fire equipment building, fire equipment house

Feuerwehrgeräteschuppen – fire equipment storage shed

Feuerwehrgurt – fireman's belt, fireman's safety belt

Feuerwehrhalle – fire station

Feuerwehr-Haltegurt – fire brigade restraint belt

Feuerwehrhandschuhe – fire-fighter gloves

Feuerwehrhaus (FwH) – fire station, station house

Feuerwehrhelm – firefighter's helmet, fireman's helmet

Feuerwehrkommandant – brigade commander

Feuerwehrleine – fire-fighter's rope, rescue rope, long line

Feuerwehrleitern – fire ladders

Feuerwehrleitern, tragbare – portable fire ladders

Feuerwehrleute – fire fighters, firemen

Feuerwehrmann – fireman

Feuerwehrmann, aktiver Dienst – serving fire-fighter

Feuerwehrmann, -frau – fire fighter

Feuerwehrmann, freiwilliger – volunteer fire-fighter, volunteer fireman

Feuerwehrmänner – firemen, fire-fighters

Feuerwehrpläne – pre-fire planes

Feuerwehrpumpe – fire pump

Feuerwehrsanitäter – fire brigade paramedic, fire brigade medical technician

Feuerwehrschlauch – fire hose

Feuerwehrschläuche – fire hoses

Feuerwehrschlüsseldepot (FSD) – fire-service key safe

Feuerwehrschlüsselkasten (FSK) – fire-service key safe

Feuerwehrschule – fire-service school, fire college

Feuerwehrschuppen – fire brigade shed

Feuerwehr-Schutzhandschuhe – fire-fighter gloves

Feuerwehrsicherheitsgurt – fireman's safety belt, fireman's waist belt, fireman's life belt, fireman's belt with safety rope

Feuerwehr-Sicherheitsstiefel – fireman's boots, fire-fighter boots

Feuerwehrstiefel – fireman's boots, fire-fighter boots

Feuerwehrtaktik – fire brigade tactics

Feuerwehrtaucher – fire brigade diver, fire diver

Feuerwehrtechnik – fire-fighting technic

Feuerwehrtrainingscenter (FTC) – fire training center (FTC)

Feuerwehrübung – fire drill, fire practice, fire-fighting exercise

Feuerwehruniform – firefighter uniform, fire brigade uniform

Feuerwehrverband – fire service association, fire brigade federation

Feuerwehrverein – firemen club, firemen association, firemen society

Feuerwehrwesen – firemanship, fire service

Feuerwehrzufahrt – fire rescue path, fire brigade access road

Feuerwehrzugang – fire brigade access point

Feuerwerk – fireworks

Feuerwerkskörper – fire crackers, pyrotechnics, pyrotechnic articles

Feuerwiderstand – fire resistance, fire integrity

Feuerwiderstandsdauer (eines Bauteils) – fire-resistance rating, fire-resistance duration

Feuerwiderstandsfähigkeit (FWF) – fire resistance

Feuerwiderstandsklasse – standard of fire resistance, fire-resistance class, fire-resistant rating

Fieber – fever, temperature, pyrexia

Fieber unbekannter Ursache (FUU) – fever of undetermined origin (FUO)

Fieber, hämorrhagisches – haemorrhagic fever, hemorrhagic fever

Fieber, virales hämorrhagisches (VHF) – viral haemorrhagic fever

filmbildender Zusatz – film forming additive

Filmdosimeter – badge meter, film dosimeter, film badge meter

Filmplakette – film badge

Filtergeräte – filter appliances

Filterleistung – filter capacity

Filterschicht – filter layer

Filzschreiber – felt-tipped pen

finanzielle Verantwortlichkeiten – financial responsibilities

First – ridge

FI-Schutzschalter – residual current operated circuit breaker (RCCB), residual current protective device (RCD), ground fault interrupter (GFI)

Fixpunkt – fixing point

Flachdach – flat roof

Flachdichtung – flat gasket, flat seal

Fläche – area

Fläche (eines Raumes) – floor area, floor space

Fläche, beschädigte – damaged area

Flächenabbrandrate – area burning rate

flächenbezogene Abbrandgeschwindigkeit – area burning rate

flächenbezogene Abbrandrate – area burning rate

flächenbezogene Brandlast – fire load density

Flächenbrand – conflagration

Flachzange – flat-nosed plier

Flamme – flame

Flamme, laminare – laminar flame

Flamme, rauschende – rustling flame

Flamme, turbulente – turbulent flame

Flamme, vorgemischte – pre-mixed flame

Flammen auf der Oberfläche – surface flash

Flammen ersticken – smother the flames

Flammen, offene – open flames

Flammenausbreitung – flame spread, propagation of flames

Flammenausbreitungsgeschwindigkeit – rate of flame spread

Flammenausbreitungszeit – flame-spread time

Flammenbildung, anhaltende – sustained flame

Flammenbrand – flaming fire

flammendichter Abschluss – fire resistant closure, fire resistant damper

Flammendurchschlag – flame penetration

Flammendurchschlagsicherung – flame arrester

Flammenfärbung – flame colour

flammenfest – flame resistant

Flammenfortpflanzung – flame propagation

Flammenfortpflanzungsgeschwindigkeit – flame propagation rate

Flammenfront – flame front

Flammengeschwindigkeit – flame velocity

Flammenhöhe – flame height

Flammenionisationsdetektor (FID) – flame ionisation detector (FID)

flammenlose Verbrennung – flameless combustion

Flammenmelder – flame detector

Flammenrückschlag – flare back, flash back

Flammenrückschlagsicherung – fire arrester

Flammensperre – flame barrier, flame arrestor

Flammentemperatur – flame temperature

Flammenzonen – flame zones

flammfest – flame resistant

flammhemmend – flame-retardant

flammhemmendes Mittel – flame retardant product

Flammpunkt – flash point

Flammschutzausrüstung – flame retardant treatment

Flammschutzbehandlung – fire protective treatment

Flammschutzbekleidung – flame-retardant clothing

Flammschutzhaube – fire-fighter hood

Flammschutzimprägnierung – fire-resistant finish, flame-retardant finish

Flammschutzmittel – flame retardant agents

Flanke – wing, flank

Flankenschutz – protection of the flank

Flansch – flange

Flanschverbindung – flange connection

Flaschenventil (Druckgasflasche) – cylinder valve

Flaschenzug – pulley block

fliegende Bauten – fairground rides

Fliehkraft – centrifugal force

Fliehkraftzerfall – centrifugal force decay

Fliehkraftzerknall – centrifugal force bursting

Fließdruck – flow pressure

Fließrichtung – direction of flow

Flucht – escape

Fluchtfiltergerät – emergency filtering apparatus

Fluchthaube – escape smoke hood, evacuation hood, fire escape hood

flüchtig (Gase, Dämpfe) – volatile

flüchtige Stoffe – volatile substances

Flüchtigkeit – volatility, fugacity

Fluchtmaske – filtering escape device, one-time use filtering escape device, emergency escape mask

Fluchtmöglichkeit – possibility for escape

Fluchtreaktion – avoidance reaction

Fluchttunnel – escape tunnel

Fluchtweg – means of escape, evacuation route, exit path, exit route, emergency escape route

Fluchtweg (Beschilderung) – escape symbol

Fluchtwege – means of escape

Fluchtweglänge – length of escape route

Fluchtzeit – movement time

Flugasche – fly ash

Flugfeldlöschfahrzeug (FLF) – airfield fire engine, airfield fire truck, airport fire-fighting vehicle

Flugfeuer – flying embers

Flughafen – airport

Flughafenfeuerwehr – airport rescue and fire-fighting service

Flughafenlöschfahrzeug – airport crash tender

Flugrettungsdienst – airborne rescue service, air rescue service

Fluor – fluorine

Fluor-Brom-Kohlenwasserstoffe (FBKW) – fluorobromocarbons (FBC)

Fluor-Brom-Kohlenwasserstoffe, teilhalogeniert (HFBKW) – hydrofluorobromocarbond (HFBC)

Fluorchlorkohlenwasserstoffe (FCKW) – chlorofluorocarbons (CFC)

Fluorchlorkohlenwasserstoffe, teilhalogeniert (H-FCKW) – hydrochlorofluorocarbons (HCFC)

Fluorkohlenwasserstoffe (FKW) – fluorocarbons, fluorinated hydrocarbons

Fluorkohlenwasserstoffe, teilhalogeniert (H-FKW) – hydrofluorocarbons (HFC)

Fluorproteinschaummittel – fluoro-protein foam compound

Fluortelomeralkohole (FTOH) – fluorotelomer alcohols (FTOH)

Flur – corridor

Fluss – river

Flussdiagramm – flow chart

flüssiger Sauerstoff – liquid oxygen (LOX)

flüssiger Stickstoff – liquid nitrogen

flüssiger Zustand – liquid state

Flüssiggas – liquified petroleum gas (LPG)

Flüssiggase – liquified gases

Flüssigkeit, inerte – inert liquid

Flüssigkeiten – liquids

Flüssigkeiten, brennbare – combustible liquids

Flüssigkeitsbrand – liquid fire, fluid fire

Flüssigkeitsschutzanzug – chemical splash suit

Flutlichtbeleuchtung – flood-lightning

Flutlichter – floods

Flutlichtmast – flood-light mast

Flutlichtscheinwerfer – floodlights

Folgebrand – secondary fire

Folgeschaden – consequential damage, secondary damage

Folie – film, membrane, foil

Folienschlauch – film tube

Folienschweißgerät – heat sealer

Förderdruck – delivery pressure

Förderhöhe – delivery height

Förderhöhe, geodätische – geodetic discharge height

Förderhöhe, manometrische – manometric discharge height

Förderleistung – delivery, delivery volume

Förderleitung – delivery line

Fördermenge – conveying capacity, discharge

Förderstrecke – delivery distance

Förderstrom – rate of delivery, flow rate

Formation – formation

Formel – formula

Formelsammlung – formulary

Formulare – forms

Forstamt – forestry office

fortpflanzungsgefährdend – toxic to reproduction

Fracht – freight, load, cargo

Frachtbrief – bill of lading

Frankfurter Schaufel – Frankfurt shovel

Freigrenze – exemption limit

Freileitung – overhead transmission line

Freileitungskabel – overhead cable

Freischaltelement (FSE) – release element

Freischalten – disconnect from the mains

Freisetzung – release

Freistellung – exemption

freiwillige – volunteers

Freiwillige Feuerwehr (FF) – volunteer fire brigade, volunteer fire company (VFC), volunteer fire department

Fremderwärmung – external heating

Fremdkörper – foreign body

Fremdzündung – piloted ignition, extraneous ignition, externally supplied ignition

Frequenz – frequency

Frequenzänderung – frequency change

Frequenzeinstellung – frequency adjustment

Frequenzmodulation (FM) – frequency modulation (FM)

Frischluft – fresh air

Frischluftzufuhr – fresh air supply

Frontmonitor – front monitor

Frostschutzmittel – anti-freeze, frost protection agent

Frühdefibrillator – automated external defibrillator (AED)

Früherkennung – early detection

Frühschicht – early shift, morning shift

Frühwarnung – early warning

Fuchsschwanz (Säge) – handsaw

Fühler – sensor

Führen durch Befehl – leading by orders

Führerhaus – driver's cab

Führung – leadership

Führungsassistent – command assistant

Führungsebene – level of command

Führungseinrichtung – command facility

Führungsgruppe (FüGr) – command-team of 9, command-group

Führungskraft der Feuerwehr – fire officer

Führungsleine (Leinensicherungssystem), (Atemschutzeinsatz) – guide line

Führungsmittel – means of command and control

Führungsorganisation – command organisation

Führungspersonal – command personnel

Führungsstab – command-staff, command, high command

Führungsstaffel (FüSt) – command-team of 5, command-squad

Führungsstil – leadership style, style of leadership

Führungsstil, autoritärer – authoritarian leadership style, authoritarian style

Führungsstil, kooperativer – cooperative style of leadership, cooperative style, transactional leadership

Führungsstufen – echelons of command

Führungssystem – command system, command and control system

Führungstrupp (FüTr) – command-team of 3, command-squadron

Führungsverhalten – command attitude

Führungsvorgang – command-process, command and control process

Fülldruck – filling pressure

Füllgrad – filling degree

Füllmaterial – filling material, spacing material

Füllmenge – charge, filling quantity, quantity per pack

Füllstand – fill level, liquid level

Füllstandsanzeige – level indicator, liquid level indication

Fundament – footing, foundation

Funkanlage – radio unit

Funken – spark

Funken, elektrische – electric sparks

Funkenenergie – spark energy

Funkenentladung – spark discharge

Funkenflug – flying sparks

Funker – radio operator, radioman (RM)

Funkgerät (FuG) – radio device, mobile radio, two-way radio

Funkmeldeempfänger (FME) – paging device

Funkmeldeempfänger, digitaler (DME) – digital paging device

Funkmeldesystem (FMS) – radio alarm system

Funkrauchmelder – wireless smoke detector

Funkruf – radio call

Funkrufname – radio call-sign

Funkschatten – radio shadow

Funkspruch – radio message

Funkstelle (FuSt) – radio station

Funkstille – radio silence

Funkstrecke – radio path

Funktion (an Geräten) – feature

Funktion (einer Person) – position, capacity

Funktionsautorität – authority of post

Funktionsbereitschaft – operational readiness

Funktionserhalt – functional endurance

Funktionskontrolle – functional control

Funktionsprinzip – operating principle, functional principle

Funktionsprüfung – functional test

funktionsunfähig – inoperative

Funktionsweise – mode of operation (MO)

Funkverbindung (aktuell) – radio link, radio contact

Funkverbindung (grundsätzliche) – radio communication

Funkverbindung, Ausfall der – radio failure

Furane – furanics

Furan-Verbindungen – furanics, furan compounds

Fuß (einer Leiter) – shoe, ladder heel

Fußbremse – foot brake, pedal brake

Fußweg – footpath

G

G-ABC-Zug – chemicals incident unit (CIU)

Gabelhubwagen – jack-lift, transpallet, pump truck, hand pallet truck

Gabelschlüssel – open-end wrench, fork spanner

Gabelstapler – forklift

Gamasche – gaiter

Gamma-Dosisleistungskonstante – gamma dose-rate constant

Gamma-Photon – gamma photon

Gamma-Quant – gamma quant

Gamma-Strahlenquelle – gamma ray source

Gamma-Strahler – gamma emitter

Gamma-Strahlung – gamma emission, gamma radiation

Gamma-Übergang – gamma transition

Gamma-Zerfall – gamma decay, gamma disintegration

GAMS-Regel (*G*efahr erkennen – *A*bsperren – *M*enschenrettung durchführen – *S*pezialkräfte alarmieren) – GAMS-rule (approx..: DBRA-rule: recognize *d*anger – *b*arricade – *r*escue people – *a*lert special forces)

Ganzkörperbestrahlung – whole-body exposure

Ganzkörperdosis – whole-body dose

Ganzkörper-Personendosimeter – whole-body personal dosimeter

Garage – garage

Gartenschlauch – garden hose

Gartenspritze – garden pump sprayer

Gärung – fermentation

Gas – gas

Gasalarm – gas alert

Gasanzünder – striker, gas lighter

Gasausbreitung – gas propagation, gas dispersion

Gasbehälter – gas tank

Gasbrenner – gas burner

Gaschromatographie (GC) – gas chromatography (GC)

gasdicht – gas-proof, gas-tight

Gasdichte – gas density

gasdichter Chemikalienschutzanzug (CSA), Typ 1a-ET – type 1a gas-tight suit for emergency teams

gasdichter Chemikalienschutzanzug (CSA), Typ 1b-ET – type 1b gas-tight suit for emergency teams

Gasdiffusion – gas diffusion

Gasdiffusionsschicht (GDS) – gas diffusion layer (GDL)

Gasdurchlässigkeit – gas permeability

Gase – gases

Gase, ätzende – caustic gases

Gase, ausströmende – escaping gases

Gase, brennbare – combustible gases

Gase, entflammbare – flammable gases

Gase, giftige – toxic gases

Gase, ideale – ideal gases

Gase, inerte – inert gases

Gase, kritische – critical gases

Gase, medizinische – medical gases

Gase, nitrose – nitrous gases

Gase, permanente – permanent gases

T. Schmiermund, *Fachwörterbuch Feuerwehr und Brandschutz*, https://doi.org/10.1007/978-3-662-64120-0_7

Gase, reale – real gases

Gase, reizende – irritant gases

Gasexplosion – gas explosion

Gasfeuerung – gas firing

Gasfilter – gas filter

Gasflasche – gas cylinder

Gasflaschenventil – gas cylinder valve

Gasflaschenwagen – gas cylinder cart, gas cylinder dolly

gasförmige Löschmittel – gaseous fire-extinguishing agents

gasförmige Stoffe – gaseous substances

gasförmiger Zustand – gaseous state

Gasgemisch – gas mixture

Gasgesetze – gas law's

Gasgleichung, allgemeine – ideal gas law, perfect gas equation, general gas law

Gasgleichung, universelle – universal gas equation

Gaskonstante, allgemeine – general gas constant

Gaskonstante, universelle – universal gas constant

Gaskonzentration – gas concentration

Gaskrieg – chemical warfare

Gasleitung – gas main, gas pipe

Gaslöschanlage – gaseous extinguishing system

Gaslöscher – gas extinguisher

Gas-Luft-Gemisch – gas-air-mixture

Gasmaske – gas mask

Gasmelder – gas detector

Gasphase – gaseous phase

Gasprüfröhrchen – gas testing tube

Gassammelröhrchen – gas sampling tube

Gasschutz – protection against toxic gases

Gassensor – gas sensor

Gasspürgerät – gas detector device, gas detector

Gasspürgerät (Lecks) – gas leak detector

Gasspürpumpe (handbetrieben) – gas detector bellows

Gasstrahl – gas jet

Gasstrahler – exhaust ejector primer

Gasundurchlässigkeit – gas impermeability

Gasverflüssigung – gas liquefaction

Gasvergiftung – gas poisoning

Gasversorgung – gas supply

Gaswarnlage (GWA) – gas detection system (GDS)

Gaswolke – gas cloud

Gaszustand – gaseous state

Gebäude – building

Gebäude, landwirtschaftliche – agricultural buildings

Gebäude, n-stöckiges – n-storey building

Gebäudeart – type of building

Gebäudeeinsturz – building collapse

Gebäudehöhe – height of building

Gebäudeklasse – building class

Gebäudekomplex – building complex

Gebäudeleittechnik (GLT) – building management system (BMS)

Gebäudenutzung – building utilisation, use of the building

Gebäudeversicherung – building insurance

Gebotszeichen – mandatory signs

Gebrauchsanleitung – instruction for use, user instructions

Gebrauchsanweisung – operating instructions, instruction leaflet

Gebühren – fees, dues

Gebührenordnung – scale of charges, schedule of fees

Gebührentabelle – table of fees, tariff schedule

Gebüsch – brush

geeignet – suitable

geeignete Art und Weise – adequate manner

Gefahr – danger

Gefahr beim Berühren – contact hazard

Gefahr der Brandausbreitung – risk of fire spread

Gefahr durch giftige Stoffe – toxic hazard

Gefahr für Leben und Gesundheit – danger to life and health

Gefahr für Leben und Gesundheit, unmittelbare – imminent danger to life and health (IDLH)

Gefahr, kleine – slight risk

gefährden – endanger, put at risk

Gefährdung – endangerment, hazard

Gefährdungsanalyse (GA) – hazard analysis (HA)

Gefährdungsereignis – hazardous event

Gefährdungsgrad – hazard level, degree of danger

Gefährdungspotential – hazard potential, risk potential

Gefahren – dangers, hazards

Gefahren, bestehende – existing dangers

Gefahrenabwehr – hazard control

Gefahrenabwehrmaßnahmen – emergency response operations

Gefahrenbereich – hazard area, hazard zone, danger area, danger zone, area of risk

Gefahrendiamant – hazard diamond

Gefahreneigenschaften – hazard characteristics

Gefahrengruppe – hazard group, hazmat class

Gefahreninformation – hazard information (HI)

Gefahrenlage – case of danger

Gefahrenmeldezentrale (GMZ) – hazard control center

Gefahrennummer (Kemler-Zahl) – hazard identification number, Kemler number

Gefahrenquelle – hazard source

Gefahrenstelle – danger point, danger spot

Gefahrenstufe – hazard level

Gefahrensymbol – hazard symbol

Gefahrenzone – danger zone, dangerous area

Gefahrenzulage – hazard bonus, danger allowance

Gefahrgut – dangerous goods (DG)

Gefahrguteinsatz – hazmat operation

Gefahrgutklassen – classes of dangerous goods, hazard classes, class of danger, DOT classes {US}

Gefahrguttransport – hazmat transportation

Gefahrguttransporter – hazmat truck

Gefahrgutumfüllpumpe (GUP) – hazmat transfer pump

Gefahrgutunfall – hazmat accident

Gefahrgutverordnung Straße (GGVS) – Hazardous Goods Ordinance – Road

Gefahrgutverordnung Straße, Eisenbahn (GGVSE) – Hazardous Goods Ordinance – Road, Rail

Gefahrgutzug – chemicals incident unit (CIU)

Gefahrklassen – classes of dangerous goods, hazard classes, class of danger, DOT classes {US}

gefährlich – dangerous, hazardous

gefährliche Einwirkung von außen – external danger, external dangerous influence

gefährliche Gegenstände – dangerous articles

gefährliche Güter – hazardous goods, dangerous goods (DG)

gefährliche Güter, Klassen – classes of dangerous goods, DOT classes {US}

gefährliche Stoffe – dangerous substances

gefährliche Stoffe und Güter (GSG) – hazardous materials, hazmat

Gefährlichkeitsmerkmale – hazardous characteristics

Gefahrnummer – hazard number

Gefahrschild – danger sign

Gefahrstoffe – hazardous substances

Gefahrstoffkennzeichnung – hazard labelling

Gefahrstoffmenge – hazmat quantity, quantity of dangerous material

Gefahrstoffverordnung (GefStoffV) – Hazardous Substances Ordinance

Gefahrstoffzug – chemicals incident unit (CIU)

Gefahrzettel – hazard labels, labels

Gefälle – slope

gefangener Raum – inner room

Gefrierpunkt – freezing point

Gegendruck – counterpressure

Gegenfeuer – back-burn, counter-fire

Gegengift – antidote, antitoxin, antivenin

Gegenstände mit Explosivstoff – explosive articles

Gegenstände, gefährliche – dangerous articles

Gegenstände, scharfkantige – sharps

Gehalt (Lohn) – salary, wages

Gehäuse – enclosure, housing

Gehörschutz – hearing protection

Gehörschutzstöpsel – earplugs, hearing protection plugs

Geiger-Müller-Zähler – Geiger-Muller counter

Geiger-Müller-Zählrohr – Geiger-Muller counter

Geigerzähler – Geiger-Muller counter

Geisterflammen – ghost flames

Gelakkumulator – gel battery

Gelände – terrain, area

Gelblicht (als RKL) – yellow rotating light, yellow strobe light

Geldbuße – fine, financial penalty

Geldstrafe – fine, financial penalty

Gelenklöscharm (GLA) – extinguishing boom

Gelenkmast (GM) – hydraulic platform

Gemeinsames Melde- und Lagezentrum (GMLZ) – fusion center of the German BOS-Organisations

Gemenge – heterogenous mixture

Gemisch – mixture, intermixture

Gemisch, brennbares – combustible mixture

Gemisch, explosibles – explosible mixture

Gemisch, explosionsfähiges – explosible mixture

Gemisch, heterogenes – heterogenous mixture

Gemisch, homogenes – homogenous mixture

Gemisch, stöchiometrisches – stoichiometric mixture

Gemisch: Wasser/Schaummittel – foam solution

Gemischbildung – mixture formation

Genehmigung – approval

Generalschlüssel – universal key

genetische Schäden – genetic damage

Gentechnik – genetic engineering (GE)

Gentechniksicherheitsverordnung (GenTSV) – Genetic Engineering Safety Ordinance

gentechnisch veränderte Organismen (GVO) – genetically modified organism (GMO)

geodätische Förderhöhe – geodetic discharge height

geodätische Saughöhe – geodetic suction height

Gerätegruppe – equipment group (EG), device group

Geräteraum – locker

Gerätewagen (GW) – tool and gear vehicle

Gerätewagen Atemschutz (GW-A) – breathing apparatus unit (BA unit)

Gerätewagen Atemschutz/Strahlenschutz (GW-AS) – BA and radioactive incident unit

Gerätewagen Gefahrgut (GW-G) – hazardous substances combat vehicle

Gerätewart (Gw) – equipment manager

gereizte Haut – irritated skin

Gericht (Justiz) – court

Geruch, angenehmer – fragrance, scent, pleasant smell

Geruch, unangenehmer – unpleasant smell

Geruchssinn – sense of smell

Geruchssinn, Lähmung dessen – paralysis of the sense of smell

Gerüst – scaffold, framework

Gesamtabsicht – overall intention

Gesamtausfall – total breakdown

Gesamtbefehl – general order

Gesamtdruckverlust – total loss of pressure

Gesamteinsatzleitung – overall command

Gesamtförderhöhe – total static head

Gesamtgewicht – gross weight, total weight

Gesamthärte – total hardness

Gesamthärte (Wasser) – total hardness of water

Gesamtlage – overall situation

Gesamtmasse – total weight

Gesamtmasse, zulässige – permissible total weight

Gesamtverantwortlicher, politischer – overall political responsibility

Geschoss (Etage) – storey, floor

Geschossanzahl – number of storeys

Geschossdecke – storey ceiling

Geschossflächenzahl (GFZ) – floor-space index (FSI)

geschützter Bereich – protected area

Geschwindigkeitsbegrenzung – speed limit, speed restriction

Gesetz – law, legislation

Gesetz (einzelnes) – act

Gesetz der Partialdrücke – law of partial pressures

Gesetz, chemisches – chemical law

Gesetz, Faraday'sches – Faraday's law

Gesetz, kubisches – cubic law

Gesetz, Ohm'sches – Ohm's law

Gesetz, physikalisches – physical law

Gesetz, unbedingtes – absolute law

Gesetz, ungeschriebenes – unwritten rule

Gesetzbuch – code of law

Gesetzgeber – law maker, law-giver, legislator

gesetzlich – legal, lawful

gesetzliche Bestimmung – legal requirement

gesetzliche Grundlage – legal basis, statutory basis

gesetzliche Regelungen – legal regulations

Gesichtspunkte – aspects, point of view

Gesichtspunkte, taktische – tactical point of view

Gesichtsschutz – face guard, face shield

Gesichtsschutzschirm – visor, face shield

Gestank – bad smell, stink, stench

Gesundheit – health

Gesundheitsamt – local health authority

Gesundheitsbehörde – public health authority, board of health

Gesundheitsgefahr – health hazard, health risk

Gesundheitsgefährdung – health hazard, health risk

gesundheitsschädlich – harmful

Gesundheitsschutz – health protection

Gesundheitsuntersuchung – health examination

Gesundheitsvorsorge – health care

Gesundheitszustand – state of health, physical condition

Getriebeöl – gear oil, gearbox oil

Gewässer – aquatic environment, waterbody

Gewässer, offenes – open water

Gewebe – tissue, fabric

Gewebeband – webbing

Gewebeschlauch – textile hose

Gewebewichtungsfaktor – tissue weighting factor

Gewerbe – trade, industry

Gewicht – weight

Gewichtskraft – weight force

Gewinde – thread

Gewitter – thunderstorm

Gewölbe – vault

Giebel – gable

Giebelwand – gable wall

Gift, Giftstoff – toxicant, toxin, toxic substance

Giftgas – poison gas

Giftgaswolke – toxic gas cloud

giftig – toxic, poisonous

giftig beim Einatmen – poisonous if inhaled

giftige Dämpfe – toxic vapours

giftige Gase – toxic gases

giftige Stoffe – toxic substances

Giftigkeit – toxicity, poisonousness

Giftwirkung – toxic effect

Giftwirkung, akute – acute toxic effect

Giftwirkung, chronische – chronic toxic effect

Giftwirkung, systemische – systemic toxic effect

Glasbaustein – glass block

glasfaserverstärkter Kunststoff (GFK) – glass fibre reinforced plastic (GRP)

Glasfasssprinkler – glass-bulb sprinkler

Glaswolle – glass-wool

Gleichgewicht – balance

Gleichgewicht, chemisches – chemical equilibrium

Gleichgewichtszustand – steady state, state of balance

Gleichstrom – direct current (dc)

Gleichung, chemische – chemical equation

Gliedermaßstab – folding ruler, folding yardstick, double meter stick

Gliederung – organisation

Gliederung eines Befehls – structure of an order

Gliederung, taktische – tactical organisation

Glimmbrand – smoldering fire, smouldering fire

glimmen – glow, smolder, gleam

Glimmen – glow

Glimmtemperatur – glow temperature

globales Positionssystem – global positioning system (GPS)

Glühbirne – light bulb

Glühen – glowing

Glühfarbe – annealing colour

Glut – embers, glow

Glutbrand – glowing fire

Glutbrandpulver – ABC powder

Glutfarbe – annealing colour

Glutnester – pockets of embers, hot spots, fire pockets

Glutwolke – ash flow

Grad – degree

Granulat – granules

Graphit – graphite

Grasbrand – grass fire

Greifzug® – come along

Grenzkonzentration – limited concentration

Grenzspaltweite – maximum experimental safe gap (MESG)

Grenzwerte – limits, threshold limit values (TLV)

Grenzwerte für Radioaktivität – radiation limits

Grenzwertüberwachung – limit value monitoring

Griff (eines Werkzeugs) – helve, handle

grob fahrlässig – grossly negligent, reckless

Grobreinigung – coarse purification

Grobstaubmaske – dust mask, dust respirator

Großbrand – blaze, large fire, major fire, conflagration

großer Zapfenstreich – great tattoo

Großfeuer – conflagration, large fire

Großraumnetzwerk – wide area network (WAN)

Großraumrettungswagen (G-RTW) – major emergency response vehicle (MERV)

Großschadensereignis – large scale disaster, major incident

Großtanklöschfahrzeug (GTLF) – large tank fire tender, large tank fire-fighting vehicle

größter anzunehmender Unfall (GAU) – worst-case accident

Grundfläche (eines Raumes) – floor area, floor space

Grundflächenzahl – site occupancy index

Grundlage – basis, ground

Grundlage, gesetzliche – legal basis, statutory basis

Grundregeln – basic rules

Grundregeln, taktische – basic tactical rules

Grundsatz der Verhältnismäßigkeit – principle of reasonableness, principle of proportionality

grundsätzlich – generally

Grundübungen – basics training

Gruppe (Gr) – group, team-of-nine

Gruppenalarm – group alarm

Gruppenalarmierung – group alert, group alerting

Gruppenführer (GrFü) – group leader

gruppenweise – in groups

Guedeltubus – oropharyngeal airway (OPA)

Gülle – liquid manure

Gully – gully, drain

Gullydeckel – gully cover, manhole cover

Gültigkeit – validity

Gummidichtung – rubber seal, rubber gasket

Gummierung – rubber lining

Gummihammer – rubber mallet

Gummischürze – rubberized apron

Gummistiefel – rubber boots, gumboots

Gurtschneider – seat belt cutter

Gusseisenspäne – cast iron chippings

Gutachten – expert assessment

H

Haager Konvention – Hague Convention

Haager Landkriegsordnung (HLKO) – Hague Land Warfare Convention

Habersches Tödlichkeitsprodukt – Haber's lethality product

Hafenfeuerwehr – harbor fire brigade

Haftpflichtversicherung – liability insurance

Haken – hook

Hakengurt – hook belt, ladder belt

Hakenleiter – pompier ladder, hook ladder

Hakenschlüssel – hook wrench, C-wrench

Halbduplex – half-duplex

halbdurchlässig – semipermeable

Halbmaske – half mask

Halbmastwurfsicherung (HMS) – crossing hitch, Italian hitch, Munter hitch, HMS

Halbschlag – half hitch

Halbwertschicht (HWS) – half-value layer (HVL)

Halbwertschichtdicke (HWS) – half-value layer (HVL)

Halbwertszeit (HWZ) – half-life period, half-time radioactive period

Halluzinogene – hallucinogens

Halogenkohlenwasserstoffe – halogenated hydrocarbons, haloalkanes

Halone – halons

Halonlöscher – halon extinguisher

Hals – neck

Halskrause – cervical collar, neck brace

Halswirbelsäule (HWS) – cervical spine

Haltebügel – retainer

Halterung – holding fixture, fixture, mounting

Haltevorrichtung – mounting device

Haltezeit (Tauchen) – decompression stop

Hämatom – haematoma, hematoma

Hammer – hammer

Hammerkopf – hammer head

Hammerstiel – hammer handle

hämorrhagisches Fieber – haemorrhagic fever, hemorrhagic fever

hämorrhagisches Fieber, virales – viral haemorrhagic fever

hämorrhagisches Krim-Kongo-Fieber – Crimean-Congo haemorrhagic fever (CCHF)

Handabsaugpumpe – hand suction pump

Handauslösung – manual release

Handbetrieb – manual mode

handbetriebene Winde – hand-operated winch

Handbewegung – hand motion

Handbürste – hand brush

Handfeger – hand broom

Handfeuerlöscher – hand-held fire extinguisher

Handfunkgerät (HFuG) – portable radio device, handheld transceiver (HT), walkie-talkie

Handgriff, lebensrettender – head-tilt and chin-lift manoeuvre (HTCL-manoeuvre)

Handhabung – handling

Handlampe – hand lamp

Handlauf (Treppe) – handrail

T. Schmiermund, *Fachwörterbuch Feuerwehr und Brandschutz*, https://doi.org/10.1007/978-3-662-64120-0_8

Handlungsablauf – process of action, course of action

Handlungsunfähigkeit – incapacitation

Handrad – handwheel

Handrad für ein Ventil – handle for valves

Handscheinwerfer – hand searchlight

Handschuhe – gloves

Handschuhe, medizinische – medical gloves

Handschutz – fingerguard

Handsirene – manual fire siren, hand-operated siren

Handsprechfunkgerät (HFuG) – portable radio device, handheld transceiver (HT), walkie-talkie

Handwerker – craftsman

Handzeichen – hand signals

Hängetrauma – suspension trauma

harte Bedachung – non-inflammable roof-covering

Härte, permanente (Wasser) – permanent hardness of water

Härte, temporäre (Wasser) – temporally hardness of water

Hartholzgriff – hardwood handle

Hasenpest – tularemia, rodent plague, rabbit fever

Haspel – reel

Hauptabsperrschieber – main stop tap, main shut-off device

Hauptalarm – master alarm

Hauptangriffsrichtung – main attack direction

Hauptangriffsweg – main attack path, main attack route

Hauptaufgabe – main objective

Hauptfeuerwache – fire brigade headquarters

Haupthahn – main tap, main valve

Hauptlöscheffekt – main extinguishing effect

Hauptschalter – main switch

Hauptschieber – main slide, main spool

Hauptschlüssel – master key, primary key

Hauptwasserleitung – main water pipe

Hausanschluss – house connection

Hausanschlusskasten – house connection box, house service connection

Hausanschlussraum (HAR) – service entrance room

Hausbrand – house fire, home fire

Haushalt – household

Hausmüll – household waste

HAUS-Regel (*H*indernisse – *A*bstände – *U*ntergrund – *S*icherheit) – HAUS-rule (approx.: ODBS-rule: *o*bstacles – *d*istances – *b*ase – *s*afety)

Haut – skin

Haut, gereizte – irritated skin

Hautblasen – skin blisters

Hautgift – skin poison

Hautkampfstoff – blister agent, vesicant agent, vesicant gas

Hautkontakt – skin contact

Hautmilzbrand – cutaneous anthrax, skin anthrax

Hautpflege – skin care

Hautreizung – skin irritation

Hautresorption – skin resorption

Havarie – sea damage (SD)

Hebebaum – lever, lifting beam

Hebekissen – air-lifting units, ‚air-bags‘, air lifting bag, lifting bag

Hebelgesetz – lever principle

Hebemittel – hoisting gears

Heber, hydraulischer – hydraulic jack, lifting jack

Hebevorrichtung – lifting device

Hecke – hedge

Heckpumpe – rear-mounted pump

Heidebrand – moor fire

Heidelberger Verlängerung – Heidelberger extension

heiße Oberfläche – hot surface

Heißluftpistole – heat gun

Heizgerät – heating equipment

Heizkraftwerk (HKW) – combined heat and power station (CHP)

Heizleistung – heat output

Heizöl – heating oil

Heizungsanlage – heating installation

Heizungsraum – boiler room

Heizwert – net heat of combustion, heat value, heating value

Heizwert, oberer – → Brennwert

Heizwert, spezifischer – specific heating value

Heizwert, unterer – → Heizwert

Helfer – helper

Heliumkern – helium core

Helm – helmet

Helm ab! – Helmets off!

Helmkennzeichnung – helmet marking

Helmlampe – helmet lamp

Helmlampe, ansteckbare – clip-on helmet lamp

Helmvisier – helmet visor, visor for helmet

Hemmstoff – inhibiting agent

Henkersknoten – hangman's knot

Hersteller – manufacturer, producer

Herz – heart

Herzdruckmassage – cardiac massage, chest compression, cardiac compression

Herzfrequenz (HF) – heart rate (HR)

Herzfrequenz, maximale – maximum heart rate (MHR)

Herzinfarkt – heart attack, cardiac infarction

Herzkammerflimmern (HKF) – ventricular fibrillation (VF)

Herz-Kreislauf-Stillstand – cardiac arrest (CA), full arrest

Herz-Lungen-Wiederbelebung (HLW) – cardiopulmonary resuscitation (CPR)

Herz-Lungen-Wiederbelebungstraining (HLW-Training) – cardiopulmonary resuscitation training (CPR-training)

Herzmassage – heart massage

Herzminutenvolumen (HMV) – cardiac output per minute (CO)

Herzschrittmacher (HSM) – pacemaker (PM)

Herzstillstand – cardiac arrest (CA)

Hess'scher Satz – Hess theorem

heterogene Inhibition – heterogeneous inhibition

heterogene Katalyse – heterogeneous catalysis

heterogenes Gemisch – heterogenous mixture

Heu – hay

Heugabel – pitchfork, hay fork

Heulton – wailing sound, wail

Heuselbstentzündung – hay self-ignition

Hilfe – help

Hilfe, nachbarliche – mutual aid

Hilfeleistung – assistance, help

Hilfeleistung, technische (TH) – rescue service

Hilfeleistungslöschfahrzeug (HLF) – auxiliary fire tender, rescue pumper

Hilfsfrist – time to assistance, help period

Hilfsgeräte – ancillary equipment

Hilfskonstruktion – auxiliary construction

Hilfskraft (Person) – temporary helper

Hilfsmittel – adjuvant

Hilfsorganisationen (HiOrg) – non-governmental sanitary organizations

Hin- und Rückweg – outward and return (route)

Hindernisse – obstacles

Hintergrund – background

Hintergrundstrahlung – background radiation

Hintergrundstrahlung, natürliche – natural background radiation

Hinweg – outward route

Hinweise – guidance

Hinweise, zusätzliche – additional guidance

Hinweisschild (z. B. für Hydrant) – direction sign

Hinweiszeichen – indication sign, information sign

Hirnhautentzündung – meningitis

Hitze – heat

Hitzeeinwirkung – heat exposure

Hitzeentwicklung – heat development

Hitzeschock – heat shock

Hitzeschutz – heat protection

Hitzeschutzanzug – heat protective suit, fire proximity suit

Hitzschlag – heat stroke

hochangereichertes Uran – highly enriched uranium

Hochbehälter – elevated tank, elevated reservoir, water tower

Hochdruck – high pressure

Hochdruckleitung – high-pressure line

Hochdruckpumpe – high-pressure pump

Hochdruckreiniger – high-pressure cleaner

Hochdruckschlauch – high-pressure hose

hochentzündlich – extremely flammable

hochentzündlicher Stoff – extremely flammable substance

hochgiftig – extremely toxic

hochgiftiger Stoff – extremely toxic substance

Hochhaus – skyscraper

Hochhausbrand – skyscraper fire, high-rise building fire

hochkonzentriert – highly concentrated

hochradioaktiver Abfall – high active waste (HAW)

hochreaktiv – highly reactive

Hochspannung – high voltage (HV)

Hochspannungsleitung – high voltage cable

Hochspannungsverteilung – high voltage distribution

Höchstkonzentration – maximum concentration

Höchstkonzentration, zulässige – maximum permitted concentration

Hochwasser – flooding

Hochwassersperre – flood barrier

Höhendifferenz – height difference

Höhenlinie (Kartenkunde) – contours

Höhenretter – height rescuer, high-altitude rescuer

Höhenrettung – height rescue, high-altitude rescue, altitude rescue

Höhensicherung – fall protection

Höhenunterschied – height difference

Höhle – cave

Hohlraum – hollow space

Hohlstrahlrohr (HSR) – fog nozzle, fog nozzle branch pipe

Holm (einer Leiter) – ladder stringer

Holz – wood

Holzaxt – felling axe, large axe

Holzbalkendecke – wooden beam ceiling

Holzfaserplatte – fibreboard

Holzfaserplatte, mitteldichte – medium-density fibreboard (MDF)

Holzgas – wood gas

Holzkohle – charcoal

Holzkohleschicht – charcoal layer

Holzkonstruktion – timber construction, wooden construction

Holzsäge – wood saw

Holzschutzmittel – timber preservative, wood preservative, wood protection agent

Holzspan – chip of wood

Holzwolle – wood wool

homogene Inhibition – homogenous inhibition

homogene Katalyse – homogeneous catalysis

homogenes Gemisch – homogenous mixture

Hörgerät – hearing aid, deaf-aid

Hubarbeitsbühne – aerial ladder platform

Hubhöhe – lifting height

Hubrettungsfahrzeuge – aerial appliances, lifting rescue vehicles

Hubschrauber – helicopter

Hubschrauber im Löscheinsatz – fire-fighting helicopter

Hubsteiger – cherry picker, hydraulic platform

Hubwagen – jack-lift

Hüftgurt – sit harness

Hupe – hooter, horn

HWS-Halskrause – cervical collar, Stiff Neck™

Hydrant (H) – hydrant, fire plug, fire hydrant

Hydrantendeckel – hydrant cover

Hydrantendeckel-Auftaugerät – thawing device for hydrants

Hydrantenpläne – maps of hydrants

Hydrantenschild – hydrant sign

Hydrantenschlüssel – fire hydrant key and bar, hydrant wrench

Hydraulikaggregat – hydraulic power unit

Hydraulikflüssigkeit – hydraulic liquid

Hydraulikleitung – hydraulic line

Hydrauliköl – hydraulic oil

Hydraulikpumpe – hydraulics pump

hydraulische Rettungsschere – hydraulic rescue cutter, Jaws of Life®

hydraulische Stützen (Drehleiter, Kran) – jacking pads, jacks

hydraulische Stützen (Drehleiter, Kran) – jacking pads, jacks

hydraulischer Heber – hydraulic jack, lifting jack

hydraulischer Rettungsspreizer – hydraulic rescue spreader

hydraulischer Spreizer – hydraulic rescue spreader

hydraulischer Wagenheber – hydraulic jack

hydraulisches Rettungsgerät – hydraulic rescue device, hydraulic rescue apparatus, Jaws of Life®

Hydrierung – hydrogenation

hydrophil – hydrophilic

hydrophob – hydrophobically

Hydrophobierung (Ergebnis) – hydrophobicity

Hydrophobierung (Vorgang) – hydrophobisation

Hydroschild – hydro-shield, water-shield

hydrostatischer Druck – hydrostatic pressure

Hygiene – hygiene

Hygienevorsorge – sanitation care

I, J

ideales Gas – ideal gas

identisch – identical

im Dienst – on duty

Imker – beekeeper, apiarist

Imkerpfeife – beekeeper's pipe, smoker

Imkerschutzanzug – bee protective suit, bee suit

Immission – immission

Immission – immission

Immissionsschutz – immission control

Immobilisation – immobilization

immobilisieren – immobilize

immun – immune

Immunisierung – immunization

Immunität – immunity

Impfausweis – vaccination certificate

Impfpass – vaccination passport, vaccine passport

Impfpass, internationaler – International Certificate of Vaccination or Prophylaxis (ICVP)

Impfung – vaccination, immunization, injection, shot, vaccine injection

Impfung gegen Corona – corona vaccination

Impfung gegen Grippe – flu vaccination, influenza vaccination

Impfung gegen Hepatitis A (HAV-Impfung) – hepatitis A vaccination (HAV vaccination)

Impfung gegen Hepatitis B (HBV-Impfung) – hepatitis B vaccination (HBV vaccination)

Impfung gegen Masern-Mumps-Röteln (MMR-Impfung) – measles-mumps-rubella vaccination (MMR vaccination)

Impfung gegen Tollwut – anti-rabies inoculation

Impfung, erste/zweite – first/second vaccination

Implosion – implosion

Implosionsgefahr – risk of implosion

Imprägnieren – impregnating

Imprägnierung – impregnation

Impuls – momentum

Impulserhaltung – momentum conservation

Impulsrate – counting rate, pulse rate

inaktiv – inactive

Inbusschlüssel – hex key, hex wrench

indirekte Kühlung – indirect cooling

indirekter Löschangriff – indirect attack

Induktionszeit – induction time

Industriebauten – industrial buildings

Industriechemikalien – industrial chemicals

Industrieklebeband – duct tape, Duck Tape™

inerte Flüssigkeiten – inert liquids

inerte Gase – inert gases

Inertgas – inert gas

inertisieren – render inert

Inertisierung – intertisation, inerting

Infektionsdosis (ID) – infectious dose (ID)

Infektionsgefahr – risk of infection

T. Schmiermund, *Fachwörterbuch Feuerwehr und Brandschutz*, https://doi.org/10.1007/978-3-662-64120-0_9

Infektionskrankheit – infectious disease

Infektionsrettungswagen (I-RTW) – special infection ambulance

Infektionsschutzanzug – infection control suit, infection protective suit

Infektionsschutzgesetz (IfSG) – Law for the Protection against Contagious Disease

infektiös – infectious

infektiöser Abfall – infectious waste

Infektiosität – infectiosity, infectivity

Information und Kommunikation (IuK) – information and communication, information and communication technology (ICT)

Information, benötigte – needed information

Information, detaillierte – detailed information

Informationsfluss – flux of information

Informationsgewinnung – gathering information

Informationspflicht – duty to inform

Informationsverarbeitung – information assessment

Infrarotmelder – infrared detector

Infrarotstrahlung (IR-Strahlung) – infrared radiation (IR-radiation)

Infraschall – subsonic noise

Infrastrukturen, kritische (KRITIS) – critical infrastructures

Infusion – infusion

Infusionsbesteck – infusion set

Infusionsbeutel – infusion bag, intravenous bag (IV bag)

Infusionsdauer – infusion period

Infusionsflasche – intravenous bottle (IV bottle), infusion bottle, drip bottle

Infusionsgeschwindigkeit – infusion rate

Infusionskatheter – infusion catheter

Infusionspumpe – infusion pump

Infusionspumpe, tragbare – ambulatory infusion pump

Infusionsschlauch – IV line, intravenous line

Infusionsständer – drip stand, intravenous pole, IV pole, infusion stand, intravenous bottle holder, IV bottle holder

Infusionstropfflasche – intravenous drip bottle

Infusionsvorrichtung – infusion device

Inhibition – inhibition

Inhibition, heterogene – heterogeneous inhibition

Inhibition, homogene – homogenous inhibition

Inhibitionseffekt (Löscheffekt) – inhibition effect

Inhibitor – inhibitor, anti-catalyst

Initialzünder – initiator

Injektion – injection

Injektionsmilzbrand – injectional anthrax

Injektorpumpe – injector pump

Inkorporation – incorporation

Inkubation – incubation

Inkubationszeit (IKZ) – incubation period (ICP)

Innenangriff – interior attack

Innenbrand – inside fire

Innengewinde – internal thread, female thread

Innenmaske – inner mask

Innenputz – plaster

Innensechskantschlüssel – hex wrench, hex key

Innenwand – internal wall

innere Bestrahlung – internal irradiation

Innerer Dienst (S1) – administration

Insektenvernichtungsmittel – insecticide

Installationsschacht – vertical duct

Instandhaltung – maintenance, upkeeping

Instandsetzung – refurbishment

Institut der Feuerwehr (IdF) – fire brigade's institute

Intensität – intensity

Intensität der Löschmittelzufuhr – intensity of extinguishing agent supply

Intensivtransporthubschrauber (ITH) – intensive care helicopter (ICH)

Intensivtransportwagen (ITW) – mobile intensive care unit (MICU)

internationales Einheitensystem (SI) – international unit system (SI)

Internationales Rotes Kreuz (IRK) – International Red Cross (IRC)

intravenös – intravenous

intravenöser Zugang – intravenous line

Intubation – intubation

Intubation, schnelle – rapid sequence intubation (RSI)

intumeszierend – intumescent

Inventar – inventory, stock

inventarisieren – make an inventory

Inversionstemperatur – inversion temperature

Inzidenz – incidence

Ion – ion

Ionenbindung – ionic bond

Ionendosis – ion dose, ion exposure dose

Ionenmobilitätsspektrometer (IMS) – ion-mobility spectrometer (IMS)

Ionisation – ionisation

Ionisationsdichte – ionization density

Ionisationskammer – ionization chamber (IC)

Ionisationsmelder – ionization detector

Ionisationsrauchmelder – ionization smoke detector

Ionisationszähler – ionization counter

ionisch – ionic

ionisierende Strahlung – ionizing radiation

Isolierband (elektr.) – insulating tape

Isoliergerät – insolating breathing apparatus, insolating BA

isoliert – insulated

Isolierung – insulation

isotonisch – isotonic

Isotop – isotope

Isotopenregel – isotope rule

Istwert – actual value

Ist-Zustand – actual state, actual condition

Jahreszeit – time of year

Jalousien – venetian blinds

Johanniter-Unfall-Hilfe (JUH) – St. Johns ambulance, St. Johns ambulance brigade

Joule-Thomson-Effekt – Joule-Thomson effect

Jugendfeuerwehr – youth fire brigade

Jugendfeuerwehrangehörige – junior firefighters

juristische Person – legal entity

justieren – adjust

Justierung – adjustment

K

Kabel – cord, cable

Kabelbinder – cable ties

Kabelbrand – cable fire

Kabelkanal – cable duct, wiring duct, wireway

Kabelschach – cable shaft

Kabeltrommel – cable reel, cable spool, cable extension reel

kalibrieren – calibrate

Kalibrierstrahler – calibration source

Kalibrierung – calibrating, calibration

Kalilauge – caustic potash lye, potassium lye

Kalium – potassium

Kalk – lime

Kalkseife – lime soap

Kalkstein – limestone

Kalorimeter – calorimeter

Kälteschutz – cold protection, cryoprotection

Kälteschutzkleidung – cold protection suit, cold protection clothes

Kamin – chimney

Kaminbrand – chimney on fire

Kamineffekt – chimney effect

Kaminwirkung – chimney effect

Kammerflimmern – ventricular fibrillation (VF)

Kampfgas – war gas

Kampfmittel – warfare agents

Kampfmittel, biologische – biological warfare agents (BWA)

Kampfmittel, chemische – chemical warfare agents (CWA)

Kampfmittelräumdienst – explosive ordnance disposal service, UXO-clearing service

Kampfmittelräumung – UXO clearing

Kampfstoff – warfare agent

Kampfstoff, biologischer – biological warfare agent (BWA)

Kampfstoffe – warfare agents

Kampfstoffe, biologische – biological warfare agents (BWA)

Kampfstoffe, chemische – chemical warfare agents (CWA)

Kampfstoffe, psychotoxische – psycho-toxic agents

Kampfstoffe, radiologische – radiological agents

Kampfstoffe, strategische – strategic agents

Kampfstoffe, taktische – tactical agents

Kampfstoffspürpapier – CWA detection-paper

kampfunfähig machendes Konzentrations-Zeit-Produkt – incapacitating concentration-time product

Kanal (Funk) – channel, radio channel

Kanal (z. B. für Wasser) – channel

Kanalabdeckung – drain seal

Kanaldichtkissen – seal bag

Kanalisation – sewage system

Kanüle – needle, tube, drain tube

Kanüle, stumpfe – blunt cannula

Kanüle: Butterfly – butterfly needle

T. Schmiermund, *Fachwörterbuch Feuerwehr und Brandschutz*, https://doi.org/10.1007/978-3-662-64120-0_10

Kapazität – capacity

Kappmesser – jack knife

Kapselgehörschutz – earmuff

Karabinerhaken – carabine, snap hook, carabine hook

kardiogener Schock – cardiogenic shock

Karten (Kartenkunde) – maps

Kartusche – cartridge

Katalysator (Fahrzeuge) – catalytic converter

Katalysator (Kat) – catalyst

Katalysatorgift – catalyst poison

Katalyse – catalysis

Katalyse, heterogene – heterogeneous catalysis

Katalyse, homogene – homogeneous catalysis

katalytisch – catalytic

Katastrophe (Kat) – disaster, catastrophe

Katastrophe, technische – technological disaster

Katastrophenalarm – disaster alert

Katastrophenfall – state of calamity, state of disaster, disaster situation

Katastrophenhilfe – disaster relief

Katastrophenschutz (KatS) – disaster control

Katastrophenschutz (Organisation) – disaster relief

Katastrophenschutz-Dienstvorschrift (KatSDV) – disaster control regulation

Katastrophenschutzhelfer – disaster relief crew, disaster relief worker

Katastrophenschutz-Organisationen – disaster relief organisations

Katastrophenschutzübung – disaster response exercise, disaster control exercise

Katheter – catheter

kathodischer Korrosionsschutz (KKS) – cathodic corrosion protection (CCP)

Kation – cation

kationisch – cationic

Katzenpfote (Knoten) – cat's paw

Kavitation – cavitation

Kehrblech – dustpan

Kehrmaschine – road sweeper

Keil – wedge

Keilriemen – V-belt

Keim – germ

Kelle – trowel

Keller – cellar, basement

Kellerbrand – cellar fire

Kemler-Zahl – hazard identification number, Kemler number

Kenngröße – characteristic value, parameter

Kenntnisstand – level of knowledge

Kennzahlen, brandschutztechnische – fire protection indexes

Kennzahlen, sicherheitstechnische – safety-related values, safety indexes

Kerma (kinetische Energie, die je Masseneinheit freigesetzt wird) – kerma (kinetic energy released per unit mass)

Kernbindungsenergie – nuclear binding energy

Kernchemie – nuclear chemistry

Kernenergie – nuclear power, nuclear energy

Kernenergie-Ausstieg – pull-out from the nuclear energy

Kernfusion – nuclear fusion

Kernkettenreaktion – nuclear chain reaction

Kernkraft – nuclear energy

Kernkraft, schwache – → Wechselwirkung, schwache

Kernkraft, starke – → Wechselwirkung, starke

Kernkraftwerk (KKW) – nuclear power plant (NPP)

Kernladung – nuclear charge

Kernladungszahl – proton number, atomic number

Kernmanteldynamikseil – dynamic kernmantle rope

Kernmantelseil – kernmantle rope

Kernphotoeffekt – photonuclear reaction

Kernreaktion – nuclear reaction

Kernreaktor – nuclear reactor

Kernschmelze – nuclear meltdown, core meltdown, core melt accident

Kernseife – curd soap, hard soap

Kernspaltung – nuclear fission

Kerntechnik – nuclear technology, nuclear power engineering

Kernzerfall – nuclear disintegration

Kesselwagen (Eisenbahn) – rail tank

Kesselwagen (Straße) – tank truck, tank vehicle, tank car

Kette – chain

Kettenabbruch – chain termination, chain breaking

Kettenfortpflanzung – chain propagation

Kettenknoten – chain sinnet

Kettenreaktion – chain reaction

Kettenreaktion, nukleare – nuclear chain reaction

Kettenreaktion, unverzweigte – unbranched chain reaction

Kettenreaktion, verzweigte – branched chain reaction

Kettenstart – chain initiation

Kettenverzweigung – chain branching

Kettenwachstum – chain propagation, chain growth

Kies – gravel

Kieselsteine – pebbles

Kinderfeuerwehr – children's fire brigade

kinetische Energie – kinetic energy

Kinnriemen – chin strap

Kippschutz – tilt protection

Kirche – church

Kirchenbrand – church fire

Kitt – putty

Klappleiter – folding ladder

Klappspaten – collapsible spade

Klapptrage – folding stretcher

Klapptritt – folding step

Klassen gefährlicher Güter – classes of dangerous goods, DOT classes {US}

Klassifizierungssystem – classification system

Klebeband – adhesive tape

Klebstoff – glue, adhesive

Kleeblatt (Trefoil) = Strahlenwarnzeichen – trefoil = basic ionizing radiation symbol

Kleiderkammer – uniform store

Kleinbrand – small fire

Kleinlöschfahrzeug (KLF) – small fire-fighting vehicle

Kleinmengenregelung – small quantity exception

Klemmbacken – clamping jaws

Klempner – plumber

Kletterausrüstung – climbing gear

Klettergurt – climbing harness

Klimaanlage – air conditioner system

Klimakanal – air conditioner duct

klinischer Abfall – clinical waste

Klumpen – clump

Knaggenteil – dog of a hose coupling

Knallgas – oxyhydrogen

Knautschzone – crush-collapsible zone

Knickschutz (z. B. Kabel) – kink protection

Knie (Rohrbogen) – elbow, pipe elbow

Knopfzelle (Batterie) – coin cell, button cell

Knoten – knots, bends

Knoten (zum Verbinden von Leinen) – bend knots

köcheln – simmering

kochen – boiling

Kochsalz – table salt

Kochsalzlösung, physiologische – physiological saline solution

Kodierung von Notfallmaßnahmen – emergency action code (EAC), dangerous goods emergency action code (DG-EA-Code) {GB}

Kohäsion – cohesion

Kohlehydrate – carbohydrates, carbs

Kohlendioxid (CO_2) – carbon dioxide

Kohlendioxidlöscher – carbon dioxide extinguisher

Kohlendioxidschnee – solid carbon dioxide, carbon-dioxide snow

Kohlenhydrate – carbohydrates, carbs

Kohlenmonoxid (CO) – carbon monoxide

Kohlensäure (H_2CO_3) – carbonic acid

Kohlensäuregas (CO_2) – carbon dioxide

Kohlensäureschnee (CO_2) – carbon-dioxide snow, solid carbon dioxide

Kohlenstoff (C) – carbon

Kohlenstoffdioxid (CO_2) – carbon dioxide

Kohlenstoffmonoxid – carbon monoxide

Kohlenwasserstoffe – hydrocarbons

Kolbenpumpe – reciprocating primer, reciprocating pump

Kolonne – convoy

Kolonnenfahrt – convoy of fire vehicles

Kombigurt – full-body harness

Kombinationsfilter – combined filter

Kombizange – all-purpose wrench, combination pliers

Kommando – command

Kommando übernehmen – assume command

Kommandowagen (KdoW) – command car

Kommunikationsmittel – means of communication

Kommunikationsstruktur – communications-structure

Komplettgurt – full-body harness

Komplexbildner – chelating agent

Kompression – compression

Kompression, adiabatische – adiabatic compression

Kompressionswärme – compression heat

Kompressor – compressor

Komprimierbarkeit – compressibility

Kondensat – condensate

Kondensatabscheider – condensate separator

Kondensation – condensation

Kondensationspunkt – condensation temperature

Kondensationswärme – heat of condensation

Konservierung – preservation

Konservierungsmittel – preservative

Kontaktgifte – contact poisons

Kontaktinfektion – contact infection

Kontamination – contamination

Kontaminationsgefahr – risk of contamination

kontaminationsgefährdet – susceptible to contamination

Kontaminationsmessgerät – contamination detector, contamination meter

Kontaminationsmonitor – contamination monitor (CoMo)

Kontaminationsnachweis – evidence of contamination, detection of contamination

Kontaminationsnachweisgerät (KNG) – contamination detector, contamination meter

Kontaminationsnachweisplatz – contamination detection area

Kontaminationsschutzanzug – contamination protective suit

Kontaminationsschutzhaube – contamination protection hood

Kontaminationsschutzkleidung – contamination protective clothing

Kontaminationsverdacht – suspicion of contamination, contamination suspected

Kontaminationsverschleppung – contamination spread, contamination carry-over

kontaminierte Verletzte – contaminated casualties

Kontrollbereich – control zone

Kontrolle – control

Kontrolleinrichtung – control equipment

kontrollierte Evakuierung – phased evacuation, controlled evacuation

Kontrollleuchte – indicator lamp, warning lamp, control lamp

Kontrollmaßnahmen – control measures

Kontrollpflicht – duty of control, inspection duty

Konvektion – thermal up current, convection

Konzentration – concentration

Konzentration, explosionsgefährliche – explosion hazard concentration

Konzentration, letale (LC) – lethal concentration

Konzentration, löschwirksame – extinguishing concentration

Konzentration, stöchiometrische – stoichiometric concentration

Konzentrationsgrenzen – concentration limits

Konzentrations-Zeit-Produkt – concentration-time product

Konzentrations-Zeit-Produkt, kampfunfähig machendes – incapacitating concentration-time product

Konzentrations-Zeit-Produkt, mittleres – mean concentration-time product

Konzentrations-Zeit-Produkt, mittleres letales – mean lethal concentration-time product

konzentriert – concentrated

Konzept – concept, plan

Kopfbebänderung – head harness

Kopf-Kiefer-Handgriff – jaw thrust manoeuvre

Kopierer – copy-machine

Korbtrage – basket stretcher

Koronaentladung – corona discharge

Körperflüssigkeiten – body fluids

körperliche Arbeit – physical work

Körperoberfläche – body surface

Körperschutz – personal protection

Körpertemperatur – body temperature

Korpuskularstrahlung – corpuscular radiation

Korrekturfaktor – corrective factor, correction factor

Korrosion – corrosion

Korrosionsgefahr – risk of corrosion

Korrosionsschaden – corrosion damage

Korrosionsschutz – corrosion protection, corrosion prevention

Korrosionsschutz, kathodischer (KKS) – cathodic corrosion protection (CCP)

Korrosionswirkung – corrosive effect

kosmische Strahlung – cosmic radiation, cosmic rays (CRs)

Kräfte, eingesetzte – involved forces

Krafteinwirkung – force effect

Kraftfahrzeug – motor vehicle

Kraftstoffkanister – gasoline canister, fuel canister

Krähenfuß (Brechwerkzeug) – crow bar

Kranausleger – jib

Kranführer – crane driver

Kranführerkabine – crane driver's cabin

Kranhaken – load hook, crane hook

Krankenbett, fahrbar – gurney

Krankenhaus – hospital

Krankenliege, mobile – gurney

Krankenschwester – nurse

Krankenschwester, examinierte – enrolled nurse

Krankentrage – stretcher

Krankentragen-Fahrgestell – (wheeled) stretcher carrier

Krankentragen-Träger – stretcher-bearer

Krankentransport – ambulance service

Krankentransportwagen (KTW) – patient transport ambulance (PTA)

Krankentransportwagen, 2 Tragen (2KTW) – twin-stretcher ambulance

Krankentransportwagen, 4 Tragen (4KTW) – four-stretcher ambulance

Krankenwagenfahrer – ambulance driver

Krankenwagenliege – wheeled stretcher

Krankheitserreger – disease-causing agent, pathogen

Krankmeldung – sick note, sick call

Kranwagen (KW) – mobile crane, crane truck

kratzfest – scratchproof

Krawattenknoten – knot of a tie

Krebs – cancer

krebserregend – cancerogenic

krebserzeugend – cancer causing, oncogenic

Kreide – chalk

Kreidetafel – blackboard

Kreiselpumpe – rotodynamic pump, centrifugal pump

Kreislaufgerät (Atemschutz) – rebreather, closed-circuit breathing apparatus

Kreislaufstillstand – circulatory arrest

Kreislaufversagen – acute circulatory failure

Kreisverkehr – roundabout, traffic rotary, traffic circle

Kreppband – masking tape

Kreuzknoten – square knot, reef knot, reef bend, crown knot, Carrick bend

Kreuzschraubendreher – cross-tip screwdriver

Kriegswaffenkontrollgesetz (KWKG) – War Weapons Control Act

Kriminalpolizei – Criminal Investigation Department

Krim-Kongo-Fieber – Crimean-Congo fever (CCF)

Krim-Kongo-Fieber, hämorrhagisches – Crimean-Congo haemorrhagic fever (CCHF)

Kriseninterventionsteam (KIT) – crisis intervention team (CIT)

Krisenstab – crisis unit, disaster management

Kristallwasser – crystal water

Kritikalitätssicherheit – criticality safety

Kritikalitätssicherheitskennzahl – criticality safety index (CSI)

kritische Infrastrukturen (KRITIS) – critical infrastructures

kritische Masse – critical mass, chain-reacting mass

kritische Temperatur – critical temperature

kritischer Druck – critical pressure

kritischer Durchmesser – critical diameter

kritischer Punkt – critical point

kritisches Gas – critical gas

Kronenkorken – crown cap

Krümmer (Rohrbogen) – elbow, pipe elbow

Kübelspritze – stirrup pump, bucket pump

kubisches Gesetz – cubic law

Küchenhandtuch – kitchen towel

Küchenrolle – kitchen paper towel

Kufen (z. B. Hubschrauber) – skids

Kugelhahn – ball cock, ball valve, spherical valve

Kugelhahn mit Hebelbetätigung – lever-operated ball valve

Kugellager – ball bearing

Kühleffekt – cooling effect

Kühlraum – cold store

Kühlung, direkte – direct cooling

Kühlung, indirekte – indirect cooling

Kühlverfahren – cooling treatment

Kunstfasern – man-made fibres, man-made fibers

Kunstharz – synthetic resin, ace resin

Kunstharzlack – synthetic resin varnish

künstliche Beatmung – artificial respiration

Kunststoffabfall – plastic waste

Kunststoffbehälter – plastic container

Kunststoffbrand – plastics fire

Kunststoffe – plastics

Kunststoffe, glasfaserverstärkte (GFK) – glass fibre reinforced plastics (GRP)

Kunststoffe, wärmebeständige – heat resistant plastics

Kupplungsschlüssel – spanner, coupling spanner

Kurzinformation – quick reference, brief information

kurzlebig (Isotope) – short-lived, short-living

Kurznachricht (Digitalfunk) – short data service (SDS)

Kurzschluss (elektr.) – electrical short, short circuit

Kurzzeitaufenthalt – short-term stay

Kurzzeitaufnahme – short-time exposure

Kurzzeitdosis – short-time dose

L

Lache (Pfütze) – pool, puddle

Lachenbrand – pool fire

Lackmus – litmus

Lackmuspapier – litmus paper

Ladegerät – charging set, charger

Laderaum – load compartments

Ladespannung – charging voltage

Ladestrom – charging current

Ladung, elektrische – electrical charge

Ladung, elektrostatische – electrostatic charge

Lage – situation

Lage, gesamte – overall situation

Lagebericht – situation report (SITREP)

Lagebesprechung – briefing meeting

Lagebeurteilung – assessment, estimate of the situation

Lagebewertung – evaluation of the situation

Lagedarstellung – representation, representation of the situation

Lageenergie – potential energy

Lageerkundung – situation reconnaissance

Lagefeststellung – establishing the situation, assessment of the situation

Lagekarte – situation map

Lagemeldung – situation report

Lageplan – ground plan

Lagerbereich – storage area

Lagergut – stored goods

Lagerhaus – warehouse

Lagerung (Personen) – position

Lagerung (Waren) – storage

Lageskizze – sketch of the situation

Lagevortrag – situation conference

Lähmung des Geruchssinns – paralysis of the sense of smell

Laibung – scuncheon

Laie – lay, lay person

Laienhelfer – lay helper, lay rescuer

laminar – laminar

laminare Flamme – laminar flame

laminare Strömung – laminar flow

Landebahnbeschäumung – foaming the landing runway, foaming the landing strip

Landebahnbeschäumungsanhänger – runway foaming tender

Landebahnbeschäumungsfahrzeug – runway foam laying device

Landrat – governor

landwirtschaftliche Gebäude – agricultural buildings

Längenausdehnung – linear thermal expansion

Längenausdehnungskoeffizient – linear thermal expansion coefficient

langlebig (Isotope) – long-lived, long-living

Lärmschutz – noise protection, noise prevention

Lassa-Fieber – Lassa fever

Last – load

Lastaufnahmemittel – load handling device, load lifting device, load lifting means

latente Wärme – latent heat

Latenzzeit – latent period, latency time

Latte – lath, slat, batten

T. Schmiermund, *Fachwörterbuch Feuerwehr und Brandschutz*, https://doi.org/10.1007/978-3-662-64120-0_11

Latthammer – roofing hammer

Lattung – lathing, lathwork

Laubwald – deciduous wood

Lauffeuer – wildfire

Laufrad (Pumpe) – rotor

Lauge – lye, basic solution, alkaline solution

laugenbeständig – alkali-proof, lye-proof

Lawine – avalanche

Lawinenhund – avalanche dog

Lawinenrettungshund – avalanche rescue dog

Lawinensuchhund – avalanche search dog

lebendbedrohend – life-threatening

lebensbedrohender Zustand – critical condition

Lebensdauer – lifetime

Lebensgefahr – mortal danger, risk of life

Lebenskraft – vitality

Lebensrettende Maßnahmen – life-saving measures

lebensrettende Sofortmaßnahmen (LSM) – basic life support (BLS), immediate life-saving measures

lebensrettender Handgriff – head-tilt and chin-lift manoeuvre (HTCL-manoeuvre)

Leck – leak

Leckage – leakage

Leckverlust – leakage loss

Lederstiefel – leather boots

Leermasse – tare weight, unloaded weight

Legende (Kartenkunde) – key, explanation of the map

Legierung – alloy

Lehrmittel – training aids, teaching aids, teaching material

Lehrplan – curriculum, lesson plan

Lehrrettungsassistent – paramedic instructor

Leichenspürhund – cadaver dog

leicht flüchtig – highly volatile

Leichtbenzin – light gasoline

Leichtbeton – lightweight concrete

leichtbrennbar – easily combustible, highly combustible

leichtentflammbar – highly flammable

Leichtentflammbarkeit – high flammability

leichtentzündbar – highly flammable, readily flammable

leichtentzündlich – easily ignitable

Leichtschaum – high expansion foam (HX)

Leichtschaumgenerator (LG) – foam generator (FG)

Leichtwasser – deuterium-depleted water

Leichtwasserreaktor (LWR) – light water reactor (LWR)

Leinensicherungssystem – line securing system

Leinöl – linseed oil

Leistung – power, engine power

Leistungsabgabe – power drain, output

Leistungsaufnahme – power consumption

Leistungsfähigkeit – efficiency

leitender Notarzt (LNA) – emergency doctor in charge

Leiter (für elektr. Strom) – conductor

Leiter (Person) – chief, leader, manager, warden

Leiter (Steiggerät) – ladder

Leiter des Brandschutzes – chief fire prevention officer

Leiter erster Ordnung – first-order conductor

Leiter zweiter Ordnung – second-order conductor

Leiteranstellwinkel – proper angle of a ladder

Leiterfahrzeug – ladder truck

Leiterfuß – ladder heel, shoe

Leitergetriebe (Drehleiter) – ladder mechanism

Leiterholm – ladder stringer

Leiterkopf – ladder head

Leiterkorb (Drehleiter) – aerial ladder platform

Leiterlänge – length of the ladder

Leitern, tragbare – portable ladders

Leiterteil – section of ladder

Leitfähigkeit (elektrische) – conductivity, electric conductivity

Leitfähigkeit (Wärme) – thermoconductivity, thermal conductivity

Leitplanken – crash barriers

Leitrad (Pumpe) – diffuser

Leitstelle – control centre, dispatch center, information and control center (ICC)

Leitstellendisponent – emergency dispatcher, dispatcher

Leitstellendisponent (Rettungsdienst) – emergency operator, 911-operator

Leitsubstanz – lead substance, lead structure

Leitung (Energie) – conduit, line, wire, pipeline, cable, conductor, cord

Leitung (Führung) – command, management, guidance

Leitungsquerschnitt – cable cross section

Leitungsstab – ‚crisis cell'

Leitungswasser – tap water

Lenzpumpe – bilge pump, light portable pump (LPP)

letale Dosis (LD) – lethal dose (LD)

letale Dosis, mittlere (LD_{50}) – median lethal dose (LD_{50})

letale Expositionsdosis – lethal exposure dose

letale Konzentration – lethal concentration

letale Konzentration (LC) – lethal concentration

letale Konzentration, mittlere (LC_{50}) – median lethal concentration (LC_{50})

letales Konzentrations-Zeit-Produkt, mittleres – mean lethal concentration-time product

Letalität – lethality

Leuchtstoffröhre – fluorescent tube

Lewisit – lewisite

Lichtäquivalent – light equivalent

Lichtbalken – light bar

Lichtbogen – electric arc, arc of light

Lichtbogenschweißen – electric-arc welding

Lichtgeschwindigkeit – speed of light, light speed

Lichthof – atrium, cortile

Lichtkuppel – light dome

Lichtmast – light mast, light pole

Lichtmastfahrzeug – light pole vehicle

Lichtquanten – light quanta, light corpuscles

Lichtschacht – light well

Lichtschranke – light barrier

Lichtschwächung (durch Rauch) – opacity (of smoke)

Lichtstrahl – light beam

Lineal – ruler

lineare Abbrandgeschwindigkeit – linear burning rate

Locheisen – hole punch

Locher – puncher, office puncher

logistische Unterstützung – logistical support

Lokalelement – local microgalvanic cell, local galvanic element

Löschangriff – fire attack

Löschangriff, direkter – direct attack, direct fire attack

Löschangriff, indirekter – indirect attack, indirect fire attack

Löschangriff, umfassender – surrounding fire attack

Löschanlage – fire-extinguishing system

Löscharbeiten – fire-fighting operations

Löschdampf – extinguishing vapour

Löschdecke – fire blanket

Löschdüse (Sprinkler) – extinguishing nozzle

Löscheffekt – extinguishing effect

Löscheffektivität – extinguishing efficiency

Löscheinrichtung – extinguishing device

Löscheinsatz – fire-fighting operation

löschen – extinguish

Löschfahrzeug – fire engine, fire tender, (fire truck), (fire-fighting truck)

Löschgas – extinguishing gas, gaseous extinguishing agent

Löschgeräte – extinguishing equipment, fire-fighting equipment

Löschgruppe – fire-fighting group

Löschgruppenfahrzeug (LF) – fire engine (group fire-fighting vehicle)

Löschhubschrauber – fire-fighting helicopter

Löschintensität – intensity of extinguishing agent supply

Löschlanze – fire-extinguishing lance

Löschmannschaft – team of fire-fighters, team of firemen

Löschmechanismus – extinguishing mechanism

Löschmethode – extinguishing method

Loschmidt-Konstante – Loschmidt number

Löschmittel – extinguishing agent, extinguishing medium, extinguishing substance

Löschmittel für Metallbrände – fire extinguishing agent for metals

Löschmittel, feste – solid fire-extinguishing agents

Löschmittel, flüssige – liquid fire-extinguishing agents

Löschmittel, gasförmige – gaseous fire-extinguishing agents

Löschmittelbedarf – need of extinguishing agent, requirement of extinguishing agent

Löschmittelbehälter – container for extinguishing agents

Löschmitteleinheit – extinguishing unit, extinguishing agent unit

Löschmittelmenge – quantity of extinguishing agent

Löschmittelrate – extinguishing rate

Löschmittelschaden – damage caused by extinguishing agent

Löschmittelstrahl – extinguishing agent jet

Löschmittelversorgung – supply of extinguishing agent

Löschmittelvorrat – stock of extinguishing agent

Löschmittelzufuhr – supply of extinguishing agent

Löschmittelzusätze – extinguishing agent additives

Löschnagel – fog nail

Löschpanzer – fire-fighting tank

Löschpulver – extinguishing powder, extinguishing dry powder

Löschpulverrate – extinguishing powder rate

Löschschaum – foam, extinguishing foam

Löschstaffel – fire squad

Löschstrategie – fire extinguishing strategy

Löschsysteme – fire-extinguishing systems

Löschtaktik – fire-fighting tactics

Löschtechnik – fire-fighting techniques

Löschteich – fire-fighting pond

Löschtrupp – fire-fighting team, extinguishing crew

Löschübung (einsatzmäßig) – fire drill

Löschübung (Unterricht) – extinguishing exercise

Löschverfahren – extinguishing method

Löschvermögen – extinguishing capacity

Löschversuch – attempt to extinguish a fire

Löschvorgang – extinguishment

Löschwasser – fire-extinguishing water, extinguishing water, water for fire-fighting

Löschwasser-Außenlastbehälter – helicopter bucket

Löschwasserbedarf – extinguishing water demand, need of extinguishing water

Löschwasserbehälter – extinguishing water tank, extinguishing water reservoir

Löschwasserbrunnen – fire well

Löschwassereinspeisung – extinguishing water supply, water supply for fire-fighting

Löschwasserentnahmestelle – water supply point

Löschwasserförderung – delivery of water

Löschwasserförderung über lange Wegstrecken – long-distance delivery of water

Löschwasserleitung – extinguishing water pipe

Löschwassermenge – quantity of water, quantity of extinguishing water

Löschwasser-Pendelverkehr – shuttle water relay

Löschwasserreserve – reserve of extinguishing water

Löschwasserrückhaltung – extinguishing-water retention

Löschwasserteich – fire pond

Löschwasserversorgung – water supply for fire-fighting, supply of extinguishing water, water supply

Löschwasserversorgung, abhängige – dependent water supply

Löschwasserversorgung, erschöpfliche – exhaustible water supply for fire-fighting

Löschwasserversorgung, unabhängige – independent water supply

Löschwasserversorgung, unerschöpfliche – inexhaustible water supply for fire-fighting

Löschwasserzuleitung (Schlauch) – extinguishing water supply pipe

Löschwasserzuleitung (Tätigkeit) – supply of extinguishing-water

Löschwasserzusatz – water additive

löschwirksame Konzentration – extinguishing concentration

Löschwirksamkeit – extinguishing efficiency

Löschwirkung – extinguishing effect

Löschzeit – extinguishing time

Löschzug (LZ) – fire-fighting unit, fire-fighting company, fire brigade

Lösemittel – solvent

Lösemittel, brennbare – flammable solvents

Lösemittel, organische – organic solvents

Lösemitteldampf – solvent vapour

Löseverhalten – dissolving behaviour

Lösevermögen – dissolving capacity

Löslichkeit – solubility

Löslichkeitskurve – mutual solubility curve, solubility curve

Lost – mustard gas

Lösung – solution

löten – solder

Lötlampe – soldering lamp, blowlamp

Luft – air

Luftbedarf – air need

Luftbedarf, theoretischer – theoretical air need

Luftbewegung – air movement

Luft-Dampf-Gemisch – air-vapor-mixture

Luftdruck – air pressure

Luftdruck, atmosphärischer – atmospheric pressure

Luftdruckprüfer (Reifendruck) – air gauge

Lufteinlass – air inlet

Lüfter – exhaust fan, exhauster, ventilator, fan

Luftfeuchtigkeit – air humidity

Luft-Gas-Gemisch – air-gas-mixture

Luftgeschwindigkeit – air velocity, airflow velocity

Lufthebekissen – air-lifting units, ‚air-bags‘

Luftheber (LH) – lifting bag

Luftkammerschiene – air splint

Luftmangel – lack of air

Luftrettung – air rescue

Luftrettungsdienst – air rescue service

Luftsauerstoff – atmospheric oxygen

Luftschaum – air foam

Luftschaumrohr – air foam branch pipe

Luftschutzraum – air raid shelter

Luftüberschuss – excess of air

Lüftung – ventilation, airing, exhaust

Lüftungsanlage – ventilation system

Lüftungskanal – ventilation duct

Lüftungsschacht – ventilation shaft

Luftverflüssigung – air liquefaction

Luftverhältnis – ratio of air

Luftverschmutzung – air pollution

Luftvolumen – air volume

Luftwechsel – exchange of air

Luftzufuhr – air supply

Lumineszenz – luminescence

Lungenautomat – lung demand valve

Lungenkampfstoffe – pulmonary agents, choking agents

Lungenmilzbrand – inhalation anthrax, lung anthrax

Lungenpest – pneumonic plague

Lutte – air-ducting system

M

Magnesiumfackel – magnesium torch

Magnetfeld – magnetic field

magnetische Feldstärke – magnetic field strength

Magnetventil – magnetic valve

Mängel – deficiencies

Mängelbericht – deficiency report

Manipulationen – tampering

manipulieren – tamper

Mannschaftsraum (Fahrzeug) – crew cab

Mannschaftstransportfahrzeug (MTF) – fire dept crew car, personnel car

Mannschaftsunterkunft – fire-fighter's quarters, firemen's quarters

Manometer – pressure gauge

Manometer für Über-/Unterdruck – pressure-vacuum gauge

manometrische Druckhöhe – manometric pressure height

manometrische Förderhöhe – manometric discharge height

manometrische Saughöhe – manometric suction height

Marburg-Virus – Marburg virus

Markierung – marking

Martinshorn – fire siren

Maschinist – pump operator, machinist, engine driver

Maske – mask

Maskenbrille – mask glasses

Masse, kritische – critical mass, chain-reacting mass

Masse, molare – molar mass

Massenabbrandgeschwindigkeit – mass burning rate

Massenanfall von Verletzten (MANV) – mass casualty incident (MCI), multiple-casualty incident (MCI)

Massenanteil – mass fraction

massenbezogene Abbrandrate – mass burning rate

Massenbrand – mass fire

Massendefekt – mass defect

Massenkonzentration – mass concentration

Massenvernichtungswaffen – weapons of mass destruction (WMD)

Massestrom – mass flow

massiv – solid, massive

Massivbauweise – solid construction

Maßnahme – action

Maßnahme, bauliche – structural measure

Maßnahme, vorbeugende – preventive action

Maßnahmen, administrativ-organisatorische – administrative-organisational-measures

Maßnahmen, ergänzende – supplementary actions

Maßnahmen, operativ-taktische – operative-tactical-measures

Maßnahmen, Reihenfolge der – sequence of measures

Maßstab (Kartenkunde) – scale

Mastwurf – clove hitch

Mastwurf, doppelter – double clove hitch

Material, spaltbares – fissionable material

T. Schmiermund, *Fachwörterbuch Feuerwehr und Brandschutz*, https://doi.org/10.1007/978-3-662-64120-0_12

Materie, Aufbau der – composition of matter

Materiewellen – matter waves

Mauerwerk – masonry

Maurerhammer – brick hammer, mason's hammer

maximale Arbeitsplatzkonzentration (MAK) – maximum permissible workplace concentration

maximale Herzfrequenz – maximum heart rate (MHR)

maximaler Druckanstieg – maximum pressure-rise

maximaler Explosionsdruck – maximum explosion-pressure

maximaler zeitlicher Druckanstieg – maximum pressure-rise at to time

Maximalmelder – maximum detector

mechanisch – mechanical, physical

mechanische Energie – mechanical energy

mechanische Festigkeit – mechanical strength

medizinische Gase – medical gases

medizinische Handschuhe – medical gloves

medizinischer Notfall – medical call

Mehrbereichsöl – multi-grade oil

Mehrbereichsschaummittel (MBS) – multi-purpose foam compound, multi-purpose foam

mehrgeschossig – multi-stored

Mehrkammerbehälter – multi-chamber container

Mehrzweckstrahlrohr – multi-use branch pipe; hand-controlled branch pipe, branch pipe, multi-purpose branch pipe

Mehrzweckzug – come-along

Meldeempfänger – pager

Meldeempfänger, digitaler (DME) – digital paging device

Meldekette – chain of reporting

Meldekopf – contact point

Melder (Brandmelder) – detector

Melder (Funktion) – communication assistant, ‚runner'

Meldergehäuse – detector housing

Meldergruppe – group of detectors

Meldesystem – report system

Meldungen – reports, messages

Meldungen abgeben – transmit of reports

Melioidose – melioidosis

Menge an Gefahrstoff – hazmat quantity, quantity of dangerous material

Mengenverhältnis – proportion, quantity ratio

Meningitis – meningitis

Menschenrechte – human right's

Menschenrettung – human rescue, rescue of people, saving human life

Menschenwürde – human dignity

menschliches Fehlverhalten – human error

Mercator-Projektion (Kartenkunde) – Mercator projection

Mercator-Projektion, universelle transversale (UTM) – universal transversal Mercator projection

Merkblätter – leafleats, explanatory leaflets, handouts

Merkmal – character, attribute, feature

Messbecher – measuring cup

Messgeräte – measuring equipment, measurement equipment

Messing – brass

Messingdrahtbürste – brass wire brush

Messkammer – measuring chamber

Messung der Strahlenbelastung – radiation-level check

Messunsicherheit – measuring uncertainty

Messwerkzeug – measuring tool

Messwert – measured value

Metallbindung – metallic bond

Metallbrand – metal fire

Metallbrandpulver – metal fire extinguishing powder

Methan – methane

METHANE-Merkschema: Einsatzdetails – exakte Ortsangabe – Art und Zeit des Vorfalls – Gefahren in der Umgebung – Anfahrtswege – Verletztenanzahl – erwartete Reaktion – METHANE-mnemonic: *m*ission details – *e*xact location – *t*ime and *t*ype of incident – *h*azards in the area – *a*pproach routes – *n*umber of casualties – *e*xpected response

metrische Pferdestärke – metric horsepower (mhp)

Milchrohrverschraubung – dairy coupling, German food coupling

Milliarde – billion (1 000 000 000, 10^9)

Millimeterpapier – metric graph paper, millimeter paper

Million – million (1 000 000, 10^6)

Milzbrand – anthrax

Mindestanforderung – minimum requirement

Mindestbetriebsdruck – minimum operating pressure

Mindestbetriebszeit – minimum operating time

Mindestdruck – minimum pressure

Mindestfeuerwiderstandsdauer – minimum fire resistance

Mindestluftbedarf – minimum air requirement

Mindestsauerstoffbedarf – minimum oxygen requirement

Mindestverbrennungstemperatur – minimum burning temperature

Mindestzünddauer – minimum ignition time

Mindestzündenergie – minimum ignition energy

Mindestzündstrom – minimum gate-trigger current

Mindestzündtemperatur – minimum ignition temperature, minimum auto-ignition temperature

Mineralfaser – mineral fibre

Mineralöl – mineral oil

mineralölbeständig – mineral-oil resistant

Mineralwolle – mineral wool

minimale Behandlung – minimal treatment (MT)

minimaler zündgefährlicher Sauerstoffgehalt – minimal ignition-hazardous oxygen concentration

Minimierungsgebot – minimization requirement

Mischbarkeit – miscibility

Mischbarkeit mit Wasser – miscibility with water

Mischelement – mixing element

Mischphasen – mixed phases

Mischung – mixture, blend

Mischungslücke – miscibility gap

Mischungsregel – dilution rule

Mischungstemperatur – mixing temperature

Mischwald – mixed wood

Missbildungen verursachend – teratogenic

Mist (Dung) – manure

Mistgabel – pitchfork, manure fork

mit dem Rettungswagen eingeliefert – brought in by ambulance (BIBA)

Mittel – means

Mittel, flammhemmendes – flame retardant product

Mittel, richtige (Einsatzmittel) – right means

Mittel, zugewiesene – assigned means

Mittelbrand – medium fire

Mitteldruck – medium pressure

Mitteldruckleitung – medium pressure line

Mitteleinbaupumpe – mid ship mounted pump

mitteleuropäische Sommerzeit (MESZ) – Central European Summer Time (CEST)

mitteleuropäische Zeit (MEZ) – Central European Time (CET)

Mittelschaum – medium expansion foam (MX)

Mittelschaumrohr – medium expansion foam branch, medium expansion foam branch pipe

mittlere Abbrandgeschwindigkeit – mean burning rate

mittlere letale Dosis (LD_{50}) – median lethal dose (LD_{50})

mittlere letale Konzentration (LC_{50}) – median lethal concentration (LC_{50})

mittleres Konzentrations-Zeit-Produkt – mean concentration-time product

mittleres letales Konzentrations-Zeit-Produkt – mean lethal concentration-time product

mobile Krankenliege – gurney

mobiler Rauchverschluss – portable smoke blocker, portable smoke curtain

Modell für Effekte mit toxischen Gasen (MET) – model for effects with toxic gases

Modulbauweise – modular construction, building-block design

Möglichkeit – possibility

Mol – mol, mole

Molalität – molality

molare Masse – molar mass

molares Volumen – molar volume

Molekül – molecule

Molekularbewegung – molecular motion

Molekülmasse – molecular mass

Molekülmasse, relative – relative molecular mass (RMM)

Molmasse – molar mass

Molotow-Cocktail – Molotov-cocktail (an incendiary composition)

Moorbrand – marsh fire

Morbidität – morbidity

Mortalität – mortality

Mörtel – mortar

Motorbremse – exhaust brake

Motordrehzahl – engine revolution

Motorkettensägenkette – chainsaw chain

Motorleistung – motor power

Motoröl – engine oil

Motorsäge – chain saw

Motorsäge, benzinbetrieben – gas chain saw, petrol-engined saw

Motorschutzschalter – motor circuit switch, motor circuit breaker

Mulde – hollow, trough

Müllhalde – dump, landfill, garbage dump

Müllhaldenbrand – dump fire, landfill fire

Multifunktionsleiter – multi-function ladder

Multifunktionswerkzeug – multitool

mündlicher Befehl – oral order

Mund-Nase-Bedeckung (MNB) – face covering, cloth face mask

Mundstück (Strahlrohr) – mouth piece, jet pipe mouth piece

Mund-zu-Mund-Beatmung – mouth-to-mouth resuscitation

Mund-zu-Nase-Beatmung – mouth-to-nose resuscitation

Münzmetalle – coin metals

Muschelvergiftung (durch STX verursacht) – paralytic shellfish poisoning (PSP)

Mützenabzeichen – cap badge

Mykotoxin – mycotoxin

N

Nachalarmierung – post-alerting

nachbarliche Hilfe – mutual aid

Nachbarraum – adjacent room, adjoining room

Nachbarschaft – neighborhood

Nachbarschaftshilfe – neighbourly help

Nachbarschaftsschutz – neighborhood protection

Nachbrennen – after flame

Nachbrennzeit – after burning time

Nachforderung – additional call

Nachfüllung – refill, replenishing

Nachkontrolle – follow-up check

nachleuchtend – phosphorescent

Nachlöscharbeiten – post-extinguishing work, follow-up extinguishing work

Nachlöschen – follow-up extinguishing

Nachmeldung – supplemental report

Nachricht – dispatch, message

Nachschlagewerke – reference books

Nachschub – ordnance

Nachsteiger (Höhenrettung) – second climber

Nachstieg (Höhenrettung) – second climbing

Nachtbeleuchtung – night lighting

Nachteil – disadvantage

Nachtschicht – nightshift

Nachtsichtgerät – night-vision device, night observation device (NOD)

Nachtübung – night drill

Nachuntersuchung – follow-up examination, check-up, reexamination

Nachweisen – detect, verify, prove

Nachweisgeräte – detection devices

Nachweisgrenze – identification limit

Nachweisgrenze, untere – minimum detection limit (MDL)

Nacken – nape

Nackenschutz – neck guard, neck protection, neck cover

Nadelwald – coniferous wood

Nagelbettprobe (NBP) – capillary refill time (CRT)

Nagelzieher – pry bar

Nahaufnahme – close-up view

Nähe – vicinity

Namensschild – name plate, name tag

Namensschildchen (an der Kleidung) – name badge

Narkose – anesthesia

Nasen-Rachen-Reizstoffe – nasopharyngeal agents

Nasopharyngealtubus – nasopharyngeal tube

Nasslöschverfahren – extinction by water (wet extinguishing process)

Natrium – sodium

Natronlauge – caustic soda lye, sodium lye

natürliche Hintergrundstrahlung – natural background radiation

Naturwissenschaften – sciences, natural sciences

Nebel – mist, fog

Nebelgerät – smoke-generating device

Nebellöschverfahren – fog extinction

T. Schmiermund, *Fachwörterbuch Feuerwehr und Brandschutz*, https://doi.org/10.1007/978-3-662-64120-0_13

Nebelstrahl – water fog

Nebenantrieb – secondary drive, power take-off (PTO)

Nebengruppe – side group

Nebengruppenelement – side-group element

Nebenlöscheffekt – secondary effect of extinguishing

Nebenwache – sub-station

neigen – depress

Neigung – angel of inclination

Neigungsmesser – inclinometer

Nennausbeute – notional yield

Nenndrehzahl – nominal speed

Nennfestigkeit – tensile strength

Nennförderhöhe – nominal discharge height

Nennförderleistung – nominal discharge, nominal displacement

Nennfördermenge – nominal delivery rate

Nennleistung – nominal power, nominal capacity

Nennöffnungstemperatur (Sprinkler) – nominal opening temperature

Nennrettungshöhe – nominal rescue height

Nennweite (NW) – nominal width, nominal diameter

Nennwert – nominal value

Nernstsche Gleichung – Nernst equation

Nervengas – nerve gas

Nervenkampfstoffe – nerve agents

Nesselstoff (Kampfstoff, Rotkreuz) – urticant, nettle agent

Netzausfall (Strom) – power failure

Netzdruck (Wasserversorgung) – system pressure

Netzersatzanlage (NEA) – emergency standby power system (ESPS)

Netzmittel – wetting agent

Netzmodus (Digitalfunk) – trunked mode operation (TMO)

Netzwasser – wet water

Netzwerk, drahtloses, lokales – wireless local area network (WLAN)

Netzwerk, drahtloses, privates – wireless private area network (WPAN)

Netzwerk, lokales – local area network (LAN)

Netzwerk, virtuelles, privates – virtual private network (VPN)

Neutralisation – neutralization

Neutralisationsmittel – neutralization agent, neutralizer

Neutralisationswärme – neutralizing heat, heat of neutralization

Neutralpunkt – neutral point

Neutron – neutron

Neutronen, thermische – thermal neutrons

Neutronenaktivierungsanalyse – neutron activation analysis

Neutronenfluss – neutron flux

Neutronenstrahlung – neutron radiation

Neutronenzähler – neutron counter

nicht betriebsbereit – inoperable

nicht brennbar – non-combustible

nicht entflammbar – non-inflammable

nicht fahrlässig – non-negligent

nicht flüchtig – nonvolatile

nicht ortsfest – non-stationary

nicht selbstentzündlich – non-self-ignitable

nicht sicher – not safe, unsafe

nicht verwendbar – non-usable

nicht wässrig – non-aqueous

nichtionisch – nonionic

nichtionisierende Strahlung – non-ionizing radiation

Nichtnotfallpatient – non-emergency patient

nichtöffentliche Feuerwehr – non-public fire brigade

Niederschlag, radioaktiver – fallout

Niederschraubventil – landing valve, screw-on valve

Niederspannung – low voltage

Niederspannungsanlage – low voltage installation

Niederspannungsverteiler – low voltage distribution

nitrose Gase – nitrous gases

Niveauregulierung – level control

Norm – standard

normal brennbar – normally combustible

normal entflammbar – inflammable, normally inflammable

normal entzündlich – normally flammable

Normalbedingungen (25 °C, 1013 mbar) – normal conditions (25 °C, 1013 mbar)

Normaldruckpumpe – normal pressure pump

Normalwasserstoffelektrode – normal hydrogen electrode (NHE)

Normalzustand – normal state

Normbedingungen (0 °C, 1013 mbar) – standard conditions (0 °C, 1013 mbar)

Normen des Brandschutzes – fire standards

Normkubikmeter (Gase) – standard cubic metre

Normzustand (0 °C, 1013 mbar) – standard conditions, standard temperature & pressure (STP)

Notarzt – emergency doctor

Notarzteinsatz – emergency doctor mission

Notarzteinsatzfahrzeug (NEF) – emergency doctor's car

Notarztwagen (NAW) – doctor's incident vehicle, doctor's emergency ambulance

Notaufnahme – casualty department, casualty

Notaufnahme (Einrichtung) – accident and emergency (A&E), emergency room (ER), emergency department (ED)

Notaufnahme (Patient) – emergency admission, emergency hospitalization

Notaufnahmeabteilung – casualty department, emergency ward (EW)

Notausgang – emergency exit, fire exit

Notausstieg – escape hatch, emergency exit

Notbeleuchtung – emergency lighting, emergency light

Notbetrieb – emergency mode

Not-Dekontamination – emergency decontamination

Notdienst – emergency services

Notdusche – safety shower, emergency shower

Notfall, medizinischer – medical call

Notfallarmband – medical ID alert bracelet

Notfalleingang – emergency entrance

Notfalleinsatz – medical call

Notfallevakuierung – emergency evacuation

Notfallfluchtmaske – emergency escape mask

Notfallkoffer – first-aid case, emergency case

Notfallmaßnahmen – emergency action

Notfallpatient – emergency patient

Notfallrucksack – emergency backpack

Notfallsanitäter (NotSan) – emergency medical technician-paramedic (EMT-P), paramedic, emergency medical technician

Notfallstation (NFS) – emergency ward (EW)

Notfallübung – emergency drill

Notfallversorgung, psychosoziale (PSNV) – psychosocial emergency care

Nothammer – life hammer, emergency hammer

Notkühlung – emergency cooling

Notnagel – emergency nail, stopgap

Notruf – emergency call, 112 call {EU}, 999 call {GB}, 911 call {US}

Notruf, falscher – bogus emergency call

Notrufabfragestelle – public-safety answering point (PSAP)

Notrufdienst – emergency call service

Notrufkanal (Funk) – emergency line, emergency channel

Notrufnummer – emergency number

Notrufsäule – roadside emergency telephone

Notrufzentrale – emergency call centre

Notrutsche – emergency chute

Notsignal – signal of distress, distress signal

Notsignalgeber (Atemschutz) – personal alert safety system (PASS), distress signal unit (DSU), automatic distress signal unit (DSU), personal distress alarm (PDA)

Notsignalgeräte – distress-signal devices

Notstand – state of calamity, state of emergency

Notstrom – emergency power

Notstromaggregat – emergency power supply unit (EPSU), emergency generator, emergency power aggregate

Notstromaggregat, mit Diesel betrieben – diesel emergency power aggregate

Notstromakku – emergency power battery

Notstromanlage – emergency power source

Notstromerzeuger – emergency power generator

Notstromversorgung – emergency power supply, backup power supply

Nottreppe – emergency staircase

Notunterkunft – emergency shelter

notwendige Treppe – necessary stair

Notwendigkeit – necessity

Notwendigkeit, zwingende – compelling necessity

Notzeichen – signal of distress, distress signal

Nowitschok – Novichok

nukleare Kettenreaktion – nuclear chain reaction

Nuklearforschung – nuclear research

Nuklearmedizin – nuclear medicine

Nuklide – nuclides

Nuklidkarte – chart of nuclides

Nullleiter – neutral wire

Nullpunkt, absoluter (Temperatur) – absolute zero

Nullrate – zero rate

Nullzeit (Tauchen) – no decompression limit (NDL)

Nummer zur Kennzeichnung der Gefahr – hazard identification number, Kemler number

Nur für Amtsgebrauch – for official use only (FOUO)

Nur für Dienstgebrauch (NfD) – for internal use only (FIUO)

Nut (Nut-Feder-Verbindung) – groove

Nut-Feder-Verbindung – tongue and groove joint (TG)

Nutzfeuer – useful fire

Nutzleistung – effective power, usable power

Nutzung – usage

O

Obdachlosenheim – homeless shelter

obere Explosionsgrenze (OEG) – upper explosion limit (UEL)

oberer Explosionspunkt – upper explosion point

oberer Heizwert – → Brennwert

Oberfläche, heiße – hot surface

Oberflächenbrand – surface fire

Oberflächenspannung – surface tension

Obergeschoss (OG) – upper floor

Oberlicht – skylight, fanlight

obstruktive Erkrankung – obstructive disease, obstructive disorder

Odoriermittel – odorant

Odorierung – odorization

offene Flammen – open flames

offene Tuberkulose – active tuberculosis

offene Wunden – open wounds

offener Brand – open fire, outside fire

offenes Gewässer – open water

öffentliche Feuerwehr – public fire brigade

öffentliche Wasserleitung – public water main

öffentliche Wasserversorgung – public water supply

Ohm'sches Gesetz – Ohm's law

Oktanzahl – octan number, octane rate

Oktettregel – octet rule

Ölablassschraube – oil drain plug

Ölabscheider – oil trap, oil separator

Ölbindemittel – oil binding agent, oil adsorbent

Ölbinder – oil binder, oil adsorbent

Ölbrenner – oil burner

Ölheizung – oil heating

Ölheizungsanlage – oil-fired heating system

Ölkanne – oil can, hand oiler

Ölnachweispaste – oil finding paste, fuel finding paste, gasoline finding paste

Ölskimmer – oil skimmer

Ölsperre – oil barrier, oil spill barrier, oil boom

Ölspur – oil on road

Ölstandkontrolle – oil level check

Ölstandkontrollleuchte – oil level warning light

Ölunfall – oil accident, accident with oils

Ölverbrauch – oil consumption

Ölverschmutzung – oil spill

Ölwechsel – oil change

Ölzentralheizung – oil central heating

operativer Dienst – operational duty

operativ-taktische Adresse (OPTA) – operative-tactical address (OPTA)

operativ-taktische Maßnahmen – operative-tactical-measures

Opfer – victim

optische Rauchdichte – optical density of smoke

optische Signale – optical signals

optischer Rauchmelder – scattered-light smoke detector

Orden – medal, decoration

Ordensband – medal ribbon, collar band

Ordensspange – service ribbon

T. Schmiermund, *Fachwörterbuch Feuerwehr und Brandschutz*, https://doi.org/10.1007/978-3-662-64120-0_14

Ordnungswidrigkeit – regulatory offence, administrative offence

Ordnungszahl – proton number, atomic number

Organdosis – organ dose, equivalent dose

Organisation für das Verbot chemischer Waffen – Organisation for the Prohibition of Chemical Weapons (OPCW)

Organisatorischer Leiter Rettungsdienst (OrgL, OLRD) – chief ambulance officer

organische Chemie – organic chemistry

organische Lösemittel – organic solvents

organische Peroxide – organic peroxides

Orientierungshilfe – guidance

Oropharyngealtubus – oropharyngeal tube

Ort – locality

Ortgang – bargeboard

ortsbeweglich – movable, portable

Ortsdosis – local dose

Ortsdosisleistung – local dose rate

ortsfest – stationary, immovable

ortsfeste Anlage – fixed installation

ortsfeste Befehlsstelle – fixed command post

Ortsfeuerwehr – local fire brigade

Öse – loop, dee

Oxidation – oxidation

oxidationsempfindlich – oxidation-sensitive

Oxidationsgeschwindigkeit – oxidation rate

Oxidationsgrad – oxidation degree

Oxidationskatalysator – oxidizing catalyst

Oxidationsmittel – oxidizing agent, oxidant

Oxidationsprozess – oxidation process

Oxidationsstufe – oxidation state

Oxidationsverhinderer – oxidation inhibitor

Oxidationsvorgang – oxidation process

Oxidationszahl – oxidation number

oxidierbar – oxidable

oxidierend wirkende Stoffe – oxidizing substances

P, Q

Paarbildung – pairing

Paarbildungsenergie – pairing energy

Paarvernichtung – annihilation

Pahlstek (Knoten) – bowline

Panikgefahr – risk of panic

Panikstange – panic bar

Panzerband – duct tape, Duck Tape™

Parallelschaltung – parallel connection

Parkposition – parking position

Partialdampfdruck – partial vapour pressure

Partialdruck – partial pressure

Partialdrücke, Gesetz der – law of partial pressures

Partikelfilter (Atemschutzmaske) – particulate respirator

Passivität – passivity

pastös – pasty

Pathogenität – pathogenicity

Patienten – patients

Patienten, ambulante – outpatients

Patienten, stationäre – inpatients

Patiententransport – patient transfer

Patienten-Transport-Einheit (PTE) – patient transport unit (PTU)

Patiententransportwagen – gurney

PECH-Regel (Pause – Eis – Compression – Hochlagern) – RICE-principle (rest – ice – compression – elevation)

Pedal – pedal

Pedalschneidgerät – pedal cutting device, pedal cutter

Pendelatmung – to-and-fro breathing

Pendelatmung – pendulum breathing

Pendelatmungssystem – to-and-fro system

Pendelverkehr (Löschwasser) – shuttle water relay

Pensionierung – retirement, act of retiring

per-/polyfluorierte Chemikalien (PFC) – per-/poly-fluorinated carbons (PFC)

Perfluoroctansäure – perfluorooctanoic acid (PFOA)

Perfluoroctansulfonsäure (PFOS) – perfluorooctanesulfonic acid (PFOS)

Perfluortenside – per-fluorinated tensids (PFT)

Periodensystem der Elemente (PSE) – Periodic Table of the Elements (PTE)

permanente Härte (Wasser) – permanent hardness of water

permanentes Gas – permanent gas

Peroxide – peroxides

Peroxide, organische – organic peroxides

Peroxyessigsäure (PES) – peroxyacetic acid (PAA)

persistent – persistent

Person, bewusstlose – unconscious person

Person, eingeschlossene – enclosed person

Person, erkrankte – ill person, sick person

Person, evakuierte – evacuated person

Person, fachkundige – qualified person, expert

Person, juristische – legal entity

Person, verschüttete – trapped person, spilled person

Personal – personnel

T. Schmiermund, *Fachwörterbuch Feuerwehr und Brandschutz*, https://doi.org/10.1007/978-3-662-64120-0_15

Personenäquivalentdosis – personal dose equivalent

Personendosimeter – individual dosimeter

Personendosimeter, elektronisches (EPD) – electronic personal dosimeter (EPD)

Personendosis – personal dose

Personenschaden – personal injury

persönliche Ausrüstung – personal equipment

Persönliche Schutzausrüstung (PSA) – personal protective equipment (PPE)

persönliche Sicherheitsleine (Atemschutzeinsatz) – personal line

Persönliche Sonderausrüstung – personal special equipment

Pest – plague

Pestsepsis – septicaemic plague

Pfahlstich – bowline

Pfahlstich, doppelter – bowline on a bight

Pfeiler – pier, pillar

Pferdestärke (PS) – horsepower (HP)

Pferdestärke, britische – British horsepower (BHP)

Pferdestärke, metrische – metric horsepower (mhp)

Pfette – purlin, stringer

pflanzenschädlich – phytotoxic

Pflanzenschädling – plant pest

Pflanzenschutz – plant protection

Pflanzenschutzmittel – pesticide, plant-protective agent

Pflichten – obligations

Pflichtenheft – specification

Pflichtfeuerwehr (PF) – statutory fire brigade

Pflichtverletzung – breach of duty

Pförtner – porter, janitor, gatekeeper

Pförtnerloge – porter's lodge, porter's office

Pfosten – mullion, stanchion

Pfütze – puddle, wet spot

Pharyngealtubus – pharyngeal tube

Phasenprüfer – mains tester, phase tester

Phlegmatisierung – desensitization

Phlegmatisierungsmittel – desensitization agent

pH-Meter – pH-meter

Phosgen – phosgene

Phosgenoxim – phosgen oxim (CX)

Phosphoreszenz – phosphorescence

Phosphorsäure (H_3PO_4) – phosphoric acid

Phosphorsäureester – phosphoric acid esters, organophosphates

Phosphorsäureestervergiftung – phosphoric acid ester poisoning

Photoeffekt (PE) – photoelectric effect (PE)

Photoionisationsdetektor (PID) – photoionization detector (PID)

Photonenäquivalent – photon equivalent

Photonenenergie – photon energy

Photovoltaik – photovoltaics (PV)

Photovoltaik-Anlage – photovoltaic system

Photovoltaikelement – photovoltaics module

pH-Skala – pH scale

pH-Wert – pH-value

pH-Wert-Messgerät – pH-meter

pH-Wert-Skala – pH scale

physikalische Explosion – physical explosion

physikalisches Brandmodell – physical fire model

physikalisches Gesetz – physical law

physiologische Kochsalzlösung – physiological saline solution

Pickel – pick axe

Piepser – pager

Pilze – mushrooms, fungi

Pilzgift – mycotoxin

Pinzette – tweezer

Pistolenstrahlrohr – pistol type branch

Pkw-Brand – vehicle fire

Planck'sches Wirkungsquantum – elementary quantum of action, Planck's quantum

Planübung – model town exercise, map exercise

Planung – planning

Planung, falsche – faulty planning

Planungsunterlagen – planning documents

Plasmaschneidgerät – plasma cutting device, plasma cutter

Platzdruck – → Zerplatzdruck

Platzverweis – (prohibition to return to a particular place)

Podest – pedestal

polieren – polishing

Poliermittel – polishing agent

politisch Gesamtverantwortlicher – overall political responsibility

Politur – polish

Polizei (POL) – police

Polizeifunk – police radio

Polytrauma – multiple trauma

Porosität – porosity

Potentialausgleich – potential equalization

potentielle Energie – potential energy

Pressluft – compressed air

Pressluftatmer (PA) – compressed air breathing apparatus, self-contained breathing apparatus (SCBA)

Primärexplosion – primary explosion

Primärkühlmittel – primary coolant

Primärkühlmittelkreislauf – primary coolant circuit

Prionen – prions

Prionenerkrankung – prion disease (PD)

Probealarm – practice alarm

Probekörper – test specimen

Probenahme – sampling, sample taking

Probenahme, Datum der – sampling date

Probenflasche – sampling jar, specimen jar, sampling vial

Probenflasche mit Schraubverschluss – screw-cap vial

Profilstahl – profiled steel

Prognose – forecast

Proportionalzähler – proportional counter

Proportionalzählrohr – proportional counter tube

Proteinschaummittel (PS) – protein foam compound

Protolyse – protolysis

Protolysegrad – degree of dissociation

Proton – proton

Prozessabschluss – closing

Prüfbedingungen – test requirements

Prüfdruck – test pressure

Prüfeinrichtung – testing facility

Prüffristen – test intervals

Prüfgas – calibration gas

Prüfgerät – testing device, tester

Prüfgrundlage – test specification

Prüfmethode – test method

Prüfplakette – inspection plate

Prüfprotokoll – test report

Prüfröhrchen – detector tube, indicator tube

Prüfspannung – test voltage

Prüfstelle – test center

Prüfstrahler – standard radioactive source

Prüfung (Abschlusstest) – examination, exam

Prüfung (Geräte, Anlagen) – test, testing

Prüfverfahren – test method

Prüfzeugnis – test certificate

Prusikknoten – prusik knot

psychosoziale Notfallversorgung (PSNV) – psychosocial emergency care

psychotoxische Kampfstoffe – psycho-toxic agents

Puffer – buffer

Pufferlösung – buffer solution

Pulaski-Axt – Pulaski

Pulsfrequenz – pulse rate, heart rate (HR)

Pultdach – pent roof, mono-pitch roof

Pulver – powder

Pulverdüse – dry powder nozzles

Pulverlöscher – powder extinguisher, dry powder extinguisher

Pulverlöschfahrzeug – dry powder appliance

Pulverlöschfahrzeug (PLF) – powder fire-fighting vehicle

Pulverlöschgerät – powder equipment

Pulverlöschverfahren – dry powder extinguishing method

Pulverstrahlrohr – powder nozzle

Pulverwerfer – powder monitor

Pumpe – pump

Pumpe, einstufig – single stage pump

Pumpe, mehrstufig – multi stage pump

Pümpel (Gummi-Saugglocke) – plunger, plumber's helper

Pumpenausgangsdruck – outlet pressure

Pumpendruck – pump outlet pressure

Pumpendruckregelung – pump-outlet pressure control

Pumpeneingangsdruck – inlet pressure, input pressure

Pumpengehäuse – pump casing

Pumpenkennlinie – pump characteristic, pump characteristic line

Pumpenleistung – pump capacity

Punkt, kritischer – critical point

Pupillen – pupils

Pupillen, erweiterte – dilated pupils

Pupillen, verengte – contracted pupils

Putztuch – cleaning rag, cleaning cloth

Putzzeug – cleaning utensils

Pyrolyse – pyrolysis

Pyrolysegase – pyrolysis gases

Pyrolyseprodukte – pyrolysis products

pyrophor – pyrophoric

pyrotechnische Erzeugnisse – pyrotechnics

Q-Fieber – Q-fever

Qualitätsfaktor – quality factor

Qualitätsfaktor (QF) – quality factor (QF)

Qualm – smoke, → Rauch

Quecksilber – mercury

Quellenaktivität – source strength

Querbalken – crossbar

Querempfindlichkeit – cross-sensitivity

Querlüftung – transverse ventilation

Querverkehr – cross traffic

R

Radarstrahlen – radar beams

Radikale – radicals

Radikalkettenreaktion – radical chain reaction

radioaktive Stoffe – radioactive material

radioaktive Strahlung – → ionisierende Strahlung

radioaktiver Abfall – active waste, radioactive waste

radioaktiver Niederschlag – fallout

radioaktiver Zerfall – radioactive decay

Radioaktivität – radioactivity

Radioaktivität, erhöhte – increased radioactivity

Radiocarbon-Methode (Altersbestimmung) – radiocarbon method, ^{14}C method

Radiographie – radiographic

radiologische Waffe – radiological weapon

radiologischer Kampfstoff – radiological agent

Radiolyse – radiolysis

Radiolyseprodukte – radiolysis products

Radiometer – radiometer

Radionuklid – radionuclide

Radionuklidgenerator – radionuclide generator

Radios – radios

Radiowellen – radio waves

Radlast (Kfz) – wheel load

Radmutternschlüssel – tyre wrench, wheel brace

Radon – radon

Radonbelastung – radon exposure

Radongas – radon gas

Radoninhalation – inhalation of radon

Radonkonzentration – radon concentration

Radstand – wheelbase

Rahmenbedingungen – general conditions

Rahmendienstplan – general duty roster, skeleton duty roster

Rahmenempfehlung – framework recommendation

Rang – echelon, rank

Rangabzeichen – insignia, badge of rank

Rangordnung – rank order, ranking, hierarchy

rasch/zügig – swift

Raspel (Werkzeug) – rasp

Raster – grid

Rauch – smoke, fume

Rauch- und Wärmeableitung – smoke and heat outlet

Rauch- und Wärmeabzugsanlage (RWA) – smoke and heat exhaust system (SHE)

Rauchabsaugung – fume extraction

Rauchabschnitt – smoke section

Rauchabzugseinrichtung – smoke exhaust venting equipment

Rauchabzugsöffnung – smoke vent opening

raucharme Schicht – low-smoke layer

Rauchausbreitung – smoke spreading

Rauchbildung – smoke formation

rauchdicht – smoke-proof

Rauchdichte – smoke density

T. Schmiermund, *Fachwörterbuch Feuerwehr und Brandschutz*, https://doi.org/10.1007/978-3-662-64120-0_16

Rauchdichte, optische – optical density of smoke

rauchdichter Abschluss – smoke-proof closure

Rauchdichtheit – denseness of smoke

Raucheinwirkung – smoke influence effect

Rauchempfindlichkeit – smoke sensitivity

Rauchentwicklung – smoke production

Rauchentwicklungsrate – smoke production rate

Raucherinsel – smoking corner

Raucherlaubnis – permission to smoke

Rauchfang – chimney, smokestack

Rauchfangkehrer – chimney sweeper

rauchfreie Schicht – smokeless layer

Rauchfreihaltung – keep free of smoke

Rauchgasausbreitung – smoke distribution, smoke propagation

Rauchgasdurchzündung – roll-over

Rauchgase – fire effluent, flue gases, smoke gases

Rauchgase, Entzündung von – flue gas ignition

Rauchgasentzündung – flue gas ignition

Rauchgasexplosion – back-draft

Rauchgasintoxikation – smoke inhalation injury, fume poisoning, smoke poisoning, smoke intoxication

Rauchgaskühlung – flue gas cooling

Rauchgasschicht – smoke gas layer, fume gas layer, flue gas layer

Rauchgasvergiftung – smoke inhalation injury, smoke poisoning, smoke intoxication, fume poisoning

Rauchgeruch – scent of smoke, smell of smoke

rauchgeschwärzt – smoke-blackened

Rauchklappe – smoke vent, smoke flap

Rauchkonzentration – smoke concentration

Rauchmelder – smoke detector

Rauchmelder, einzelner – single-point smoke detector

Rauchmelder, kabellos – wireless smoke detector

Rauchmelder, optischer – scattered-light smoke detector

Rauchpartikel – smoke particles

Rauchpunkt (Öle) – smoke point

Rauchsäule – pillar of smoke, column of smoke

Rauchschaden – smoke damage

Rauchschleier – veil of smoke

Rauchschürze – fire curtain, smoke barrier

Rauchschutz – smoke protection

Rauchschutztür – smoke control door

rauchschwach – smokeless

Rauchschwaden – billows of smoke, clouds of smoke

Rauchverbot – no smoking

Rauchverbotsbereich – no-smoking area

Rauchvergiftung – smoke intoxication, smoke poisoning, fume poisoning

Rauchverringerung – smoke reduction

Rauchverschluss, mobiler – portable smoke blocker, portable smoke curtain

Rauchwarnmelder – smoke alarm

Rauchwolke – cloud of smoke, smoke cloud

Raum – room, chamber

Raum, angrenzender – adjoining room

Raum, explosionsgefährdeter – explosion hazardous room

Raum, gefangener – inner room

Raum, geschlossener – enclosure

Raum, umbauter – cubature

Raumabschluss – room enclosure

Räumdienst (im Winter) – snow ploughing service

räumen (einen Bereich) – vacate (an area)

Raumexplosion – space explosion

räumlich – three-dimensional

Raumluft – room air

Raumluftqualität – indoor air quality (IAQ)

Raumtemperatur – room temperature

Räumung, eines Blindgängers – UXO clearance

Räumungssignal – evacuation signal

Räumungsübung – evacuation exercise

Räumungszeit – evacuation time

rauschende Flamme – rustling flame

Rauschunterdrückung – noise suppression

Rautek-Griff – Rautek grip

Reaktion – reaction

Reaktion, chemische – chemical reaction

Reaktion, endotherme – endothermic reaction

Reaktion, exotherme – exothermic reaction

Reaktion, heftige – vigorously reaction

reaktionsfreudig – highly reactive

Reaktionsgefahr – risk of reaction

Reaktionsgeschwindigkeit – reaction rate

Reaktionsgeschwindigkeit – reaction velocity

Reaktionsgleichung – chemical equation, reaction equation

Reaktionsmechanismus – reaction mechanism

Reaktionswärme – heat of reaction

Reaktionszone – reaction zone

Reaktordruckbehälter (RDB) – reactor pressure vessel (RPV)

Reaktorhülle – containment shell

Reaktorlaufzeit – reactor life

Reaktorschnellabschaltung (RESA) – SCRAM (safety cut rope axe man)

Reaktorunfall – nuclear reactor accident

reale Abbrandgeschwindigkeit – real burning rate

reales Gas – real gas

Reanimation – resuscitation, reanimation

Reanimationsraum (im Krankenhaus) – resuscitation room

Rechen – rake

Rechte – rights

Rechte und Pflichten – rights and obligations

Rechtsaufsicht – legal supervision

Rechtsgrundlage – legal basis

Rechtskraft – legal force, force of law

rechtskräftig – legally valid

Redox-Gleichung – redox equation

Redox-Paar – redox pair

Redox-Reaktion – redox reaction

Redox-System – redox system

Reduktion – reduction, diminution

Reduktionsmittel – reducing agent, reducer, reductant

Redundanz – redundancy, redundance

Reduzierstück – reducing adapter

Regallager – rack storage

Regalsprinkler – in-rack sprinkler

Regalstapellager – pallet store

Regel – rule, norm, regulation

regelmäßig – regular

Regelung (Steuerung) – control, regulation

Regelungen (Regeln) – rules, regulations, arrangements

Regelungen, gesetzliche – legal regulations

Regelventil – control valve

Regenbekleidung – rain clothes

Regenerationsgerät (RG) (Atemschutz) – regenerative breathing apparatus, regenerator

Regenhose – rain trousers

Regenjacke – rain jacket

Regenrinne – gutter

Reibung – friction

Reibungsbremse – friction brake

Reibungsenergie – friction energy

Reibungsverlust – friction loss

Reibungswärme – frictional heat

Reibungswiderstand – frictional resistance

Reichweite – range

Reifen (Kfz) – tyres

Reihenfolge – sequence

Reihenfolge der Maßnahmen – sequence of measures

Reihenfolge, zeitliche – temporal order

Reihenschaltung (elektr.) – series circuit

Reihenschaltung (von Feuerlöschpumpen) – series connection of fire pumps

rein – pure, clean

Reinelement – mononuclidic element

Reinheit – purity

Reiniger – cleaner

Reinigung – cleaning

Reinigungsarbeiten – cleaning work

Reinigungsmethode – cleaning method

Reinigungsmittel – detergent, cleaning agent, cleaning compound

Reinstoff – pure substance, mono-constituent substance

Reiz – irritation, stimulus

reizend – irritant

Reizgase – irritant gases

Reizstoff – irritant

Reizstoffe – irritants, riot control agents

Relaisbetrieb – relay operation

Relaisfunkstelle – radio relay station

Relaisstelle – relay station

relative Abbrandgeschwindigkeit – relative burning rate

relative Atommasse – relative atomic mass (RAM)

relative Molekülmasse – relative molecular mass (RMM)

Reparatur – repair

Reserve – reserve, reserve material

Reservekräfte – relief-forces

Reserveluft – breathing air reserve

Resistenz – resistance

Restrisiko (RR) – residual risk (RR)

Reststrahlung – residual radiation

Restwärme – residual heat

Resublimation – desublimation

Retentionsfläche (Hochwasserschutz) – washland

retten – rescue

Rettung – rescue

Rettung aus fließendem Gewässer – swift water rescue

Rettung von Menschenleben – saving lives, saving human lives

Rettung, technische – extrication

Rettungs- und Bergungsarbeiten – rescue and salvage operations

Rettungsaktion – rescue operation, rescue mission

Rettungsarbeiten – rescue operations, rescue work

Rettungsassistent (RettAss, RA) – mergency medical technician-paramedic (EMT-P), paramedic, emergency medical technician

Rettungsaxt – fireman's axe

Rettungsboot (RTB) – lifeboat

Rettungsbrett – spine board (SB)

Rettungsdecke (Folie, gold/silber) – heat insulating foil

Rettungsdecke (Stoff-/Wolldecke) – rescue blanket, space blanket

Rettungsdienst (RD) – emergency medical services (EMS), accident ambulance, ambulance corps, ambulance service, rescue service

Rettungseinheit – rescue unit

Rettungseinrichtung – rescue device

Rettungseinsatz – rescue operation, rescue mission, rescue work

Rettungsfahrzeug – ambulance

Rettungsfloß – life raft, live-saving raft

Rettungsflug – rescue flight

Rettungsflugzeug – rescue aircraft, rescue airplane, rescue plane

Rettungsgasse – emergency lane

Rettungsgerät, hydraulisches – hydraulic rescue device, hydraulic rescue apparatus, Jaws of Life®

Rettungsgeräte – rescue equipment

Rettungsgürtel – life belt, cork belt

Rettungshelfer (RH) – emergency medical technician-basic (EMT-B)

Rettungshöhe – rescue height

Rettungshubschrauber (RTH) – medivac chopper, rescue helicopter, emergency rescue helicopter

Rettungshund – rescue dog

Rettungshundeführer – rescue dog handler

Rettungshundestaffel – rescue dog squadron, rescue dog team

Rettungsinsel – inflatable life raft

Rettungskette – rescue chain

Rettungskorb (Drehleiter) – rescue cage

Rettungskosten – rescue costs

Rettungskräfte – rescue forces

Rettungsleine – lifeline, rescue line

Rettungsleiter – rescue ladder

Rettungsleitstelle – emergency call center, 911 call center

Rettungsmannschaft – rescue crew, rescue party, rescue team

Rettungsmaßnahme – rescue measure, rescue effort

Rettungsmesser – emergency rescue knife, rescue knife

Rettungsmittel – rescue resources, rescue equipment

Rettungsorganisation – rescue organization

Rettungsplattform – rescue platform

Rettungspuppe – rescue dummy, dummy

Rettungsring – life ring, life saver, life belt

Rettungsrutsche – evacuation chut, rescue chut

Rettungssanitäter (RettSan, RS) – emergency medical technician-intermediate (EMT-I)

Rettungssatz – hydraulic rescue device, Jaws of Life®

Rettungsschere (med.) – rescue scissors

Rettungsschere, hydraulische – hydraulic cutter, hydraulic rescue cutter, Jaws of Life®

Rettungsschlinge – bowline

Rettungsschlitten – rescue sledge

Rettungsschwimmer – lifeguard, life-safer

Rettungsseil – rescue rope

Rettungsseilwinde – rescue winch

Rettungsspreizer – rescue spreader

Rettungsspreizer, hydraulischer – hydraulic rescue spreader

Rettungstafel – rescue panel

Rettungstaucher – rescue diver

Rettungstechnik – rescue technique, rescue technology

Rettungstrage – rescue stretcher

Rettungstrupp – rescue team

Rettungsversuch – attempt to save, rescue bid

Rettungswache (RetW) – ambulance station

Rettungswagen (RTW) – ambulance, emergency ambulance

Rettungsweg – emergency escape route

Rettungsweg, erster – first escape route, first fire escape

Rettungsweg, zweiter – second fire escape, second escape route

Rettungswerkzeug – rescue tool

Rettungswesen – rescue services, emergency services

Rettungsweste – flotation suit, life jacket

Rettungswinde – rescue winch

Rettungszeichen – emergency sign

Rettungszylinder – hydraulic rescue cylinder

Richtfunk – directional radio

richtige Mittel (Einsatzmittel) – right means

Richtlinie (RL) – guideline, directive

Richtungspfeil – direction arrow

Rickettsien – rickettsia

Riegel – bolt

Riegelstellung – blocking position

Rieselfähigkeit – pourability

Ringer-Lösung – Ringer solution, physiological saline solution

Ringleitung – loop

Ringleitungssystem – ring-pipe system

Ringschlüssel – ring wrench

Risiko – risk

Risikoabschätzung (RA) – risk estimation (RE)

Risikoanalyse – risk analysis

Risikoberechnung – risk calculation

Risikobewertung – risk evaluation

Risikobewertung – risk assessments

Risikofaktor – risk factor

Risikogruppen – risk groups

Risikomanagement – risk management

Risikovorsorge – risk provisions

Rizin – ricin

Rizin-Vergiftung – ricin poisoning

Rohr – pipe, tube, duct

Rohr, erdverlegtes – buried pipework

Rohr, frei verlegtes – pipework above ground

Rohraufhängung – pipe suspension

Rohrbogen – elbow, pipe elbow

Rohrdurchmesser – pipe diameter

Rohrkarabiner – large spring hook

Rohrnetz – piping network

Rohrquerschnitt – pipe cross-section, tube cross-section

Rohrschneider – pipe cutter

Rohrsteckschlüssel – tubular hexagon box wrench

Rohrzange – pipe wrench

Rollcontainer – roll cage

Rolle – reel

Rollgabelschlüssel – crescent wrench, adjustable wrench

Rollschlauch – rolled fire hose

Rolltor – roller door

Rolltreppe – moving stairway, moving staircase

Röntgenstrahlen – X-rays

Röntgenstrahlung – X-rays, X-ray radiation

Röntgenuntersuchung – X-ray examination

Rost – rust

rostfrei – rust-free

Rostlöser – rust dissolver

Rostschutzfarbe – anti-rust paint

Roter Halbmond – red crescent

rotglühend – red-glowing

Rotz – maliasmus, glanders

RTW-Besatzung – ambulance crew

Rückfallebene – fallback level

Rückhaltevermögen – retainment capacity, retainability

Rückhaltung – retainment, retention

Rücklaufsperre – backstop

Rückmeldung – informative message

Rückschlagklappe – clapper valve

Rückschlagventil – backstop valve

Rückweg – return route, way back

Rückweg-Führungsleine (Leinensicherungssystem), (Atemschutzeinsatz) – guide line

Rückzug – retreat

Rückzugsweg – retreat way, retreat path

Rückzugsweg sichern – secure the retreat way

Rückzündung – reignition, flash-back

Ruheenergie – rest energy

Ruhemasse – proper mass, rest mass

Ruhezeit – resting time

Ruhigstellung – immobilization, fixation

Rundfunkgeräte – radios

Rundumkennleuchte (RKL) – rotating beacon, revolving light

Ruß – soot, carbon black

Rußbildung – soot formation

Rußbrand – soot fire

Rüstwagen (RW) – equipment tender

Rüstwagen Öl (RW-Öl) – oil-spill tender

Rüstwagen Schiene (RW Schiene, RW-S) – railway rescue vehicle

Rüstzeit – setting-up time

Rüstzug – rescue squad

Rutschstange – sliding pole, fireman's pole

S

Säbelsäge – sabre saw

Sabotage – sabotage

Sabotagegift – sabotage toxin

Sachgebiet (S 1 – S 6) – staff section (S 1 – S 6)

Sachgebiet S 1 – Personal/Innerer Dienst – staff section S 1 – personnel/administration

Sachgebiet S 2 – Lage – staff section S 2 – information gathering and assessment

Sachgebiet S 3 – Einsatz – staff section S 3 – operation

Sachgebiet S 4 – Versorgung – staff section S 4 – logistics

Sachgebiet S 5 – Presse- und Medienarbeit – staff section S 5 – media and press

Sachgebiet S 6 – Information und Kommunikation – staff section S 6 – communications and transmission

Sachkundiger – expert

sachlich – factual

Sachschaden – damage to property

Sachverständigengutachten – expert evidence

Sachverständiger – expert witness, official expert

Sachwert – property, property value, material asset

Sachwertschutz – property protection, property value protection

Sachwertversicherung – material assets insurance

Sackkarre – hand truck

Sackstich (Knoten) – European death knot (EDK), offset overhand bend (OOB), thumb knot, open-hand knot

Safar-Tubus – Safar airway

Säge – saw

Sägeblatt – saw blade

Sägekette – saw chain

Sägeschiene – saw guide

Salbe – ointment

Salpetersäure – nitric acid

Salpetersäure, rotrauchende – red-fuming nitric acid (RFNA)

Salpetersäureester – nitric acid ester

Salpetersäureestervergiftung – nitric acid ester poisoning

Salze – salts

Salzsäure – hydrochloric acid

Sammelplatz – assembly point

Sammelruf – collective call, ‚banner cry‘

Sammelstück – collecting head, suction collecting head

Sand – sand

Sandeimer – sand bucket

Sandpapier – sandpaper, glasspaper, emery paper

Sandsack – sandbag

Sandsackbarrikade – sandbag barricade

Sanierung (Gebäude) – renovation, reconstruction

Sanierung (Grund, Boden, Umwelt) – clean-up

T. Schmiermund, *Fachwörterbuch Feuerwehr und Brandschutz*, https://doi.org/10.1007/978-3-662-64120-0_17

sanitäre Einrichtungen – sanitary facilities, sanitary installations

Sanitärraum – sanitary room

Sanitäter (Sani) – emergency medical technician-basic (EMT-B)

Sanitätsdienst – medical service

Sanitätshelfer (Sanhelfer) – emergency medical technician-basic (EMT-B)

Sanitätskasten – first aid kit

Sanitätskraftwagen (Sanka) – ambulance, medical vehicle

Sanitätspersonal – medical personnel

Sanitätsstation – first-aid station

Sarin – sarin (GB)

Satteldach – saddle roof

Sättigungskonzentration – saturation concentration

Satz von der Erhaltung der Energie – energy theorem

sauer – acid

Sauerstoff – oxygen

Sauerstoff, flüssiger – liquid oxygen (LOX)

Sauerstoffäquivalent – oxygen equivalent

sauerstoffarm – oxygen starved; low in oxygen

Sauerstoffbeatmung – oxygen resuscitation

Sauerstoffbedarf – oxygen demand, oxygen requirement

Sauerstoffbehälter – oxygen tank, oxygen cylinder

Sauerstoffentzug – oxygen deficiency

Sauerstoffflasche – oxygen cylinder

Sauerstoffgrenze – oxygen limit

Sauerstoffgrenzkonzentration – limiting oxygen concentration

Sauerstoff-Index (OI) – oxygen index (OI)

Sauerstoffintoxikation – oxygen intoxication

Sauerstofflanze – thermal lance

Sauerstoffmangel – anoxia, oxygen starvation, oxygen deficiency, lack of oxygen

Sauerstoffmaske – oxygen mask

Sauerstoffmessgerät – oxygen meter, oxmeter

Sauerstoffschutzgerät (SSG) – closed-circuit oxygen breathing apparatus

Sauerstofftoxikose – oxygen intoxication, oxygen toxicity syndrome

Sauerstoffträger (chem.) – oxidizing agent

Sauerstoffträger (med.) – oxygen carrier

Sauerstoffvergiftung – oxygen intoxication, oxygen toxicity

Sauerstoffzufuhr – oxygen supply

Saugbetrieb – suction operation

Saugeingang – pump inlet

Saughöhe – suction height

Saughöhe, geodätische – geodetic suction heigh

Saughöhe, manometrische – suction heigh manometric

Saugkorb – strainer

Saugkupplung – suction coupling

Saugleitung – suction line, suction pipe

Saugprüfung – suction check

Saugrohr – suction pipe

Saugschlauch – suction hose

Saugschutzkorb – strainer basket

Saugstelle – suction point

Saugstutzen – suction connection, suction hose connection

Säule – column, pillar

Säure – acid

Säure-Base-Reaktion – acid-base reaction

säurebeständig – acid-proof, acid-resistant, acid-stable

säurefest – acid-fast

Säuregrad – acidity

Säurehalogenide – acid halides

Säurekonstante – acid dissociation constant

Säureschürze – acid apron

Säureschutzhandschuhe – acid gloves, acid-resistant gloves

Säureschutzschürze – acid apron

Säurestärke – acid strength, acidity

Säureverätzung – acid burn

Saxitoxin – saxitoxin (STX), shellfish poison

schachbrettartig – chequered

Schacht – shaft

Schaden – damage

Schäden, genetische – genetic damage

Schäden, somatische – somatic damages

Schadensabwehr – damage control

Schadensanalyse (i. S. v. Fehlersuche) – failure analysis

Schadensanalyse (z. B. nach Sturm) – damage analysis

Schadensart – type of damage

Schadensbehebung – damage repair, remedial action

Schadensbekämpfung – damage control

Schadensereignis – damaging event

Schadensersatz – indemnity

Schadensfeuer – destructive fire

Schadensgebiet – damage territory

Schadensklasse – damage class

Schadenslage – situation of impact and damage, case of damage, incident

Schadenslage, Bewältigung der – mastering of the disaster

Schadenstelle – damage area, area of disaster

Schadensumfang – size of damage

Schadensursache – source of damage

Schadenswirkung – damage-impact

Schädigung, thermische – thermal degradation

schädlich – harmful, damaging, causing damage

Schadstoff – harmful substance, pollutant

Schadstoffbelastung – pollution level

Schäkel – shackle

Schale (Atom) – shell

Schalenmodell – shell model

Schalldruck – acoustic pressure, sound pressure

Schalldruckpegel – sound pressure level (SPL)

Schalldruckpegel, äquivalenter – equivalent sound pressure level (ESPL)

Schallgeschwindigkeit – acoustic velocity, sonic speed, sound speed, sound velocity

Schallpegel – sound level

Schallschutz – noise prevention

Schaltfunken – switching spark

Schaltuhr – time switch, switch clock, timer

scharfkantige Gegenstände – sharps

Schätzung – estimation

Schaufel – shovel

Schaufel, Frankfurter – Frankfurt shovel

Schaufeltrage – clamp shell, folding scoop stretcher, scoop stretcher

Schaum – foam

Schaumanhänger – foam trailer (FTr)

Schaumbeständigkeit – foam constancy, foam steadiness

Schaumbildner – foam concentrate, foaming agent

Schaumdecke – foam blanket

Schaumerzeuger – foaming device

Schaumerzeugung – foaming

Schaumgenerator – foam generator (FG)

schaumhemmend – anti-foaming, foam-inhibiting

Schaumkanone – foam gun

Schaumlöscher – foam extinguisher

Schaumlöschfahrzeug (SLF) – foam tender (FoT), foam fire-fighting vehicle

Schaumlöschgerät – foam equipment

Schaumlöschverfahren – foam extinguishing method

Schaummittel – foam concentrate

Schaummittel, alkoholbeständiges – alcohol-resistant foam compound

Schaummittel, wasserfilmbildendes – aqueous film forming foam (AFFF)

Schaummittelbehälter – foam concentrate container

Schaummittel-Wasser-Gemisch – foam solution

Schaumrohr – foam branch pipe (FBP)

Schaumstabilisator – foam stabilizer

Schaumstabilität – foam stability, foam constancy

Schaumstoff – foamed plastic, plastic foam

Schaumstrahlrohr – foam branch pipe (FBP)

Schaumtankfahrzeug – foam tank tender

Schaumteppich – foam blanket

schaumverträglich – foam-compatible

Schaumverträglichkeit – foam compatibility

Schaumwerfer – foam monitor

Schaumzerfall – foam decomposition, foam extinction

Scheibenzertrümmerer – glass breaker

Scherwirkung – shearing action

Scheuerbürste – scrubbing brush

Scheuermittel – scouring agent

Schicht (Arbeitszeit) – shift

Schicht (Oberfläche) – layer

Schicht, raucharme – low-smoke layer

Schicht, rauchfreie – smokeless layer

Schiebetür – sliding door

Schiebleiter – extension ladder

Schiefer – slate

Schieferdach – slated roof

Schiene (med.) – splint

schlaff (z. B. Seil) – loose

Schlaganfall – stroke, apoplexy, apoplexia cerebri stroke, cerebrovascular accident (CVA)

Schlagbaum – turnpike, barrier

Schlagbohrmaschine – percussion drilling machine

schlagempfindlich – sensitive to shocks

Schlagfunken – impact sparks

Schlagvolumen (SV) – stroke volume (SV)

Schlauch – hose

Schlauchanhänger – hose trailer

Schlauchbinde – hose gaiter

Schlauchbrücke – hose ramp

Schlauchdurchmesser – hose diameter

Schlauchgeräte (Atemschutz) – air-line system

Schlauchhalter – hose-holder, hose strap

Schlauchhaspel – hose reel

Schlauchhaspel, fahrbar – mobile hose reel

Schlauchhaspel, tragbar – portable hose reel

Schlauchkorb – hose carrier basket, hose basket

Schlauchkupplungen – hose couplings

Schlauchlager – fire hose depot

Schlauchleitung – hose line, fire hose pipe

Schlauchleitung verlegen – advancing a hose line, lay a hose line

Schlauchpaket – hose package

Schlauchplatzer – hose crack, hose burst

Schlauchpumpe – peristaltic pump

Schlauchschelle – hose clamp

Schlauchsperre – flexible barrier

Schlauchtragekorb – hose basket, hose carrier basket

Schlauchtrupp (S-Tr) – hose crew, hose team, hose squadron

Schlauchtruppführer, -in (S-TrFü, S-TrFü'in) – hose crew leader, hose team leader, hose squadron leader

Schlauchtruppmann, -frau (S-TrM, S-TrFr) – hose crew member, hose team member, hose squadron member

Schlauchturm – hose tower, hose drying tower

Schlauchverbindung – hose connection

Schlauchwagen (SW) – hose-layer (HL)

Schlauchwechsel – hose replacement

Schlaufenknoten – European death knot (EDK), offset overhand bend (OOB), thumb knot, open-hand knot

Schleiffunken – grinding sparks

Schleifkorbtrage – dragging basket

Schleuse (im Gebäude) – double door system

Schleuse (im Gewässer) – water-gate, flood-gate

Schleusenventil – sluice valve

Schließdruck (Pumpe) – closing pressure

Schließzylinder – key cylinder, locking cylinder

Schlitzschraubendreher – slot screwdriver

Schloss (in der Tür) – lock

Schlüssel (für Schrauben) – wrench, spanner

Schlüssel (für Türen) – key

Schlüsselschalter – key-operated switch

Schlüsselweite (SW) (Werkzeug) – wrench width, wrench size

Schmelzbereich – melting range

schmelzen – melting

Schmelzflusselektrolyse – fused salt electrolysis

Schmelzlot – fusible link, fusible solder

Schmelzlotsprinkler – fusible-link sprinkler

Schmelzpunkt (Fp) – melting point (mp)

Schmelzsicherung (elektr.) – safety fuse

Schmelzverbindung – fusible link

Schmelzverhalten – melting behaviour

Schmelzvorgang – melting process

Schmelzwärme – heat of melting

Schmelzwärme, spezifische – specific heat of melting

Schmierfett – grease

Schmiernippel – grease nipple

Schmieröl – lubricating oil

Schmierseife – soft soap

Schmierung – lubrication

Schmutzfänger (Kfz) – mud-flap

schmutzige Bombe – dirty bomb

Schmutzwasser – sewage

Schneelast – snow load

Schneerohr – snow pipe

Schneeschaufel – snow shovel

Schneeschieber – snow shovel

Schnellangriffseinrichtung – quick attack, booster-line

Schnellangriffsfahrzeug – rapid intervention vehicle (RIV)

Schnellangriffsschlauch – quick attack hose

Schnell-Intubation – rapid sequence intubation (RSI)

Schnellkupplung – quick coupling

Schnellkupplungsgriff – quick coupling handle

Schnellschlussventil – quick-acting valve

Schnellstraße – expressway, speedway

Schnitt – cut

Schnittmodell – sectional model

Schnittschutz – cut protection

Schnittschutzhandschuhe – cut protection gloves

Schnittschutzhose – cut protection trouser

Schnittschutzjacke – cut protection jacket

Schnittstelle – interface

Schnittwunde – cut, incision

Schnur – cord, lanyard, string

Schock – shock

Schock, allergischer – allergic shock

Schock, anaphylaktischer – anaphylactic shock

Schock, kardiogener – cardiogenic shock

Schock, septischer – septic shock

Schock, toxischer – toxic shock (TS)

Schocklagerung – passive leg-raising (PLR)

Schockwelle – shock wave

Schornstein – chimney, smokestack

Schornsteinbrand – chimney fire

Schornsteinfeger – chimney sweeper

Schornsteinfegermeister – master chimney sweeper

Schornsteinfegerwerkzeug – chimney sweeping tools, chimney sweeper set

Schotenstich – sheet bend, weaver's hitch, becket bend, weaver's knot

Schotenstich, doppelter – double sheet bend

Schotstek (Knoten) – sheet bend, weaver's hitch, becket bend, weaver's knot

Schraubendreher – screw driver

Schraubkarabiner – screw carabiner, screw karabiner

Schraubkupplung – screw-on coupling

Schraubstock – jaw vice

Schraubzwinge – clamp, screw clamp

Schreibgeräte – writing tools

Schreibmaschine – typewriter

schriftlicher Befehl – written order

Schrittspannung – step voltage

Schrott – scrap, scrap metal

Schubkarre – wheelbarrow

Schubkarre – wheel barrow

Schulterklappe – shoulder board, shoulder strap

Schüttdichte – bulk density

Schuttmulde – debris trough

Schutz – protection

Schutz von Personen – protection of persons

Schutz von Personen und Sachwerten – protection of persons and material assets

Schutz von Personen, Tieren und Sachwerten – protection of persons, animals and material assets

Schutz von Sachwerten – protection of material assets

Schutz von Tieren – protection of animals

Schutzabdeckung – protection cover, protective hood

Schutzanzug – protective suit

Schutzausrüstung – protective equipment, protective gear

Schutzausrüstung – safety equipment

Schutzausrüstung, persönliche (PSA) – personal protective equipment (PPE)

Schutzbereich – protection area

Schutzbrille – goggles, safety glasses, protective spectacles

Schutz-Erdung – protective grounding

Schutzgas – inert gas

Schutzhandschuhe – protective gloves

Schutzhaube – protective hood

Schutzhelm – safety helmet

Schutzimpfung – protective immunization, vaccination

Schutzklasse – protective class, protection category

Schutzkleidung – protective clothing

Schutzmaßnahme – precautionary measure, protective measure

Schutzschirm – protective shield

Schutzsystem – protective system

Schutzumfang – scope of protection

Schutzvorhang (Theater) – protective curtain

Schutzvorkehrungen – protective arrangement, safeguards

Schutzvorschriften – safety precautions

Schutzwert – protective value

Schutzwirkung – protective effect

Schutzziele – protection goals

Schutzzone – protection zone

schwach radioaktiver Abfall – low active waste (LAW)

schwache Kernkraft – → schwache Wechselwirkung

schwache Wechselwirkung – weak interaction

Schwachstrom – low current

Schwamm – sponge

Schwarzbereich – black area

schwarzer Körper – black body

Schwebstofffilter (SSF) – high-efficiency particulate absorbing (air) filter (HEPA filter)

Schwefel – sulfur, sulphur

Schwefeldioxid – sulfur dioxide, sulphur dioxide

Schwefel-Lost (S-Lost) – sulphur mustard gas, mustard gas

Schwefelsäure – sulfuric acid, sulphuric acid

Schwefelwasserstoff – hydrogen sulfide, hydrogen sulphide

Schweiß – sweat, perspiration

schweißen – weld

Schweißen – welding

Schweißen (Acetylen-Sauerstoff) – autogenous gas welding, autogenous welding, oxygen-acetylene welding

Schweißerbrille – welder's goggles

Schweißerhelm – welder's helmet

Schweißfunken – welding sparks

Schweißperlen – welding beads

Schwelbrand – smouldering fire, smoldering fire

Schwelen – smouldering combustion, smoldering combustion

Schwelen, fortschreitendes – progressive smouldering, progressive smoldering

Schwelle – threshold

Schwellenwerte – threshold values

Schwellung (med.) – swelling

Schwelpunkt – smouldering point, smoldering point

schwer abbaubar – persistent

schwer brennbar – hardly combustible

schwer brennbarer Stoff – self-extinguishing material

schwer entflammbarer Stoff – material flammable with difficulty

schwer entzündbarer Stoff – material difficult to ignite

schwer entzündlich – hardly flammable

schwer flüchtig – less volatile

Schwerbenzin – heavy gasoline

schweres Wasser – deuterium oxide

Schweres-akutes-Atemwegssyndrom-Coronavirus Typ 2 – severe acute respiratory syndrome coronavirus type 2 (SARS-CoV-2)

Schwergasausbreitung – heavy-gas propagation

Schwergaseffekt – heavy-gas effect

Schwergaswolke – heavy-gas cloud

Schwerschaum – low expansion foam (LX)

Schwerschaumrohr – low expansion foam branch, low expansion foam branch pipe

Schwerwasserreaktor – heavy water reactor (HWR)

Schwimmdachtank – floating roof tank

Schwimmerschalter – float switch

Schwimmerventil – float valve

Schwimmweste – flotation suit, life jacket

schwingungsarm – low-vibration

Schwingungsdämpfer – vibration dampener

Schwingungsdämpfer (im Kfz) – engine mount damper

schwitzen – sweat, perspire

Schwitzen – sweating, perspiration

Sechskantschlüssel – hex wrench

Seenot – distress at sea

Seenotrettungskreuzer – rescue cruiser

Segeltuch – canvas

Segeltucheimer – canvas bucket

sehr giftig – very toxic

sehr giftiger Stoff – very toxic substance

Seife – soap

Seil – rope, cord

Seil ausgeben – give out rope

Seil einholen – take in rope

Seilschlauchhalter – rope hose holder

Seilspannung – rope tension

Seilstoppgerät – rope control device

seilunterstützte Zugangstechniken (SZT) – rope access

Seilverkürzung – shank

Seilwinde – winch

Seitenairbag – side airbag

Seitenlage, stabile (Erste Hilfe) – lateral recumbent position

Seitenschneider – diagonal cutter, wire cutter

Sekundärbrand – secondary fire

Sekundärinfektion – secondary infection

Sekundenkleber – superglue, crazy glue

selbstansaugend – self-priming

selbstbeschleunigt – self-acceleration

selbstentflammend – self-inflaming

selbstentzündlich – auto-ignitable, self-ignitable, spontaneously ignitable, self-inflaming

selbstentzündliche Stoffe – substances liable to spontaneous combustion

Selbstentzündlichkeit – auto-ignitability, self-ignitability

Selbstentzündung – auto-ignition, self-ignition

Selbstentzündungstemperatur – self-ignition temperature, self-ignition point, auto-ignition temperature

selbsterhaltend – self-sustaining

Selbsterwärmung – self-heating, spontaneous heating

Selbsterwärmungstemperatur – self-heating temperature, spontaneous heating temperature

selbstlöschend – self-extinguishing

Selbstrettung – self-rescue

Selbstschutz – self-protection

Selbstüberwachung – internal checking

selbstverlöschend – self-extinguishing

Selektierung (med.) – triage

Seltene Erden – rare earths, rare earth elements

Senfgas – mustard gas

Senke – basin, depression, well

Sepsis – sepsis, septicaemia

septischer Schock – septic shock

sesshaft (z. B. Kampfstoff) – persistent

Sesshaftigkeit – sedentariness

Seuchenschutz – disease control

Seuchenschutzbehörde – disease control center, Center for Disease Control (CDC)

Sheddach – sawtooth roof

Sicherheit – safety, security

Sicherheitsabstand – safety distance, safety clearance

Sicherheitsbeauftragter – safety officer

Sicherheitsbehälter – containment

Sicherheitsbeleuchtung – safety lighting

Sicherheitsbestimmungen – safety regulations, security regulations

Sicherheitsdatenblatt (SDB) – safety data sheet (SDS) {GB}, material safety data sheet (MSDS) {US}

Sicherheitseinrichtung – safety installation

Sicherheitsgefährdung – safety hazard

Sicherheitsgeschirr (Absturzsicherung) – fall arrest harness, fall protection harness, full-body fall arrest harness, safety harness

Sicherheitsglas – safety glass

Sicherheitsgurt – safety belt

Sicherheitsgurt (Kfz) – seat belt

Sicherheitshelm – safety helmet

Sicherheitskanne – safety can

Sicherheitskenndaten – safety criteria

Sicherheitskennzeichnung – safety marking

Sicherheitslabor – biohazard containment laboratory (L1-L4)

Sicherheitsleine, persönliche (Atemschutzeinsatz) – personal line

Sicherheitsleuchte – safety lamp

Sicherheitsprüfsiegel – security seal

Sicherheitsrisiko – safety risk

Sicherheitsschleuse – security door system

Sicherheitsschloss – tamper-proof lock

Sicherheitsschuhe – safety shoes

Sicherheitsseil – safety rope

Sicherheitssiegel – security seal

Sicherheitsstufe (Labor) – safety level, physical containment level (L1-L4)

sicherheitstechnische Kennzahlen – safety-related values, safety indexes

Sicherheitstreppe – safety stair

Sicherheitstreppenraum – safety stair

Sicherheitstrupp (SiTr) – protection team

Sicherheitsventil – safety valve

Sicherheitsvorschrift – safety regulation

Sicherheitsweste – fluorescent jerkin, high-visibility vest (HV-vest)

Sicherheitszeichen – safety signs

Sicherstellung – ensuring

Sicherung – protection

Sicherung der Einsatzstelle – protection of place of operation

Sicherungsbereich – protected area

Sicherungsbereich, abgesetzter – separated protected premises

Sicherungsmann/-frau – belayer

Sicherungsmaßnahme – protective measure

Sicherungsmaßnahme, bauliche – physical safeguarding measure

Sicherungspunkt (Höhenrettung) – belay point

Sicherungsseil – belay rope

Sichtbereich – visual range

sichten (med.) – triage

Sichtkontrolle – visual inspection

Sichtminderung (durch Rauch) – obscuration (by smoke)

Sichtprüfung – visual testing (VT), visual check

Sichtscheibe (Atemschutzmaske) – visor

Sichtung (med.) – triage

Sichtweite – visibility

Siebanalyse – sieve analysis

Siedepunkt (Kp) – boiling point (bp)

Siedeverzug – delay in boiling, defervescence

SI-Einheit – SI unit

Signalanlage – signalling installation

Signale, akustische – acoustic signals

Signale, optische – optical signals

Signalgeber – signal generator, signaller

Signallaterne – signal lantern

Signalpfeife – signal whistle, signal pipe

Signalwandler – signal transductor, signal converter

Silicium – silicon

Silikon – silicone

Sims – sill, ledge

Sinneswahrnehmungen – sensory perception

Sirene – alarm siren

Sirenensignale – siren signals

Situation – situation

Situation, ausweglose – hopeless situation

Sitzgurt – sit harness

Sitzordnung – seating order

Skala – scale

Skalierung – scaling

Skelett – skeleton

Skelettbauweise – frame construction

Skelettformel – skeleton formula

Skizze – sketch, draft

S-Lost – mustard gas

Sofortbehandlung – immediate treatment (IT)

Sofortbildkamera – instant-picture-camera

Sofortinformation – immediate information, instant information

Sofortmaßnahmen, lebensrettende (LSM) – basic life support (BLS), immediate lifesaving measures

Sofortmeldung – immediate report, instant report

Solaranlage – solar collector, solar plant

Solarenergie – solar energy

Solarpaneel – solar panel

Solarstrom – solar electricity

Solarzelle – solar cell

Sollbruchstelle – predetermined breaking point

Sollwert – nominal value, set point

Sollzustand – nominal condition, specified condition

Soman – soman (GD)

somatische Schäden – somatic damages

Sonderausrüstung – special equipment, optional equipment

Sonderausrüstung, vollständige – complete special equipment, complete optional equipment

Sonderbauten – special constructions

Sondereinheit – special unit

Sonderfahrzeuge – special vehicles

Sonderfall – special case

Sondergenehmigung – special permission

Sondergenehmigung – special permit, special license

Sonderlöschfahrzeug – special fire-fighting vehicle

Sonderlöschmittel – special extinguishing media

Sondermüll – hazardous waste

Sonderrechte – special rights

Sonderregelung – special regulation, special arrangement

Sondersignal – blue flashing light and siren

Sonnenbrand – sunburn

Sonneneinstrahlung – insolation

Sonnenstich – sunstroke, insolation encephalopathy

Sonnenstrahlung – solar radiation

Spaltaxt – splitting axe

spaltbare Stoffe (Kerntechnik) – fissile material

spaltbares Material – fissionable material

Spalthammer – splitting maul

Spaltkeil – splitting wedge

Spaltprodukte (chemische Reaktion) – decomposition products

Spaltprodukte (Kerntechnik) – fission products

Spaltreaktion – fission reaction

Spaltung (Kerntechnik) – fission, fissioning

Spaltung (mechanisch) – cleavage, breakage, splitting

Spannbeton – prestressed concrete

Spannung (elektrische) – electrical voltage, voltage

Spannung (mechanische) – tensions, frictions

Spannungsabfall – voltage drop

Spannungsreihe – electrochemical series

Spannungstrichter – voltage funnel, voltage crater

Spannweite – span width

Spanplatte – particleboard, chipboard

Sparren – cheveron, rafters

Spaten – spade

Spätfolgen – late sequelae

Spätschäden – long-term damages, delayed damages

Spätschicht – late shift, evening shift

Speicherring – storage ring

Speichertank – storage tank

Speicherung – storage

Spektrum, elektromagnetisches – electromagnetic spectrum

Sperrbereich – prohibited area

Sperrgebiet – no-go area, no-go zone, restricted area, restricted zone, exclusion area

Sperrholz – plywood

spezifische Aktivität – specific activity

spezifische Schmelzwärme – specific heat of melting

spezifische Sublimationswärme – specific heat of sublimation

spezifische Verbrennungswärme – specific heat of combustion

spezifische Verdampfungswärme – specific heat of vaporization

spezifische Wärmekapazität – specific heat capacity

spezifischer Brennwert – specific caloric value

spezifischer Heizwert – specific heating value

Spielraum (bei Entscheidungen) – latitude

Spierenstich (Knoten) – fisherman's knot, English knot

Spineboard – spine board (SB)

Spitzhacke – pickax, pickaxe

Spitzzange – long-nose pliers

Splitter – splinter

Spontanentzündung – spontaneous ignition

Spontanentzündungstemperatur – spontaneous ignition temperature

Sportboot – pleasure craft

Sprechanlage – intercom system

Sprechfunkverkehr – radio communication

Sprechmembran – speech diaphragm

Spreizer (SP) – rescue spreader

Spreizer, hydraulischer – hydraulic rescue spreader

Sprengkraft – explosive force, explosive power

Sprengmittel – exploder

Sprengstoff – explosive

Sprinkler – sprinkler

Sprinkleranlage – fire sprinkler, fire sprinkler system

Sprinklerdüse – sprinkler nozzle

Sprinklerkopf – sprinkler head

Sprinklerpumpe – sprinkler pump

Sprinklerzentrale (SPZ) – sprinkler control center

Spritze – syringe

Spritze aufziehen – fill a syringe, draw up into a syringe

Spritze erhalten – have an injection

Spritze verabreichen – give a shot, give an injection

Spritzenhaus – fire engine house

Sprosse – rung

Sprossenabstand – rung spacing

Sprühbild – spray pattern

Sprühdose – spray can

Sprühdüse – spray nozzle

Sprühnebel – drizzling fog

Sprühstrahl – spray jet

Sprühstrahlrohr – diffuser nozzle

Sprühwasser – water spray

Sprühwasser-Löschanlage – water spray system

Sprungkissen – rescue cushion

Sprungpolster – jumping cushion

Sprungtuch – jumping blanket, jumping sheet

Spuckbeutel – vomit bag

Spule – reel

Spundwand – sheet pile wall

Spüren (Nachweisen) – sensing

Spürpapier – detection paper

Spürpulver – detection powder

Stab (Einsatzstab) – command staff, ‚crisis cell'

Stab für außergewöhnliche Ereignisse – staff for extraordinary events

Stabdosimeter – pocket dosimeter

stabile Seitenlage (Erste Hilfe) – lateral recumbent position

Stabilität, thermische – thermal stability

Stadtgas – city gas, mains gas, town gas

Stadtplan – city map, town map

Staffel (St) – squad, team-of-five

Staffelführer (StFü) – squad leader

Stahl – steel

Stahlbeton – steel concrete

Stahlkonstruktion – steel construction

Stahlrohr – steel pipe

Stahlstütze – stanchion

Stahlstütze, leichte – light steel support

Stahlträger – steel girder

Stahl-Trapezblech-Dach – steel deck

Standard-Dekontamination – standard decontamination

Standard-Einsatz-Regel (SER) – standing order

Standard-Temperatur-Zeit-Kurve – standard-temperature-time-curve

Standardverbrennungsenthalpie – standard combustion enthalpy

Standardvorgehen – standard intervention process

ständig – permanent, constantly, all the time

ständig alarmierbar – permanently available to intervence

ständig besetzt – permanently manned

Standrohr – standpipe

Standsicherheit – stability

Staphylokokken – staphylococci

Staphylokokken, multiresistente – multiresistant staphylococci

Staphylokokken-Enterotoxin B (SEB) – Staphylococcal Enterotoxin B (SEB)

Staphylokokken-Infektion – staphylococcus infection

starke Kernkraft – → starke Wechselwirkung

starke Wechselwirkung – strong interaction

Starklichtfackel – high light torch

Starkstrom – heavy current, power current

Starkstromanlage – power installation, high voltage system

Starkstrom-Werkzeugkasten – power current tool box

Starkwind – gale

Starthilfe – jump start

Statik – statics

stationär – inpatient

stationärer Zustand – steady state

statischer Druck – static pressure

Status – status, state

Statusmeldung – status message

Staub – dust

Staub nicht einatmen. – Do not breathe dust.

Staub, brennbarer – combustible dust

Staub, entzündbarer – flammable dust

Staubbrand – dust fire

Staubexplosion – dust explosion

Staubexplosionsgefährdung – dust-explosion hazard

staubexplosionsgeschützt – dust ignition proof

Staub-Luft-Gemische – dust-air-mixtures

Staubmaske – dust mask, dust respirator

Staudamm – dam, reservoir dam

Staudruck – ram pressure

Stauraum – stowage place

Staustelle – dam point

Stauwehr – weir

Steckleiter – scaling ladder

Steckleiter, A-Einschubteil – attachment module (2-runged) of a scaling ladder

Steckleiter, A-Teil – base module (9-runged) of a scaling ladder

Steckleiter, B-Teil – attachment module (7-runged) of a scaling ladder

Steckschlüssel – socket wrench

Steigeisen – pole climber

Steighöhe – working height

Steigklemme – ascender

Steigleiter – ascending ladder

Steigleitung – rising main

Steigleitung, nasse – wet rising main

Steigleitung, trockene – dry rising main

Steinwolle – rock wool

Stellungswechsel (räumlich) – positional change

Stemmeisen – pry bar

Sterberate – mortality

Sterilisation (Gegenstände) – sterilization

Sterilisierung – sterilization

Steuerleitung – pilot line, control line

Steuerventil – control valve, pilot valve

Stiche (Knoten) – hitches

Stichflamme – flash, darting flame, jet of flame, explosive flame

Stichleitung – isolated line

Stichprobe – random sample

Stichsäge – jigsaw

Stickeffekt – smothering effect

Stickgase – asphyxiant gases

Stickoxide (NO_x) – nitrogen oxides

Stickstoff – nitrogen

Stickstoff, flüssiger – liquid nitrogen

Stickstoff-Lost (N-Lost) – nitrogen mustard gas

Stiefel – boots

Stiefelspanner – boot tree

Stiel – shaft, stick

Stirnlampe – head lamp, headtorch

stochastische Strahlenwirkung – stochastic radiation effect

Stöchiometrie – stoichiometry

stöchiometrische Ausbeute – stoichiometric yield

stöchiometrische Konzentration – stoichiometric concentration

stöchiometrische Verbrennung – stoichiometric combustion

stöchiometrisches Gemisch – stoichiometric mixture

Stockpunkt – pour point

Stockwerk – floor

Stoffaustausch – mass exchange, material exchange

Stoffe, ansteckungsgefährliche – infectious substances

Stoffe, ätzende – corrosive substances

Stoffe, ausgelaufene – spilled substances

Stoffe, brennbare – combustible substances

Stoffe, die in Berührung mit Wasser entzündbare Gase entwickeln – substances which, in contact with water, emit flammable gases

Stoffe, dispergierte – dispersed substances

Stoffe, erwärmte – elevated temperature substances (HazMat)

Stoffe, explosionsfähige – → Explosivstoffe

Stoffe, explosive – explosives

Stoffe, feste – solid substances

Stoffe, feste, entzündbare – flammable solids

Stoffe, feste, explosive – solid explosives

Stoffe, flüchtige – volatile substances

Stoffe, flüssige – liquid substances

Stoffe, gasförmige – gaseous substances

Stoffe, gefährliche – dangerous substances

Stoffe, giftige – toxic substances

Stoffe, hochentzündliche – extremely flammable substances

Stoffe, hochgiftige – extremely toxic substances

Stoffe, leicht flüchtige – highly volatile substances

Stoffe, nicht flüchtige – nonvolatile substances

Stoffe, oxidierend wirkende – oxidizing substances

Stoffe, polymerisierbare – polymerizing substances

Stoffe, radioaktive – radioactive material

Stoffe, schwer flüchtige – less volatile substances

Stoffe, sehr giftige – very toxic substances

Stoffe, selbstentzündlich – substances liable to spontaneous combustion

Stoffe, selbstzersetzliche – self-reactive substances

Stoffe, spaltbare (Kerntechnik) – fissile material

Stoffe, tiefkalte – cryogenic substances

Stoffe, umweltgefährdende – environmentally hazardous substances

Stoffe, wasserdampfflüchtige – steam-volatile substances

Stoffeigenschaften – substance properties

Stoffgemisch – mixture of substances

Stoffmenge – amount of substance

Stoffmengenanteil – amount-of-substance fraction, mole fraction

Stoffmengenkonzentration – molar concentration, amount concentration

Stoffmengenverhältnis – amount-of-substance ratio

Stoffnummer – substance key, substance number

Stopfbuchse – gland seal, stuffing-box seal

Stopperknoten – backup knot, stopper knot

Stopptaste – stop button

Störeinfluss – confounding effect

Störfall – abnormal occurrence, incident

Störfallanalyse – accident analysis

Störung (Zustand) – fault, malfunctioning

Störungen – disturbances

Störungsdienst – fault-clearing service

Störungsmeldung – fault report, trouble report

Störungsursache – cause of a malfunction

Storz-Kupplungen – Storz couplings

stoßempfindlich – sensitive to impact

Stoßwelle – impact wave

Strahlen – rays, radiancy, beams

Strahlenabschirmung – radiation shielding

Strahlenbelastung – radiation load, radiation exposure

Strahlenbelastung, Messung der – radiation-level check

Strahlenbiologie – radiation biology

strahlenbiologisch – radiobiological

Strahlendosis – radiation dose, radiation dosage

strahlendurchlässig – radiolucent

Strahleneinwirkung – radiation exposure

Strahlenempfindlichkeit – radiosensitivity, radiation sensitivity

Strahlenerkrankung, akute – acute radiation syndrome (ARS)

Strahlengefährdung – radiation hazard

Strahlengrenzwerte – radiation limits

Strahlenkater – radiation sickness

Strahlenkrankheit – radiation sickness

Strahlenmessgerät, tragbares – cutie-pie

Strahlenquelle – radiation source

Strahlenquelle, abgeschirmte – shielded radiation source

Strahlenquelle, nicht abgeschirmt – unshielded radiation source

Strahlenschäden – radiation damage, radiation injuries

Strahlenschutz – radiation protection

Strahlenschutzanzug – radiation suit, radiation protection suit

Strahlenschutzbeauftragter (SSB) – radiation protection officer (RPO), radiation safety officer (RSO)

Strahlenschutzbereich – radiation protection area

Strahlenschutzmaßnahmen – radiation protection measures

Strahlenschutzverantwortlicher – radiation protection supervisor

Strahlenschutzverordnung (StrlSchV) – German Radiation Protection Ordinance

Strahlenspürgerät – radiation detector

strahlenundurchlässig – radiopaque, radiodense

Strahlenunfall – radiation accident

Strahlenvergiftung – radiation poisoning, radiation toxicity

Strahlenwarnzeichen – basic ionizing radiation symbol

Strahlenwichtungsfaktor – radiation weighting factor

Strahlenwirkung – radiation effect

Strahlenwirkung, deterministische – deterministic radiation effect

Strahlenwirkung, stochastische – stochastic radiation effect

Strahlrohr – fire-fighting nozzle, fire nozzle, branch

Strahlrohrführer – pipe leader, hose operator

Strahlrohrmundstück – mouth piece, jet pipe mouth piece

Strahlung – radiation

Strahlung, elektromagnetische – electromagnetic radiation

Strahlung, ionisierende – ionizing radiation

Strahlung, kosmische – cosmic radiation, cosmic rays (CRs)

Strahlung, nichtionisierende – non-ionizing radiation

Strahlung, radioaktive – → Strahlung, ionisierende

Strahlung, terrestrische – terrestrial radiation

strahlungsarm – low-radiation

strahlungsbeständig – radiation-resistant

Strahlungsdosimeter – radiation dosimeter

Strahlungsenergie – radiation energy, beam energy

strahlungsgefährdeter Bereich – high radiation area

Strahlungsleistung – radiation power, radiant flux

Strahlungsmessgeräte – radiation measuring devices

Strahlungswärme – radiation heat

Strahlungswärmestromdichte – radiant heat flux

Strahlungswolke – radiation plume, radioactive plume

strapazierfähig – durable

Straßenkarte – road map, street map

Straßenverkehrsordnung (StVO) – traffic regulations, road traffic regulations

Strategie – strategy, generalship

strategische Kampfstoffe – strategic agents

Strebe – brace, strut, shore

Streichholz – match

streng verboten – strictly forbidden, strictly prohibited

Streulicht – flare light, stray light

Streulichtmelder – stray light detector, scattered-light detector

Streulichtrauchmelder – scattered-light smoke detector

Streuversuch – scattering experiment, spread test

Strichliste – tally chart

Strickleiter – corded ladder, rope ladder

Stroh – straw, litter

Strohdach – thatched roof, straw roof

Strom, elektrischer – electric current

Stromausfall – power failure

Stromerzeuger – generator, portable generator

Stromkreis – electrical circuit

Strommarke – electric burn

Stromstärke – current strength

Stromunfall – electric accident, accident caused by electric current

Strömung, laminare – laminar flow

Strömung, turbulente – turbulent flow

Strömungsgeschwindigkeit – velocity of flow, flow speed

Stromverteiler – current distributor

Stromverteilerkasten – circuit breaker panel

Stromverteilung – current distributor

Strukturformel – structural formula

Sturm – storm, strong gale

Sturmlaterne – hurricane lamp

Sturmschäden – storm damages

Sturmstreichhölzer – storm matches

Sturmzündhölzer – storm matches

Sturz (Bauteil) – lintel

Sturzfaktor – fall factor

Stütze (Bauteil) – linchpin

Stützen, hydraulische (Drehleiter, Kran) – jacking pads, jacks

Stützfeuer – base-on fire

Stützkrümmer – hose support

Stützmauer – retaining wall

Stützpfeiler – supporting pillar

Stützpunktfeuerwehr – support fire station, main point fire brigade

Sublimation – sublimation

Sublimationsdruck – sublimation pressure

Sublimationspunkt – sublimation point

Sublimationswärme – enthalpy of sublimation, heat of sublimation

Sublimationswärme, spezifische – specific heat of sublimation

sublimieren – sublime

Substanz, betäubend wirkende – narcotic

Substanz, erstickend wirkende – asphyxiant

Such- und Rettungshubschrauber – search and rescue helicopter (SAR helicopter)

Such- und Rettungsteam – search and rescue team (SAR-team)

Suchtrupp – search party, search crew

Sulfonamide – sulfonamides

Sulfonsäuren – sulphonic acids

Summenformel – empiric formula

Summer – buzzer

Suspension – suspension, slurry

Synonym – synonym

Synthese – synthesis

systemische Giftwirkung – systemic toxic effect

Szintigraphie – scintigraphy

Szintillationszähler – scintillation counter

T

Tabun – tabun (GA)

Tachometer – speedometer

Tafelkreide – blackboard chalk

Tageslichtprojektor – overhead-projector, OH-projector

Tageszeit – time of day

tägliche Einsätze – day-to-day operations

Tagschicht – day shift, daytime shift

taktische Bezeichnung – tactical designation

taktische Einheit – tactical unit

taktische Gesichtspunkte – tactical point of view

taktische Gliederung – tactical organisation

taktische Grundregeln – basic tactical rules

taktische Kampfstoffe – tactical agents

taktische Zeichen – tactical signs

taktischer Auftrag – tactical cast, tactical mission cast

taktischer Rückzug – tactical withdrawal

Tankbrand – tank fire

Tanklager – tank farm

Tanklöschfahrzeug (TLF) – tank tender, fire-extinguishing tender, tank fire-fighting vehicle, pump water tender, water tender, tender truck

Tankstelle – filling station, petrol station, service station

Tanktasse – safety trough for tank

Tankwagen – tank car

Tara – tare

Taschenlampe – torch, flash light

Taschenrechner – pocket-calculator

tasten (fühlen) – feel, touch, palpate

Tauchausrüstung – diving equipment

tauchen – dive

Tauchen – diving

Taucherdruckkammer – decompression chamber, diver's decompression chamber

Taucherkrankheit – diver's disease

Tauchermesser – diving knife

tauchfähig (Pumpe) – submersible

Tauchgeräte – diving apparatuses

Tauchpumpe (TP) – submersible pump

Taupunkt – dew point

Täuschungsalarm – delusive alarm

technische Hilfeleistung (TH) – rescue service

technische Katastrophe – technological disaster

technische Rettung – extrication

Teer – tar, bitumen

Teerpappe – bitumized paper

Teilausfall – partial failure, partial malfunction

Teilchenstrahlung – corpuscular radiation, particles radiation

Teilflutung – partial discharge

Teilkörperbestrahlung – partial-body exposure

Teilkörperdosis – partial-body dose

Teillagerfläche – store section area

Teilschutz – partial protection

Teilüberwachung – partial monitoring

Telekommunikation – telecommunication

T. Schmiermund, *Fachwörterbuch Feuerwehr und Brandschutz*, https://doi.org/10.1007/978-3-662-64120-0_18

Telekommunikation, digital verstärkte schnurlose – digital enhanced cordless telecommunication (DECT)

Teleskopdrehleiter – extending turntable ladder

Teleskopkran – extending jib crane

Teleskopsonde – telescopic probe

Temperatur – temperature

Temperatur, kritische – critical temperature

Temperatur, selbstbeschleunigte Zersetzung – self-acceleration decomposition temperature (SADT)

temperaturabhängig – temperature-dependent

Temperaturabhängigkeit – temperature dependence

Temperaturanstieg – temperature rise

Temperaturbereich – temperature range, range of temperature

Temperatureinfluss – influence of temperature

Temperaturklasse – temperature class

Temperaturleitfähigkeit – temperature conductivity

Temperaturskala – temperature scale

Temperaturverlauf eines Brandes – temperature process by a fire

temporäre Härte (Wasser) – temporally hardness of water

Tenazität – tenacity

Tenside – surfactants, tensids

Terrasse – terrace

Terrassendach – terrace roof

terrestrische Strahlung – terrestrial radiation

Terroranschlag – terror attack, terrorist attack

terroristischer Hintergrund – terroristic background

Testbenzin – white spirit

theoretische Verbrennungstemperatur – theoretical combustion temperature

theoretischer Luftbedarf – theoretical air need

thermisch – thermal

thermische Beanspruchung – heat stress

thermische Dissoziation – thermal dissociation

thermische Eigenschaften – thermal properties

thermische Neutronen – thermal neutrons

thermische Schädigung – thermal degradation

thermische Stabilität – thermal stability

thermische Zersetzung – thermal decomposition

Thermodifferentialmelder – rate-of-rise detector

Thermoelement – thermocouple

Thermolumineszenzdosimeter (TLD) – thermoluminescent dosimeter (TLD)

Thermomelder – heat detector

Thermometer – thermometer

Thermoplaste – thermoplastics

Tiefbrunnen – deep well

tiefkalte Stoffe – cryogenic substances

tiefliegende Bereiche – low areas

Tiegel – crucible

Tierhebegerät – animal rising device

Tierheim – animal shelter

Tierrettung (Einheit) – animal rescue team

Tierrettung (Tätigkeit) – animal rescue, rescue of animals

Tochternuklid – daughter nuclide

tödliche Verletzung – fatal injury

Tödlichkeitsprodukt, Habersches – Haber's lethality produkt

Tollwut – rabies

Torx-Schraubendreher – star screwdriver

Totalausfall – total breakdown

Totenkopf (Symbol) – skull and crossbones (symbol)

Totmanneinrichtung (Atemschutz) – personal alert safety system (PASS), distress signal unit (DSU), automatic distress signal unit (ADSU), personal distress alarm (PDA)

toxischer Schock – toxic shock (TS)

Toxizität, akute – toxicity, acute

tragbare Feuerlöscher – portable fire extinguishers

tragbare Feuerlöschgeräte – portable fire-extinguishers

tragbare Feuerlöschkreiselpumpe (TS) – portable fire pump normal pressure (PFPN)

tragbare Feuerwehrleitern – portable fire ladders

tragbare Infusionspumpe – ambulatory infusion pump

tragbare Leitern – portable ladders

tragbare Schlauchhaspel – portable hose reel

tragbare Warnblinkleuchte – portable flashing light

tragbares Beleuchtungsgerät – portable lightning apparatus

Trageband – carrier strap

Träger (Tragbalken) – girder

Träger einer Krankentrage (Person) – stretcher-bearer

Tragestuhl – stretcher chair

Tragfähigkeit – load bearing capacity

Traghaken – load hook, crane hook

Tragkraftspritze (TS) – portable pump, portable fire pump, portable fire pump normal pressure (PFPN)

Tragkraftspritzenanhänger (TSA) – portable pump trailer

Tragkraftspritzenfahrzeug (TSF) – small fire engine with portable pump, portable fire pump vehicle

Tränengas – lacrimator, tear-gas

Tränkung – dipping, steeping

Transport – transfer

Transportanlage – transport system

Transportfähigkeit – movability, transportability

Transportfahrzeug – transporter, transport vehicle

Transportkategorie – transport class, transport category

Transportkennzahl (TKZ) – transport index (TI)

Transportliege – wheeled stretcher

Transportrecht – transportation law

Transport-Unfall-Informations- und Hilfeleistungssystem (TUIS) – Transport-Accident-Information-and-Assistance--System

Trapezblech – trapezoidal sheet metal

Trapezblechdach – trapezoidal roof

Traufe – eave

Trefoil = Strahlenwarnzeichen – trefoil = basic ionizing radiation symbol

Treibgas – propellant gas

Treibmittel – expellant

Trenneffekt (Löscheffekt) – separating effect

Trennscheibe – cutting disk

Trennschleifen – cut off by grinding

Trennschleifer – grinder

Trennschneiden – cutting

Trennwand – partition wall

Treppe – stairway

Treppe, notwendige – necessary stair

Treppenauge – well hole

Treppenraum – staircase

Triageabteilung – triage ward

Triagieren – triage

Trichter – funnel

Triebwerksbrand (Flugzeug) – engine fire

Trinkwasser – drinking water

Trinkwasserleitung – drinking water pipe

Trinkwasserleitungsanlage – drinking water pipework system

Tripelpunkt (TP) – triple point (TP)

Tritium (^{3}H, T) – tritium

Trittleiter – step ladder

Trittstufe (Fahrzeug) – footstep

trocken – dry, arid

Trockenbatterie – dry cell, dry cell battery

Trockeneis (CO_2) – dry-ice

Trockenlöscher – dry powder extinguisher, → Pulverlöscher

Trockenlöschmittel – → Löschpulver

Trockenlöschverfahren – extinction by dry powder

Trockensaugprüfung – dry suction test

Trockensteigleitung – dry rising main, dry riser

Trommel – drum

Trompetenknoten – sheepshank, cat-shank

Tröpfchen – droplets

Tröpfcheninfektion – droplet infection

Tropfen – drop

tropfen – drip

Trümmer – wreckage, rubble, debris

Trümmerbereich – reach of ruins

Trümmerlast – load of ruins

Trümmerschatten – collapse zone, reach of ruins

Trummsäge – two-man saw

Trupp (Tr) – squadron, team-of-two

Truppführer (TrFü) – leading fire-fighter (LFf), squadron leader

Truppmann (TrM) – fire-fighter (Ff)

Tscherenkow-Strahlung – Cherenkov radiation

Tuberkulose (Tb) – tuberculosis (TB)

Tuberkulose, offene – active tuberculosis

Tunnelbrand – tunnel fire

Tür eines Brandabschnitts – fire compartment door

Tür, feuerfest – fire resisting door

Tür, feuerhemmende – fire-retardant door

Tür, nach außen öffnend – outward opening door

Tür, nach innen öffnend – inward opening door

Tür, rauch- und flammendicht – fire resisting smoke door

Türband – door hinge

Türblatt – door leaf

turbulente Flamme – turbulent flame

turbulente Strömung – turbulent flow

Turbulenz – turbulence

Türschild – name plate, escutcheon

Türschließer – door closer

Türsturz – lintel

Typenprüfung – type testing

Typenschild – type label, type plate

U

überbewertet – overstated

Überdruck – positive pressure

Überdruck-Pressluftatmer – positive pressure breathing apparatus

Überdruckventil – pressure valve, pressure relief valve

Überflurhydrant – pillar hydrant

Überflurhydrantenschlüssel – pillar hydrant key

Übergangsstück – connector, adaptor

Übergangsstück für Feuerwehrdruckschläuche – delivery hose adaptor

Übergangsstück für Feuerwehrsaugschläuche – suction hose adaptor

Übergangszustand – transition state

Überhandknoten – overhand knot

überhitzen – overheat, superheat

Überhitzung – overheating, superheating

überkochen – boil over

überkritisch (Gas, Flüssigkeit) – super-critically

Überlastung – overload

Überlaufen – overflow, overrun

Überlebensrate – survival rate

überprüfen – check, confirm, inspect, verify

Überprüfung – check-up, examination, inspection, verification

überschreiten – exceed

Überschuhe – shoe covers, shoe protectors

Überschwemmung – flooding, flood

Überspannung – overvoltage, overpotential

Überspannungsschutz – surge suppressor

übersteuern – overshoot

Übertragungsanlage – transmission system

Übertragungseinrichtung (ÜE) – transmission device

Übertragungsweg – transmission route

Überwachung (von Geräten) – monitoring

Überwachung (von Personen) – surveillance

Überwachung, besondere – special surveillance

Überwachungsbereich – supervised area

Übungshaus – fire drill house

Übungspuppe – training dummy, dummy

Übungsturm – drill tower, practice tower

Ultraschall – ultrasound

umbauter Raum – cubature

umbauter Raum – cubature

Umdrehungen – rate of rotation, revolutions

Umdrehungsgeschwindigkeit – rate of rotation

Umfang – girth

umfassender Löschangriff – surrounding fire attack

Umgebung – surrounding, ambience

Umgebung, nahe/nähere – vicinity

Umgebungsäquivalentdosis – ambient dose equivalent

Umgebungsäquivalentdosisleistung – ambient dose equivalent rate

Umgebungsatmosphäre – ambient atmosphere

T. Schmiermund, *Fachwörterbuch Feuerwehr und Brandschutz*, https://doi.org/10.1007/978-3-662-64120-0_19

Umgebungsbrand – surrounding fire, ambient fire

Umgebungsdruck – ambient pressure

Umgebungstemperatur – ambient temperature

Umlenkrolle – pulley

Umlenkung – deflection

umluftunabhängiges Atemschutzgerät – self-contained breathing apparatus (SCBA)

umrechnen – convert

Umriss – outline, contour

Umstände – circumstances

Umstellung – change-over

Umwandlung – conversion

Umwelt – environment

Umweltbedingungen – environment conditions

Umweltbehörde – environmental authority

Umweltbelastung – environmental burden, environmental load

umweltgefährdende Stoffe – environmentally hazardous substances

Umweltradioaktivität – environmental radioactivity

Umweltrecht – environment law

Umweltschutz – environmental protection, pollution control

Umweltschutzzug – chemical incident unit (CIU)

Umweltverschmutzung – environmental pollution

umweltverträglich – environmentally compatible, environmentally friendly

Umweltverträglichkeit, elektromagnetische (EMUV) – electromagnetic environmental compatibility (EMEC)

Umweltverträglichkeitsprüfung (UVP) – environmental impact assessment (EIA)

Umweltzerstörung – environmental degradation

unabgeschirmte Strahlenquelle – unshielded radiation source

unabhängige Löschwasserversorgung – independent water supply

unbedenklich – without risk, unrisky

unbewusstes Verhalten – unconscious behaviour

undicht – leaky

undurchlässig – impermeable

unerschöpfliche Löschwasserversorgung – inexhaustible water supply for fire-fighting

Unfall – accident

Unfallambulanz – casualty department, casualty

Unfallbericht – accident report, accident log

Unfallopfer – accident victim

Unfallverhütung – prevention of accidents

Unfallversicherung – accident insurance

ungebraucht – unused, fresh, virgin

ungefährlich – not dangerous, harmless

ungeschriebenes Gesetz – unwritten rule

ungeschützter Bereich – unprotected area

Universalklebeband – duct tape, Duck Tape™

Universallöschfahrzeug (ULF) – universal fire-fighting vehicle

universelle Gaskonstante – universal gas constant

universelle transversale Mercator-Projektion (UTM) – universal transversal Mercator projection

Unkrautvernichtungsmittel – herbicide, weed killer

unmittelbare Gefahr für Leben und Gesundheit – imminent danger to life and health (IDLH)

unregelmäßig – irregular

unrein – impure, unclean

unschädlich – harmless, not harmful

unsicher – unsafe

Unterabschnitt – sub-sector

Unterbau – foundation, substructure

unterbewertet – understated

unterbrechen – interrupt

Unterdruck – negative pressure

untere Explosionsgrenze (UEG) – lower explosion limit (LEL)

untere Nachweisgrenze – minimum detection limit (MDL)

untere Wasserbehörde – lower water authority

unterer Explosionspunkt – lower explosion point

unterer Heizwert – → Heizwert

Unterflurhydrant – ground hydrant

Unterflurhydrantenschlüssel – ground hydrant key

Untergrund – substrate

Untergruppen (der Gefahrklassen) – hazard divisions

Unterkühlung – hypothermia, undercooling

Unterkunft – shelter

Unterkunft, vorübergehende – temporary shelter

Unterlegkeil – wheel chock, drag shoe

Unterlegscheibe – washer

unterstellte Einsatzkräfte – units under command, subordinated forces

Unterstellungsverhältnis – subordination

Unterstützung – support

Unterstützung, logistische – logistical support

unterteilen – subdivide

Unterweisung – instruction(s), teaching, briefing

unvermischbar – immiscible

unverzüglich – immediately

unverzweigte Kettenreaktion – unbranched chain reaction

Unwetter – tempests, severe weather, thunderstorm

Unwetter, heftiges – violent storm

Unwetter, starkes – severe thunderstorm

Unwucht – unbalanced state

unzerbrechlich – unbreakable

Uran – uranium

Uran, angereichertes – enriched uranium

Uran, hochangereichertes – highly enriched uranium

Urananreicherung – uranium enrichment

Uranbombe – uranium bomb

Uranisotop – uranium isotope

Uranstrahlung – uranium radiation

Ursprung – origin

UV-empfindlich – UV-sensitive

UV-Strahlung – UV radiation

V

Vakuum – vacuum

Vakuumpumpe – vacuum exhauster

Vakuumtrage – vacuum-stretcher

Valenzelektronen – valence electrons

Van-der-Waals-Anziehung – van der Waals attraction

Vaseline – Vaseline, petroleum jelly

Vene – vein

Ventil – valve

Ventilationsanlage – ventilation system

Ventilator – ventilator, fan

Ventil-Handrad – handle for valves

Ventilklappe – valve flap, clapper valve

Ventilscheibe – valve disk

Verabschiedung (aus dem Dienst) – retirement, act of retiring

Verankerung – anchorage

Verantwortlichkeiten, finanzielle – financial responsibilities

Verantwortung – responsibility

Verantwortungsbereich – area of responsibility

Verantwortungsebene – level of responsibility

Verästelungsnetz – ramified water main

Verätzung – cauterization, chemical burn

Verätzungsgefahr – risk of burns by corrosion

Verband (Vb) – brigade (Bde)

Verbände (bei Verletzungen) – bandages, dressings

Verbandpäckchen – roll of bandage, dressing roll

Verbandplatz – dressing station

Verbandschere – first-aid scissors, bandage scissors

Verbandsführer (VbFü) – brigade leader

Verbandskasten – first-aid box

Verbesserung – improvement

Verbindung (Material, Substanz) – compound

Verbindungsmittel (Absturzsicherung) – fastener, lanyards

Verbindungsoffizier (VO) – liaison officer (LNO)

Verbindungspersonal (z. B. zu anderen Dienststellen) – liaison-personnel

Verbot – prohibition, ban

verboten – prohibited, forbidden

Verbotszeichen – prohibition sign

Verbrauchsmaterial – consumable goods, consumables

Verbrennung – combustion

Verbrennung (medizinisch) – burn

Verbrennung dritten Grades – third-degree burn

Verbrennung ersten Grades – first-degree burn

Verbrennung zweiten Grades – second-degree burn

Verbrennung, brennstoffarme – fuel-lean combustion

Verbrennung, brennstoffreiche – fuel-rich combustion

Verbrennung, explosionsartige – explosive combustion

T. Schmiermund, *Fachwörterbuch Feuerwehr und Brandschutz*, https://doi.org/10.1007/978-3-662-64120-0_20

Verbrennung, flammenlose – flameless combustion

Verbrennung, stöchiometrische – stoichiometric combustion

Verbrennung, vollständige – complete combustion

Verbrennungsart – type of combustion

Verbrennungsenergie – energy of combustion, combustion energy

Verbrennungsgase – combustion gases

Verbrennungsgeschwindigkeit – combustion rate

Verbrennungsgleichung – combustion equation

Verbrennungsmechanismus – combustion mechanism

Verbrennungsprodukte – combustion products

Verbrennungsprozess – combustion process

Verbrennungstemperatur – combustion temperature

Verbrennungstemperatur, theoretische – theoretical combustion temperature

Verbrennungsvorgang – combustion process

Verbrennungswärme – calorific value, heat of combustion

Verbrennungswärme, effektive – actual calorific value

Verbrennungswärme, spezifische – specific heat of combustion

Verbrennungszone – zone of combustion

Verbrühung – scald

Verbund (Bauelemente) – junction, link up, bond

Verbundbauweise – composite method of building, composite construction

Verbundwerkstoff – composite material

Verdacht – suspicion

Verdachtsstoff – suspected toxin

Verdampfung – evaporation

Verdampfungseffekt – vaporization effect

Verdampfungswärme – heat of vaporization, enthalpy of vaporization, heat of gasification

Verdampfungswärme, spezifische – specific heat of vaporization, specific heat of evaporation

Verdichtung (von Gasen) – compression

Verdickung – thickening

verdrängen – dislodge, displace

Verdrängungseffekt (Löscheffekt) – crowding-out effect, displacing effect

Verdünnung – dilution

Verdünnungseffekt (Löscheffekt) – dilution effect, diluting effect

Verdünnungsmittel – diluent, diluting agent

verdunsten – evaporize

Verdunstung – evaporation

Verdunstungsfläche – evaporation surface

Verdunstungskälte – evaporative cooling, evaporation chill

Verdunstungszahl – evaporation number, evaporation rate

Vereinbarung – agreement, arrangement

Verfahren – procedure, process

verflüssigtes Erdgas – liquefied natural gas (LNG)

Verformen – deform

Verformung – deformation

Verformungsenergie – deformation energy

Verformungsgefahr – risk of deformation

Verformungswiderstand – deformation resistance

Verfügbarkeit – availability

Vergasen – gasify

Vergasung – gasification

Vergiftung – poisoning

Vergiftungsgefahr – risk of intoxication

Vergiftungszentrale – poison control center

Verglasung (Baukunde) – glazing

Verhalten – behaviour

Verhalten bei Brandeinwirkung (Bauteile) – fire performance

Verhalten, unbewusstes – unconscious behaviour

Verhaltensregeln – rules of conduct, behavioral rules

Verhaltensregeln im Brandschutz – behavioral rules in fire protection

Verhältnismäßigkeit – proportionality

Verhältnismäßigkeits-Grundsatz – principle of reasonableness, principle of proportionality

Verhütung – prevention

Verkehrsfläche – circulation area, traffic area

Verkehrsleitkegel – traffic cone

Verkehrsunfall (VU) – road traffic accident (RTA), traffic accident, motor vehicle accident (MVA)

Verkehrsunfall, schwerer – car crash

Verkehrsweg – traffic route

Verkleidung (Baukunde) – cladding

verkohlen – carbonize

Verkohlung – charring, carbonisation

Verkokung – coking

Verlängerungskabel – extension cord, extender cable, extension lead

Verlegungsbefehl – movement order

Verletzte – casualties

Verletzte, kontaminierte – contaminated casualties

Verletztendarsteller – role play actor

Verletzung – hurt, injury

Verletzung, tödliche – fatal injury

Verlustfaktor – loss factor

Verlustmeldung – casualty report

Vermeidung – avoidance

Vermutung – speculation, assumption

Vernichtung – destruction

Veröffentlichung – publication

Verordnung – regulation, ordinance, by-law, order

Verpackung – packaging

Verpackungsmaterialien – packing materials

Verpflichtung – obligation

Verpuffung – deflagration

Verputz – plaster

verqualmt – smoky, smoke-filled

verraucht – smoky, smoke-filled

Verriegeln – bolting

Versagen – break down, give out

Versagen, menschliches – man-made disaster

Versammlungsraum – assembly room, assembly hall

Versammlungsstätte – public assembly room, assembly centre

Versand – delivery

Verschalung – casing, panelling

Verschäumung – foaming

Verschäumungsbereich – foaming range

Verschäumungszahl (VZ) – foaming index, foaming ratio, expansion ratio

Verschieden – various, different, miscellaneous

Verschiedenartig – variously, variedly

Verschiedenartigkeit – diversity

verschiedenes – miscellaneous

Verschleiß – attrition

Verschleißteile – expendable parts

Verschleppung (von Substanzen) – carryover

Verschließen – locking

Verschluss – locking mechanism

Verschlüsselung – encryption

Verschlusssache (VS) – classified material, classified document

verschmoren – scorch

verschmutzen – pollute, contaminate

Verschmutzung – pollution, contamination

verschüttete Person – spilled person, avalanched person

Verschüttetenrettung – avalanche victim rescue

Verschütteten-Suchgerät – sound detector

Verschüttetes (Flüssigkeiten) – spillage

Versengen – to scorch

verseucht (durch Mikroorganismen) – infested

Verseuchung (durch Mikroorganismen) – infestation

Versicherer – insurer

Versicherter – insured person

Versicherung – insurance

Versicherungswert – insurance value, insurable value

Versorgung – logistics

Versorgungsdruck – supply pressure

Versorgungsleitung – supply conduit

Verstärker (Funktechnik) – repeater

Verstärkerpumpe – booster pump

Verstärkung – reinforcement, back-up forces, enhancement forces

Verteidigung – defence

Verteiler (Empfänger-Liste) – distribution list

Verteiler (Feuerwehr-Gerät) – dividing breeching, distributor, triple-head distributor

Verteiler mit Überdruckventil – controlled dividing breeching

Verteiler, dreifach – three-way distributor

Verteilung – distribution

Verteilungsfaktor – factor of distribution

Verträglichkeit, elektromagnetische (EMV) – electromagnetic compatibility (EMC)

Verträglichkeitsgruppe – compatibility group, compatibility class

vertraulich – confidential

Vertraulichkeit – confidentiality

Verwaltungsakt – administration act, administrative deed

Verwaltungsvorgang – administrative procedure

Verwaltungsvorschrift (VwV) – administrative regulation

verwendbar – usable

verwerfen (verformen) – warp

Verwirklichung – realization

Verzeichnisse – registers

Verzögerer – retarder

verzögerte Behandlung – delay treatment (DT)

Verzögerung – delay

Verzögerungsdauer – delay period

Verzögerungseffekt – delayed effect

verzweigte Kettenreaktion – branched chain reaction

Vielfachmessgerät – multimeter

Vier-Augen-Prinzip – four-eyes principle

Vierkantschlüssel – square spanner, square wrench

Vierkantsteckschlüssel – square-box wrench

virales hämorrhagisches Fieber – viral haemorrhagic fever

Viren – viruses

Virulenz – virulence

viruzid – virucidal, viricidal

Viskosität – viscosity

Vitalfunktionen – vital functions

Vollalarm – full alarm

Vollbrand – fully developed fire

Vollduplex – full-duplex

Vollflutung – total discharge

Vollflutungsanlage – total discharge system

Vollgeschoss – full storey

vollgestellt – cluttered

Vollgummi – solid rubber

Volllast – full load

Vollmaske (Atemschutz) – full facepiece respirator, full face mask

Vollnarkose – general anesthesia

Vollschutz – full protective clothing

vollständig – complete

vollständige Verbrennung – complete combustion

vollständiger Aktivitätsverlust – loss of activity

vollständiger Durchbruch – full penetration

Vollstrahl – full jet, solid jet, solid stream

Vollstrahlrohr – full jet pipe, smoothbore nozzle pipe

Vollüberwachung – full monitoring

Vollwärmeschutzsystem – total heat protection system

Vollzugriff – full access

Volumen, molares – molar volume

Volumenabbrandgeschwindigkeit – volume burning rate

Volumenanteil – volume fraction

Volumenausbeute – volume yield

Volumenausdehnung – thermal volume expansion

Volumenausdehnungskoeffizient – coefficient of volume expansion

Volumenkonzentration – volume concentration

Volumenmangelschock – hypovolemic shock

Volumenstrom – volumetric flow rate, volume flow

Vorausrüstwagen (VRW) – immediate rescue vehicle

Voraussage – prediction

Voraussetzung, bauliche – building requirement

Voraussetzung, betriebliche – production requirement, business requirement

Vorbaupumpe – front mounted pump

Vorbefehl – preselect command

Vorbehandlung – pretreatment

vorbereitet – prepared

vorbeugend – preventive

vorbeugende Maßnahme – preventive action

vorbeugender Brandschutz (VB) – fire prevention (FP)

vorbeugender Explosionsschutz – preventive explosion protection

Vordach – canopy, porch

Vorderseite – front side, preface

Vordrucke – forms

vorgemischte Flamme – pre-mixed flame

vorgemischte Flamme – premixed flame

Vorgesetzter – superior

Vorhängeschloss – padlock

vorherrschend – prevailing

Vorkehrung – precaution

Vorkehrungen treffen – take precautions

Vorrat – supply, stock, storage

Vorratsbehälter – supply tank, storage tank

Vorratsmenge – stored quantity

Vorreinigung – precleaning

Vorsatz (Absicht) – malice

vorsätzlich – deliberate, nonnegligent

vorsätzliche Brandstiftung – deliberate arson, nonnegligent arson

Vorschlaghammer – sledgehammer

Vorschrift – provision, rule, regulation

Vorsicht – caution

Vorsichtsmaßnahme – precautional measure, cautionary measure, precaution

Vorsteiger (Höhenrettung) – lead climber

Vorstieg (Höhenrettung) – lead climbing

Vorteil – advantage

Vorwarndauer – pre-warning duration

Vorwarnung – pre-warning

Vorwarnzeit – pre-warning time

Vorwort – preface, foreword

W

Wachablösung – end of duty

Wachabteilung – one-shift-personnel of a fire station

Wachabteilungsleiter – senior officer

Wachbereich – station turnout area

Wachsfackel – wax torch

Wackelkontakt – loose connection

Waffe, biologische – biological weapon

Waffe, chemische – chemical weapon

Waffe, radiologische – radiological weapon

Wagenheber – jack, automobile jack, lifting jack

Wagenheber, hydraulischer – hydraulic jack

Wahrnehmungen – perceptions

Wahrnehmungen, eigene – personal perceptions

Wahrscheinlichkeit – probability

Wahrscheinlichkeit des Eintretens (Ereignis) – probability of occurrence

Waldbrand – forest fire, wild fire, wildland fire

Waldbrandabwehr – forest fire defence

Waldbrandbekämpfung – forest firefighting

Waldbrandgefahrenstufe – forest fire risk level, forest fire warning level

Waldbrandprognose – forest fire forecast

Waldbrandursache – forest fire cause

Waldbrandwarnstufe – forest fire warning level

Waldbrandwarnung – forest fire warning

Walmdach – hip roof

Wand – wall

Wandeffekt (Löscheffekt) – wall effect

Wandhydrant – fire cabinet, hose reel cabinet, wall hydrant

Wandtafel (Schaubild) – wall chart

Wandverkleidung – panelling

Wandwasseranschluss – wall hydrant, hose bib

Wanne – pan, tray

Warenzeichen – trademark

Wärme – heat

Wärme abstrahlen – radiate heat

Wärme, latente – latent heat

Wärmeabfuhr – heat removal

wärmeabgebend – liberating heat, exothermic, exothermal

Wärmeableitung – heat outlet, heat dissipation

Wärmeabsorption – heat absorption

Wärmeabzugsanlage – heat exhaust venting system

wärmeaufnehmend – absconding heat, endothermic, endothermal

Wärmeauftrieb (Konvektion) – thermal up current, thermal uplift

Wärmeausbreitung – heat spread

Wärmeausdehnung – thermal expansion

Wärmeausdehnungskoeffizient (WAK) – coefficient of thermal expansion (CTE), temperature coefficient of expansion (TCE)

Wärmeaustausch – heat exchange

wärmebeständige Kunststoffe – heat resistant plastics

Wärmebeständigkeit – heat resistance

T. Schmiermund, *Fachwörterbuch Feuerwehr und Brandschutz*, https://doi.org/10.1007/978-3-662-64120-0_21

Wärmebilanz – heat balance, heat budget

Wärmebildkamera (WBK) – thermal imager, thermal imaging camera (TIC)

Wärmedämmfolie – heat-insulation foil

Wärmedämmstoff – thermal insulation material

Wärmedämmung – heat insulation

Wärmedifferentialmelder – rate-of-rise temperature detector

Wärmedurchgang – heat transfer, heat passage, heat transmission

Wärmedurchgangskoeffizient – heat transmission coefficient

Wärmeeintrag – heat input

wärmeempfindlich – sensitive to heat

Wärmeenergie – thermal energy, caloric energy

Wärmeentwicklung – heat generation

Wärmeexplosion – thermal explosion

Wärmefluss – heat flux, heat flow rate, heat flow

wärmefreisetzend – liberating heat, exothermic exothermal

Wärmefreisetzung – thermal release, heat release

Wärmefreisetzungsrate – heat release rate (HRR), rate of heat release

Wärmefühler – heat sensor

Wärmegleichgewicht – thermal equilibrium

Wärmekapazität – heat capacity, thermal capacity

Wärmekapazität, spezifische – specific heat capacity

Wärmekonvektion – thermal convection, heat convection

Wärmelehre – thermodynamics

wärmeleitend – thermo-conductive

Wärmeleiter – heat conductor

Wärmeleitfähigkeit – thermoconductivity, thermal conductivity

Wärmeleitfähigkeitsdetektor (WLD) – thermal conductivity detector (TCD)

Wärmeleitung – thermal conduction, heat conduction

Wärmeleitzahl – thermal diffusivity

Wärmemaximalmelder – maximum temperature detector

Wärmemelder – heat detector

Wärmemenge – heat quantity, amount of heat

Wärmemitführung (Konvektion) – heat convection

Wärmequelle – source of heat

Wärmerückgewinnung – heat recovery

Wärmeschutz – heat protection

Wärmeschutzanzug – heat protective suit (against thermal radiation)

Wärmeschutzhandschuhe – heat protection gloves

Wärmeschutzhaube – heat protective hood

Wärmeschutzkleidung – heat protective clothing

Wärmestau – heat accumulation, heat build-up

Wärmestrahlung – thermal radiation, heat radiation

Wärmestrahlungsübergang – radiative heat transfer

Wärmestrom – heat flux, heat flow rate, heat flow

Wärmestrom, konvektiver – convective heat flux

Wärmestromdichte – heat flow density

Wärmeströmung – heat convection

Wärmetauscher – heat exchanger

Wärmeträger – heat transfer medium

Wärmetransport – heat transfer

Wärmeübergang – heat transfer, heat transition

Wärmeüberträger – heat-transmission agent, heat transfer medium

Wärmeübertragung – heat transfer

Wärmeübertragung – heat transmission, heat transfer

Wärmeverbrauch – heat consumption

Wärmeverlust – heat loss

Wärmeversorgung – heat supply

Wärmezersetzung – thermal degradation

Wärmezündung – heat ignition

Warmwasserheizung – warm water heating

Warnblinkleuchte – flashing light

Warnblinkleuchte, tragbare – portable flashing light

Warndreieck – hazard triangle, hazard warning triangle

Warneinrichtung – alarm device

Warnflagge – signal flag, fire brigade signal flag

Warnkleidung – high-visibility clothing

Warnleuchte – warning light

Warnsignal – warning device

Warntafel (Lkw-Kennzeichnung) – hazard warning panel (on trucks)

Warnung – warning

Warnweste – warning vest, fluorescent jerkin, high-visibility vest (HV vest)

Warnzeichen – warning sign

Wartung – maintaining, maintenance

wartungsfrei – maintenance-free

Waschbenzin – cleaner's naphtha, cleaner's solvent

Wäscherei – laundry

Waschlauge – wash solution

Waschraum – washing room

Wasser – water

Wasser, Gesamthärte – total hardness of water

Wasser, permanente Härte – permanent hardness of water

Wasser, schweres – deuterium oxide

Wasser, temporäre Härte – temporary hardness of water

Wasserabscheider – water trap, water separator

Wasseraufbereitung – water treatment

Wasserbeaufschlagung – density of discharge

Wasserbedarf – water demand, water requirement

Wasserbehälter – water tank

Wasserbehörde – water authority

Wasserbehörde, untere – lower water authority

Wasserdampf – steam, water steam

wasserdampfflüchtig – steam-volatile

wasserdampfflüchtige Stoffe – steam-volatile substances

Wasserdruck – water pressure

Wasserdurchfluss – water throughput

Wasserenthärtung – water softening

Wasserentnahmestelle – water supply point

wasserfilmbildendes Schaummittel – aqueous film forming foam (AFFF)

Wasserförderung – delivery of water

wasserführende Armaturen – water carrying fittings

Wassergefährdungsklasse (WGK) – water hazard class (WHC)

Wassergehalt – water content

Wasserglas – water glass

Wasserhahn – faucet

Wasserhalbzeit (Löschmittel Schaum) – water half-life period

wasserhaltig – aqueous

Wasserhärte – hardness of water

Wasserhochbehälter – high-level water tank

Wasserleitung – water main

Wasserleitung, öffentliche – public water main

Wasserlöscher – water extinguisher

Wasserlöscher (Dauerdruck-Wasserlöscher) – pressurized water fire-extinguisher

Wasserlöschverfahren – method of extinguishing by water

wasserlöslich – soluble in water

Wasserlöslichkeit – water solubility

wassermischbar – water-miscible

Wassermischbarkeit – water miscibility

Wassernachweispaste – water finding paste

Wassernebel – water fog

Wasserpumpenzange – water pump pliers

Wasserrettung – water rescue

Wasserrettungsgeräte – water rescue apparatuses

Wasserringpumpe – water ring primer

Wassersäule – water column

Wasserschaden – water damage

Wasserschadstoffe – water pollutants

Wasser-Schaummittel-Gemisch – foam solution

Wasserschleier – water curtain

Wasserspeicher – water reservoir

Wassersperre – water barrier

Wasserspiegel – water level, water table

Wasserstand – water level

Wasserstandsglas – water gauge glass

Wasserstoff – hydrogen

Wasserstoffbombe – fusion bomb, H-bomb

Wasserstoffbrückenbindung – hydrogen bridge linkage

Wasserstrahl – water jet

Wasserstrahlpumpe – water-jet pump

Wassertankwagen – water tanker

Wassertrupp (W-Tr) – water crew, water team, water squadron

Wassertruppführer, -in (W-TrFü, W-TrFü'in) – water crew leader, water team leader, water squadron leader

Wassertruppmann, -frau (W-TrM, W-TrFr) – water crew member, water team member, water squadron member

Wasserunfall – water accident

wasserunlöslich – insoluble in water

Wasserverbrauch – water consumption

Wasserverschmutzung – water pollution

Wasserversorgung – water supply

Wasserversorgung, dezentrale – decentralized water supply

Wasserversorgung, öffentliche – public water supply

Wasserversorgung, zentrale – central water supply

Wasserversorgungsanlage – water supply installation

Wasserverteilung – water distribution

Wasserwaage – water level, mechanic's level

Wasserwerfer – monitor, fire monitor

Wasserzähler – water meter

wässrig – aqueous

Wathose – chest wader

Weberknoten – reef knot

Webleinenstek – clove hitch

Wechselladerfahrzeug (WLF) – roller container vehicle, vehicle for transportation of roller containers (ACTS)

Wechselstrom – alternating current (ac)

Wechselwirkung – interaction

Wechselwirkung, schwache – weak interaction

Wechselwirkung, starke – strong interaction

Wecker – alarm clock

Wegerecht – trespass right

Wehr (Stauwehr) – weir

Wehrersatzdienst – alternative national service

Wehrführer (der Freiwilligen Feuerwehr) (WeFü) – officer in charge (OiC) of the volunteer fire brigade

Wehrleiter (der Freiwilligen Feuerwehr) – officer in charge (OiC) of the volunteer fire brigade

Wehrpflicht – conscription, draft service

Wehrpflichtige – conscripts

weich – weak

Weißbereich – white area

Weißblech – tin plate

Weißblechdose – tinplate can

weißglühend – white-glowing, incandescent

Weisungsbefugnis – superiority

Weiterbildung – supplementary education, continuing education

Weiterbrennen – burn on

Weiterbrennen, selbstständiges – self-propagation of flame

Wellblech – corrugated sheet iron

Welle – wave

Wellenbereich (Funk) – waveband

Wellenbereichsschalter – band switch

Wellenlänge – wave length

Wellenstrahlung – wave emission of radiation

Welle-Teilchen-Dualismus – wave-corpuscle duality

Wellpappe – corrugated board

Wendehaken – turning hook, cant hook, peavey

Wendeltreppe – spiral staircase

Wendestrahlrohr – turntable monitor

Wendl-Tubus – nasopharyngeal tube

Werkbank – workbench

Werkfeuerwehr (WF) – work fire brigade (WTF), plant fire brigade, site fire brigade

Werkzeuge – tools

Werkzeugkasten – tool-box, tool-chest

Werkzeugkoffer – tool case

Wetterschutz – weather shelter

widersprüchlich – contradictory

Widerstand, elektrischer – electric resistance

Wiederaufbereitungsanlage – reprocessing plant

Wiederbelebung – resuscitation, revitalisation

Wiederherstellung der Einsatzbereitschaft – restoration of operational readiness

Wiederherstellungskosten – restoration costs

wiederverwendbar – reusable

Wiederverwendung – reuse

Wiederverwertung – recycling

Wiese – grassland

Wildwasser – white-water

Wildwasserrettung – white-water rescue

Willenskraft – volition

Wind- und Wetterschutz – shelter against wind and weather

windabgewandt – downwind

windabgewandte Seite – downwind side

Winde (Gerät) – winch

Winde, handbetriebene – hand-operated winch

Windeinfluss – wind effect

Windfahne – wind vane

Windgeschwindigkeit – wind speed

Windgeschwindigkeitsmesser – wind speed meter

Windmessgerät – wind gauge, anemometer

Windrichtung – wind direction

Windrichtungsanzeiger – wind-direction indicator

Windschutz – wind shelter

Windstärke – wind force, storm force, wind strength

Windwurf – windfall, windthrow

windzugewandt – windward

windzugewandte Seite – windward side

Winkelmesser – goniometer

Winkelschleifer – angle grinder

Winkerkelle – traffic paddle

Wipfelbrand – treetop fire

Wipfelfeuer – treetop fire

Wirbelsäulenbrett – spine board (SB)

Wirbelstrombremse – eddy (current) brake

Wird ausgeführt! – Wilco!; Will comply!

Wird erledigt! – Can do!

Wirkfläche (von Sprinklern) – sprinkler activation area

Wirksamkeit – efficiency, effectiveness

Wirkungsgrad – efficiency

wirkungslos – inoperative

Wischprobe – wipe test

Wochenendschicht – weekend shift

Wohngebiet – residential area

Wohnhausbrand – house fire

Wohnungsbrand – apartment fire, flat fire, home fire

Wundbenzin – medical-grade petroleum spirit

Wunde – wound

Wunden, offene – open wounds

Wunderkerzen – sparklers

Wundhaken – tissue retractor, retractor

Wundheilsalbe – wound healing ointment

Wundsepsis – wound sepsis

Wurfhöhe (z. B. Wasserstrahl) – height of jet

Wurfleinenknoten – bumper knot

Wurfweite (z. B. Wasserstrahl) – range of jet

X, Y, Z

x-fach – x-way

Yperit – mustard gas

Y-Schlauchanschluss – Siamese connection, Y-shape

zäh – tough, rigid

zäh (klebrig) – tacky

zähflüssig – viscous, viscid

Zähigkeit – toughness, rigidity

Zählrohr – counting tube, counter

Zange – plier, pliers

Zapfenstreich – tattoo

Zapfenstreich, großer – great tattoo

Zarge (Tür, Fenster) – frame, case

Zehenschutzkappe – protective toecap

Zeichen, taktische – tactical signs

Zeit bis zum Eintreffen – attendance time

zeitkritisch – time-critical

zeitliche Reihenfolge – temporal order

zeitlicher Druckanstieg – rate of pressure, pressure rise at to time

zeitlicher maximaler Druckanstieg – maximum rate of pressure

Zeitrahmen – time frame

Zeitschaltuhr – time switch, switch clock, timer

Zeitschlitz – time slot

Zeitüberwachung – time monitoring

zellschädigend – cytopathic, cytotoxic

zelltötend – cytocidal

Zelt – tent

Zeltdorf – tent village

Zeltkrankenhaus – tent hospital

Zement – cement

zentrale Wasserversorgung – central water supply

Zentralnervensystem (ZNS) – central nervous system (CNS)

Zentrifugalkraft – centrifugal force

zerbrechlich – fragile

zerbrechlicher Behälter – breakable container

Zerfall (Zersetzung) – disintegration, decay

Zerfall (Zusammenbruch) – breakdown

Zerfall, radioaktiver – radioactive decay

Zerfallsarten – types of decay

Zerfallsenergie – decay energy

Zerfallskonstante – decay constant

Zerfallsreihe – decay series, transformation series, decay chain

Zerfallszeit – decay period

Zerknall – bursting

Zerplatzdruck – burst pressure

zersetzen – disintegrate, decompose, degrade, decay

Zersetzung – decomposition

Zersetzung, exotherme – exothermic decomposition

Zersetzung, selbstbeschleunigte – self-acceleration decomposition

Zersetzung, thermische – thermal decomposition

Zersetzungseffekt – decomposition effect

Zersetzungsgeschwindigkeit – decomposition rate

Zersetzungsprodukte – decomposition products

T. Schmiermund, *Fachwörterbuch Feuerwehr und Brandschutz*, https://doi.org/10.1007/978-3-662-64120-0_22

Zersetzungstemperatur – decomposition temperature

zerstäuben – atomize, spray

Zerstäuber – diffuser, ‚atomizer'

Zerstäuberdüse – spray nozzle

Zerstörung – destruction

Zertifizierung – certification

Ziegelstein – brick

Ziel des Einsatzauftrags – objective of the mission

Ziele – goals, targets, aims

zielgerichtet – target orientated

zielorientiert – goal-orientated

ZIKADE-Merkschema: Zeit begrenzen – Inkorporation verhindern – Kontaminationsverschleppung vermeiden – Abschalten – Deckung ausnutzen – Entfernung groß halten – ZIKADE-mnemonic (approx.: TICDCD-mnemonic: limit time– prevent incorporation – avoid contamination carry-over – shut down – exploit cover – keep distance large)

Zimmermannsbeil – carpenter's hatched

Zimmermannshammer – carpenter's hammer

Zimmermannsstich (Knoten) – timber hitch, countryman's knot, lumberman's knot

Zimmermannsstich mit Halbschlag – killick hitch, timber hitch with a half stitch

Zisterne – cistern

Zivil-militärische Zusammenarbeit (ZMZ) – civil-military co-operation (CIMIC)

Zivilschutz (ZS) – civil protection, civil defense

Zivilschutzsirene – civil defense siren

Zivilverteidigung – civil defense

Zollstock – → Gliedermaßstab

Zubringerleitung (Löschwasserversorgung) – feeder water conduit

Zufahrten für die Feuerwehr – driveways for the fire brigade

Zufuhr – feed, supply

Zug (Einheit) – company

Zug (Fahrzeug) – train

Zug (Zugluft) – draught, draft

Zugang, intravenöser – intravenous line

Zugänge für die Feuerwehr – entrances for the fire brigade

Zugänglichkeit – accessibility

Zugangspunkt – access point

Zugangspunkt, drahtloser – wireless access point (WAP)

Zugangstechniken, seilunterstützte (SZT) – rope access

Zugbrand – train fire

zugewiesene Mittel – assigned means

Zugfahrzeug – traction vehicle

Zugfestigkeit – tensile strength

Zugführer (ZFü) – company leader

zügig/rasch – swift

Zugkraft – tractive force

Zugsäge – tractive saw, crosscut saw, pit saw

Zugseil – pull-rope, pulling cable

Zugtrupp (ZTr) – company command crew

zulässige Achslast – permissible axle load

zulässige Gesamtmasse – permissible total weight

zulässige Höchstkonzentration – maximum permitted concentration

Zulassung – authorisation, permission, licence

Zuluft – air supply

Zuluftöffnung – air supply opening, fresh air vent

Zumischer – inline inductor

Zumischer, regelbarer – adjustable inline inductor

Zumischerschlauch – pick-up hose for extinguishing agent

Zumischrate – admixture rate, admixing rate

Zumutbarkeit – reasonableness

Zündbereich – ignition range

Zünddurchschlag – explosive strike

Zündeigenschaften – ignition characteristics

zünden – ignite, fire

Zündenergie – ignition energy

zündfähig – ignitable

Zündfähigkeit – ignitability

Zündflamme – ignition flame, pilot flame

Zündfunke – ignition spark, trigger spark

Zündgefahr – ignition danger

Zündgrenzdruck – ignition limit pressure

Zündgrenze – flammable limit

Zündholz – match

Zündkerzen – spark plugs

Zündphase – ignition phase

Zündpunkt – ignition point

Zündquelle – ignition source

Zündschutzart – ignition protection type

Zündtemperatur – ignition temperature, kindling temperature, spontaneous-ignition temperature (SIT)

Zündung – ignition

Zündursache – cause of ignition

Zündverhalten – ignition characteristics

Zündverzug – ignition lag

Zündverzugszeit – ignition delay time

Zündvorgang – process of ignition

Zündwilligkeit – combustibility

Zündzeit – ignition time

Zusammenarbeit – cooperation, collaboration

Zusammenarbeit, auftragsbezogene – mission-concentrated cooperation

zusammenbrechen – collapse

zusammenziehen – contract

Zusatz, filmbildender – film forming additive

zusätzliche Ausrüstung – supplementary equipment

zusätzliche Hinweise – additional guidance

Zusatzstromversorgung – auxiliary power unit (APU)

Zusatzzeichen – additional signs

Zuschauer – bystander

Zustand, angeregter – exited state

Zustand, fester – solid state

Zustand, flüssiger – liquid state

Zustand, gasförmiger – gaseous state

Zustand, gleichbleibender – steady state

Zustand, lebensbedrohender – critical condition

Zustand, stationärer – steady state

Zuständigkeit – jurisdiction

Zuständigkeitsbereiche – areas of jurisdiction

Zustandsänderung – state change

Zustandsgröße – parameter of state, variable of state

Zustandsgrößen – parameters of state

Zuteilung – allocation

Zutrittsbeschränkung – restricted access

Zutrittsverweigerung – denial of access

Zwangslüftung – forced-air system

Zwangsmittel – means of coercion

Zweck – purpose, goal

Zweifel – doubt

Zweifelsfall, im – in case of doubt

Zwei-Helfer-Methode (Erste Hilfe, HLW) – two-helper method

Zwei-Komponenten-Kleber – two-component adhesive

Zweikreisbremse – dual-circuit brake

zweiter Abmarsch – second alarm

zweiter Rettungsweg – second fire escape, second escape route

Zwinge – clamp

zwingende Notwendigkeit – compelling necessity

Zwischenbehälter – intermediate tank

Zwischenfall – incident

Zwischengeschoss – mezzanine

zwischenmenschlicher Bereich – interpersonal area

zwischenmolekulare Anziehung – intermolecular attraction, molecular attraction

Zwischensicherung – interim security

Zylinder – cylinder

Zylinderprojektion (Kartenkunde) – cylinder projection

Zylinderschloss – cylinder lock

English-German

Inhaltsverzeichnis

A

AB powder – AB-Pulver

ABC powder – ABC-Pulver, Glutbrandpulver

ABCDE-mnemonic: attention, area, approach – break, barrier (self-protection), barrier (external protection) – count, communicate – develop, direct, delegate – evaluate, emergency departments – ABCDE-Merkschema: Achtung, Areal, Ansatz – Bedenkzeit, Barriere (Selbstschutz), Barriere (Fremdschutz) – Zählen, Kommunizieren – durchdachte Grundstrukturen, Dirigieren, Delegieren – Evaluieren, Erstkontakt Notaufnahme

abnormal occurrence – Störfall

A-bomb – Atombombe

abrasion – Abrieb

abroll container – Abrollbehälter (AB)

absconding heat – wärmeaufnehmend

abseiling – Abseilen

absolute law – unbedingt einzuhaltendes Gesetz

absolute zero – absoluter Nullpunkt (Temperatur)

absorbance agent – Absorptionsmittel

absorbance granules – Absorptionsgranulat

absorbed dose – Energiedosis

absorbed dose rate – Energiedosisleistung

absorbent – Absorptionsmittel, Aufsaugmittel

absorbent carbon – Aktivkohle

absorbing capacity – Absorptionsvermögen

absorbing rod – Absorberstab (Kerntechnik)

absorption – Absorption

absorption capacity – Absorptionsvermögen

absorption cross-section – Absorptionsquerschnitt

acceleration lane – Beschleunigungsstreifen

acceptance test – Abnahmeprüfung

access point – Zugangspunkt

accessibility – Zugänglichkeit

accident – Unfall

accident ambulance – Rettungsdienst (RD), Unfallambulanz

accident analysis – Unfallanalyse, Störfallanalyse

accident and emergency (A&E) – Notaufnahme (Einrichtung)

accident caused by electric current – Elektrounfall, Stromunfall

accident insurance – Unfallversicherung

accident log – Unfallbericht

accident report – Unfallbericht

accident victim – Unfallopfer

accident with oils – Ölunfall

accumulator – Akkumulator

accumulator acid – Akkumulatorsäure, Batteriesäure

ace resin – Kunstharz

acetic acid – Essigsäure

acetylcholinesterase-inhibitor (AChEI) – Acetylcholinesterase-Hemmer

T. Schmiermund, *Fachwörterbuch Feuerwehr und Brandschutz*, https://doi.org/10.1007/978-3-662-64120-0_23

acetylene (ethine) – Acetylen (Ethin)

acetylene black – Acetylenruß

acid – sauer; Säure

acid apron – Säureschürze, Säureschutzschürze

acid burn – Säureverätzung

acid dissociation constant – Dissoziationskonstante, Säure(dissoziations)konstante

acid gloves – Säureschutzhandschuhe

acid halides – Säurehalogenide

acid strength – Säurestärke, Acidität

acid-base reaction – Säure-Base-Reaktion

acid-fast – säurefest

acidity – Säurestärke, Acidität

acidosis – Blutübersäuerung

acid-proof – säurebeständig

acid-resistant – säurebeständig

acid-resistant gloves – Säureschutzhandschuhe

acid-stable – säurebeständig

acoustic pressure – Schalldruck

acoustic signals – akustische Signale

acoustic velocity – Schallgeschwindigkeit

act – einzelnes Gesetz

act of retiring – Pensionierung, Verabschiedung aus dem Dienst

action – Aktion, Maßnahme, Einsatz, Vorgang

action drill – Einsatzübung

action order – Einsatzbefehl, Aktionsreihenfolge

action run – Einsatzfahrt

activated atom – angeregtes Atom, aktiviertes Atom

activated carbon – Aktivkohle

activated-carbon filter – Aktivkohlefilter

activation time – Aktivierungszeit

active division – aktive Abteilung, Einsatzabteilung

active tuberculosis – offene Tuberkulose

active waste – radioaktiver Abfall

activity – Aktivität

activity concentration – Aktivitätskonzentration

activity loss – Aktivitätsverlust

actual calorific value – effektive Verbrennungswärme

actual condition – Ist-Zustand

actual state – Ist-Zustand

actual value – Ist-Wert

acute circulatory failure – Kreislaufversagen

acute disease – akute Erkrankung

acute illness – akute Erkrankung

acute radiation syndrome (ARS) – akute Strahlenerkrankung, schwere Strahlenerkrankung

acute respiratory disease (ARD) – akute respiratorische Erkrankung (ARE)

acute toxic effect – akute Giftwirkung

acute toxicity – akute Toxizität

adaptor – Adapter, Übergangsstück

additional call – Nachforderung

additional guidance – zusätzliche Hinweise

additional signs – Zusatzzeichen

adequate manner – in geeigneter Art und Weise

adhesive – Klebstoff

adhesive tape – Klebeband

adiabatic compression – adiabatische Kompression

adjacent room – Nachbarraum, Nachbarzimmer

adjoining room – Nebenzimmer, angrenzender Raum

adjust – justieren

adjustable inline inductor – regelbarer Zumischer

adjustable wrench – Rollgabelschlüssel

adjustment – Justierung

adjuvant – Hilfsmittel, Adjuvans

administer first aid – Erste Hilfe leisten

administration – Verwaltung, auch: Innerer Dienst (S1)

administration act – Verwaltungsakt

administrative assistance – Amtshilfe

administrative deed – Verwaltungsschreiben

administrative offence – Ordnungswidrigkeit

administrative procedure – Verwaltungsvorgang

administrative regulation – Verwaltungsvorschrift (VwV)

administrative-organisational-measures – administrativ-organisatorische Maßnahmen

admixing rate – Zumischrate

admixture rate – Zumischrate

adsorption – Adsorption

adsorption capacity – Adsorptionsvermögen

advancement – Förderung, Beförderung (im Dienstgrad)

advancing a hose line – eine Schlauchleitung verlegen

advantage – Vorteil

aeration – Belüftung

aerial appliances – Hubrettungsfahrzeuge

aerial ladder platform (ALP) – Drehleiter mit Korb (DLK)

aerosol – Aerosol

aerosol mist – Aerosolnebel

aerosol particle – Aerosolteichen

aerosol spray – Aerosolspray

affliction – Erkrankung, Gebrechen

after burning time – Nachbrennzeit

after flame – Nachbrennen

agency – Geschäftsstelle, Behörde

agent outlet – Auslassöffnung (z. B. für Löschmittel)

aggregates – Aggregate

aggregates of the fire brigade – Aggregate der Feuerwehr

aging – Alterung

agreement – Abkommen, Übereinkommen, Vereinbarung

agricultural buildings – landwirtschaftliche Gebäude

aim – Ziel, Zweck, Absicht

air – Luft

air brake – Druckluftbremse

air brake system – Druckluftbremsanlage

air compressor – Druckluftkompressor

air conditioner duct – Klimakanal

air conditioner system – Klimaanlage

air cylinder – Druckgasflasche mit Atemluft

air foam – Luftschaum

air foam branch pipe – Luftschaumrohr

air gauge – Luftdruckprüfer (Reifendruck)

air humidity – Luftfeuchtigkeit

air inlet – Lufteinlass

air lifting bag („air-bags") – Lufthebekissen, Hebekissen

air liquefaction – Luftverflüssigung

air movement – Luftbewegung

air need – Luftbedarf

air pollution – Luftverschmutzung

air pressure – Luftdruck

air raid shelter – Luftschutzraum, Luftschutzbunker

air rescue – Luftrettung

air rescue service – Flugrettungsdienst, Luftrettungsdienst

air splint – Luftkammerschiene

air supply – Luftzufuhr, Zuluft

air supply opening – Zuluftöffnung, Luftzufuhröffnung

air velocity – Luftgeschwindigkeit

air volume – Luftvolumen

airbag – Airbag

airbag cover – Airbag-Abdeckung

airborne rescue service – Flugrettungsdienst

air-ducting system – Lutte, Luftkanalsystem

airfield fire engine – Flugfeldlöschfahrzeug (FLF)

airfield fire truck – Flugfeldlöschfahrzeug (FLF)

airflow velocity – Luftgeschwindigkeit, Luftstromgeschwindigkeit

air-gas-mixture – Luft-Gas-Gemisch

airing – Lüftung, Belüftung

air-lifting units („air-bags") – Lufthebekissen, Hebekissen

air-line system – Schlauchgeräte (Atemschutz)

airport – Flughafen

airport crash tender – Flughafenlöschfahrzeug

airport fire-fighting vehicle – Flugfeldlöschfahrzeug (FLF)

airport rescue and fire-fighting service – Flughafenfeuerwehr

air-vapor-mixture – Luft-Dampf-Gemisch

airway – Atemwege

ALARA-principle: as low as reasonably achievable – ALARA-Prinzip: so niedrig wie vernünftigerweise erreichbar

alarm – Alarm

alarm and response regulations – Alarm- und Ausrückeordnung (AAO)

alarm bell – Alarmglocke

alarm clock – Wecker

alarm condition – Alarmzustand

alarm device – Warneinrichtung

alarm drill – Alarmübung

alarm equipment – Alarmierungseinrichtung

alarm grades – Alarmstufen

alarm horn – Alarmhorn

alarm key word – Alarmierungsstichwort

alarm organisation – Alarmorganisation

alarm receiver – Alarmempfänger

alarm regulations – Alarmordnung

alarm response – Alarmfahrt

alarm signal – Alarmsignal, Alarmzeichen

alarm siren – Sirene, Alarmsirene

alarm system – Alarmanlage

alarm threshold – Alarmschwelle

alarm time – Alarmierungszeit

alarm transmission – Alarmübermittlung, Alarmübertragung

alarm transmission equipment – Alarmübertragungseinrichtung

alarm valve – Alarmventil

alarming – Alarmierung

alcohol-resistant – alkoholbeständig

alcohol-resistant foam compound – alkoholbeständiges Schaummittel

alert – Alarm

alert dosimeter – Alarmdosimeter

alight – brennend

alkali cartridge – Alkalipatrone

alkaline – alkalisch

alkaline solution – alkalische (basische) Lösung, Lauge

alkaliproof – alkalibeständig

alkanes – Alkane

alkenes – Alkene

alkines – Alkine

all the time – ständig, die ganze Zeit

all-clear – Entwarnung

all-clear signal – Entwarnungssignal

allergic shock – allergischer Schock

allocation – Zuteilung, Zuweisung

alloy – Legierung

alpha capture – Alpha-Einfang

alpha decay – Alpha-Zerfall

alpha disintegration – Alpha-Zerfall

alpha emitter – Alpha-Strahler

alpha particle – Alpha-Teilchen

alpha radiation – Alpha-Strahlung

alternating current (ac) – Wechselstrom

alternative national service – Wehrersatzdienst

altitude rescue – Höhenrettung

aluminum wedge – Aluminiumkeil

amalgams – Amalgame

ambience – Umgebung

ambient atmosphere – Umgebungsatmosphäre

ambient dose equivalent – Umgebungsäquivalentdosis

ambient dose equivalent rate – Umgebungsäquivalentdosisleistung

ambient fire – Umgebungsbrand

ambient pressure – Umgebungsdruck

ambient temperature – Umgebungstemperatur

ambulance – Rettungsfahrzeug

ambulance corps – Rettungsdienst (RD)

ambulance crew – RTW-Besatzung, Krankenwagenbesatzung

ambulance driver – Krankenwagenfahrer

ambulance service – Krankentransport; auch: Rettungsdienst (RD)

ambulance station – Rettungswache (RetW)

ambulatory infusion pump – tragbare Infusionspumpe

Ambu™ bag ventilation – Beatmung mit Beatmungsbeutel

ammonia – Ammoniak (NH_3)

ammonia solution – Ammoniaklösung

ammonia-resistant – ammoniakfest

amount concentration – Stoffmengenkonzentration

amount of heat – Wärmemenge

amount of substance – Stoffmenge

amount-of-substance fraction – Stoffmengenanteil

amount-of-substance ratio – Stoffmengenverhältnis

amplitude modulation (AM) – Amplitudenmodulation (AM)

anaphylactic shock – anaphylaktischer Schock

anchor bend – Ankertauknoten

anchor point – Anschlagpunkt

anchorage – Verankerung

anchorage point – Festpunkt, Befestigungspunkt

ancillary equipment – Hilfsgeräte, Zusatzgeräte

anemometer – Anemometer, Windmessgerät

anesthesia – Narkose

angel of inclination – Neigung

angle grinder – Winkelschleifer

angular momentum – Drehimpuls

animal rescue – Tierrettung (Tätigkeit)

animal rescue team – Tierrettung (Einheit)

animal rising device – Tierhebegerät

animal shelter – Tierheim

anion – Anion

anionic – anionisch

annealing colour – Glühfarbe, Glutfarbe

annex – Anbau, Nebengebäude, Anhang

annihilation – Paarvernichtung, Annihilation

anomaly of water – Anomalie des Wassers

anoxia – Sauerstoffmangel

antenna – Antenne

anthrax – Milzbrand

antibacterial – antibakteriell

anti-catalyst – Inhibitor, Antikatalysator

antidote – Gegengift

anti-foaming – schaumhemmend, schaumzerstörend

anti-freeze – Frostschutzmittel

anti-rabies inoculation – Impfung gegen Tollwut

anti-rust paint – Rostschutzfarbe

antitoxin – Gegengift

antivenin – Gegengift (insbesondere nach Schlangenbissen)

aorta – Aorta

apartment fire – Wohnungsbrand

apiarist – Imker, Bienenzüchter

apnea – Atemstillstand

apoplexia cerebri stroke – Schlaganfall

apoplexy – Schlaganfall

appearance – Aussehen, Erscheinungsbild (z. B. Gefahrstoff)

appliance room – Fahrzeughalle

application – Anwendung

application limits – Anwendungsgrenzen

apprehension – Festnahme

approval – Genehmigung, Zulassung, Zustimmung

aquatic environment – Gewässer, aquatische Umwelt

aqueous – wässrig, wasserhaltig

aqueous film forming foam (AFFF) – wasserfilmbildendes Schaummittel

arc of light – Lichtbogen

arch – Bogen

area – Fläche, Gebiet, Gegend, Bereich, Gelände

area burning rate – Flächenabbrandrate, flächenbezogene Abbrandgeschwindigkeit, flächenbezogene Abbrandrate

area of disaster – Einsatzstelle (ESt), Schadensstelle

area of responsibility – Verantwortungsbereich

area of risk – Gefahrenbereich

areas of jurisdiction – Zuständigkeitsbereiche

arid – trocken

arithmetic mean – Mittelwert, Durchschnitt

arrangement – Vereinbarung, Regelung; auch: räumliche Anordnung

arrival note – Eintreffmeldung

arson – Brandstiftung

arson attack – Brandanschlag

arson investigator – Brandermittler, Brandursachenermittler

arsonist – Brandstifter

artery – Arterie

artery clamp – Arterienklemme

artery forceps – Arterienklemme

artificial respiration – künstliche Beatmung

artificial respiration mask – Beatmungsmaske für künstliche Beatmung

as matter of experience – erfahrungsgemäß

asbestos – Asbest

ascender – Steigklemme

ascending ladder – Steigleiter

ascent speed – Aufstiegsgeschwindigkeit (Tauchen)

ascertain the situation – Erkundung, Situationsfeststellung, Erkennen der Lage

ash – Asche

ash flow – Glutwolke

ash tray – Aschenbecher

ash wood – Eschenholz

aspects – Gesichtspunkte

asphyxiant – erstickend wirkende Substanz

asphyxiant gases – Stickgase

asphyxiation – Erstickung (bei Lebewesen)

aspiration – Ansaugen, Einatmen

assembly area – Bereitstellungsraum (BSR); auch: Montagebereich

assembly centre – Versammlungsstätte

assembly hall – Aula, Versammlungshalle

assembly point – Sammelplatz

assembly room – Versammlungsraum

assessment – Beurteilung, Lagebeurteilung

assessment of fire hazard – Beurteilung der Brandgefährdung, Einschätzung der Brandgefahr

assessment of the situation – Lagebeurteilung, Situationseinschätzung

assessment values – Beurteilungswerte

assigned means – zugewiesene Mittel

assistance – Hilfe, Hilfeleistung

assume command – das Kommando übernehmen

assumption – Annahme, Vermutung

atmosphere – Atmosphäre

atmospheric discharge – atmosphärische Entladung

atmospheric oxygen – Luftsauerstoff

atmospheric pressure – atmosphärischer Luftdruck

atom – Atom

atomic bomb – Atombombe

atomic bond – Atombindung, kovalente Bindung

atomic disintegration – Atomzerfall (Radioaktivität)

atomic energy – Kernenergie, Atomenergie

atomic mass – Atommasse, absolute Atommasse

atomic mass unit (amu) – Atommasseneinheit (amu)

atomic model – Atommodell

atomic nuclei – Atomkerne

atomic number – Ordnungszahl, Kernladungszahl

atomic physics – Atomphysik

atomic reactor – Atomreaktor

atomic shell – Atomhülle (Atomschale)

atomic size – Atomgröße

atomic species – Atomart

atomic structure – Atombau

atomic, atomical – atomar

atomize – zerstäuben

atomizer – Zerstäuber

atrium – Atrium

attachment module (2-runged) of a scaling ladder – A-Teil-Einschub einer Steckleiter

attachment module (7-runged) of a scaling ladder – B-Teil einer Steckleiter

attack – Angriff

attack equipment – Angriffsgeräte

attack path – Angriffsweg

attack route – Angriffsweg

attack-crew – Angriffstrupp (A-Tr)

attack-crew leader – Angriffstruppführer, -in (A-TrFü, A-TrFü'in)

attack-crew member – Angriffstruppmann, -frau (A-TrM, A-TrFr)

attack-squadron – Angriffstrupp (A-Tr)

attack-squadron leader – Angriffstruppführer, -in (A-TrFü, A-TrFü'in)

attack-squadron member – Angriffstruppmann, -frau (A-TrM, A-TrFr)

attack-team – Angriffstrupp (A-Tr)

attack-team leader – Angriffstruppführer, -in (A-TrFü, A-TrFü'in)

attack-team member – Angriffstruppmann, -frau (A-TrM, A-TrFr)

attempt to extinguish a fire – Löschversuch

attempt to save – Rettungsversuch

attendance – Teilnahme, Anwesenheit, Bereitschaft (Dienst)

attendance list – Anwesenheitsliste

attendance time – Anwesenheitszeit; auch: Dauer bis zum Eintreffen

attic – Dachgeschoss, Dachboden

attraction – Anziehung

attractive force – Anziehungskraft

attribute – Eigenschaft, Merkmal

attrition – Verschleiß

audible – akustisch

authentication – Authentifizierung

authorisation – Genehmigung, Erlaubnis, Zulassung

authoritarian leadership style – autoritärer Führungsstil

authoritarian style – autoritärer Stil/ Führungsstil

authorities and organisations with security tasks – Behörden und Organisationen mit Sicherheitsaufgaben (BOS)

authority – Behörde

authority of post – Funktionsautorität (Autorität aufgrund der Funktion)

autocatalysis – Autokatalyse

autocatalytic agent – Autokatalysator

autogenous (gas) cutting – Autogenschneiden, Brennschneiden (Acetylen-Sauerstoff)

autogenous (gas) welding – Autogenschweißen, Schweißen (Acetylen-Sauerstoff)

auto-ignitability – Selbstentzündlichkeit

auto-ignitable – selbstentzündlich

auto-ignition – Selbstentzündung

auto-ignition temperature – Selbstentzündungstemperatur

automated external defibrillator (AED) – Frühdefibrillator, automatischer, externer Defibrillator (AED)

automatic distress signal unit (DSU) – Notsignalgeber, Totmanneinrichtung (Atemschutzeinsatz)

automatic fire alarm (AFA) – automatischer Feuermelder

automatic fire detector – automatischer Brandmelder

automatic fire protection installation – automatische Brandschutzanlage

automatic fire signal – automatischer Brandalarm

automatic fire-fighting installation – automatische Feuerlöschanlage

automobile jack – Wagenheber

autooxidation – Autooxidation, Selbstoxidation

autoradiography – Autoradiographie

auxiliary construction – Hilfskonstruktion

auxiliary fire tender – Hilfeleistungslöschfahrzeug (HLF)

auxiliary power unit (APU) – Zusatzstromversorgung

availability – Verfügbarkeit, Einsatzbereitschaft

avalanche – Lawine

avalanche dog – Lawinenhund

avalanche rescue dog – Lawinenrettungshund

avalanche search dog – Lawinensuchhund

avalanche victim rescue – Verschüttetenrettung

avalanched person – verschüttete Person

average – Mittelwert, Durchschnitt

Avogadro's constant, Avogadro's number – Avogadro-Konstante, Avogadro-Zahl

Avogadro's law – Avogadro-Gesetz

avoidance – Vermeidung

avoidance reaction – Fluchtreaktion

away on duty (A.O.D.) – Dienstreise, auf einer sein

axe – Axt

axial load – Achslast

B

BA and radioactive incident unit – Gerätewagen Atemschutz/Strahlenschutz (GW-AS)

back cut – Fällschnitt

back-burn – Gegenfeuer

backdraft – Rauchgasexplosion, Rauchgasdurchzündung

backfire – Brandstreifen

background – Hintergrund

background radiation – Hintergrundstrahlung

backstop – Rücklaufsperre

backstop valve – Rückschlagventil

back-up forces – Verstärkung, Unterstützungskräfte

backup power supply – Notstromversorgung

bacterial disease – bakterielle Erkrankung

bactericidal – antibakteriell

bacteriosis – bakterielle Erkrankung

bacterium, bacteria – Bakterium, Bakterien

bad smell – schlechter Geruch, Gestank

badge – Abzeichen, Anstecknadel

badge meter – Filmdosimeter

badge of rank – Rangabzeichen

badging – Bezettelung

bag valve mask (BVM) – Beatmungsbeutel

bag valve mask ventilation – Beatmung mit Beatmungsbeutel

balance – Gleichgewicht, Ausgleich, Ausgewogenheit

ball bearing – Kugellager

ball cock – Kugelhahn

ball valve – Kugelhahn, Kugelhahnventil

ban – Verbot, Bann, Sperre

band switch – Wellenbereichsschalter, Bandschalter

bandage scissors – Verbandschere

bandages – Verbände (bei Verletzungen)

bannerman – Bannerträger

bargeboard – Ortgang

barotrauma – Barotrauma, Druckverletzung

barrel hitch – Fassstek (Knoten)

barrel pump – Fasspumpe

barricade – Absperrung, Sperre, Barrikade

barricade tape – Absperrband

barrier – Absperrung, Sperre; auch: Schlagbaum

barrier border – Absperrgrenze

barrier limit – Absperrgrenze

barrier material – Absperrmaterial

base – Boden, Sockel, Basis, Grundlage; auch: Base (Lauge)

base module (9-runged) of a scaling ladder – A-Teil einer Steckleiter

base unit – Basiseinheit

basement – Keller, Kellergeschoss, Untergeschoss

T. Schmiermund, *Fachwörterbuch Feuerwehr und Brandschutz*, https://doi.org/10.1007/978-3-662-64120-0_24

base-on fire – Stützfeuer

basic – basisch (alkalisch)

basic ionizing radiation symbol – Strahlenwarnzeichen

basic life support (BLS), immediate life-saving measures – lebensrettende Sofortmaßnahmen (LSM)

basic rules – Grundregeln

basic solution – basische (alkalische) Lösung, Lauge

basic tactical rules – taktische Grundregeln

basics training – Grundübungen

basin – Senke, Becken, Talmulde

basis – Grundlage, Basis

basket stretcher – Korbtrage

batten – Latte

battery – Batterie

battery tester – Batterietester

BC-powder – BC-Pulver

beam – Strahl; auch: Balken, Träger

beam energy – Strahlungsenergie

beard decree – Barterlass

bearded axe – Bartaxt

becket bend – Schotenstich, Schotstek

becquerel – Becquerel

bee protective suit – Imkerschutzanzug

bee suit – Imkeranzug, Imkerschutzanzug

beekeeper – Imker, Bienenzüchter

beekeeper's pipe – Imkerpfeife

begin of the mission – Einsatzbeginn

behavioral rules – Verhaltensregeln

behavioral rules in fire protection – Verhaltensregeln im Brandschutz

behaviour – Verhalten

belay point – Sicherungspunkt (Höhenrettung)

belay rope – Sicherungsseil

belayer – Sicherungsmann/-frau

belt-type fall impact absorber – Bandfalldämpfer

bend knots – Knoten (zum Verbinden von Leinen)

bends – Bunde (Knoten)

beta decay – Beta-Zerfall

beta disintegration – Beta-Zerfall

beta emitter – Beta-Strahler

beta particle – Beta-Teilchen

beta radiation – Beta-Strahlung

bight – Bucht (z. B. eines Seils)

bilge pump – Lenzpumpe

bill of lading – Frachtbrief

billion – Milliarde (1 000 000 000 = 10^9)

billows of smoke – Rauchschwaden

binary weapons – Binärwaffen

binder – Bindemittel

binding agent – Bindemittel

binding cord – Bindestrick

binding socket piece – Einbindestutzen

binoculars – Fernglas

biodegradability – biologische Abbaubarkeit

biodegradation – biologischer Abbau

biohazard containment classes (S1-S4) – Biosicherheitsstufen (S1-S4)

biohazard containment laboratory (L1-L4) – Biosicherheitslabor (Stufen L1-L4)

biohazardous waste – biogefährdender Abfall

Biological Agents Ordinance – Biostoffverordnung (BioStoffv)

biological assignment – B-Einsatz, Einsatz mit Biostoffen

biological substances – Biostoffe

biological warfare agents (BWA) – biologische Kampfstoffe, biologische Kampfmittel

biological weapon – biologische Waffe, Biowaffe

biological weapons convention (BWC) – Biowaffenkonvention

biological working substances – biologische Arbeitsstoffe

biomaterials – Biostoffe

biosafety level (S1-S4) – Biosicherheitsstufen (S1-S4)

birds' mouth – Fällkerbe

bite protection mouth wedge – Beißkeil

bitumen – Bitumen, Teer

bitumized paper – Teerpappe

black area – Schwarzbereich

black body – schwarzer Körper

blackboard – Kreidetafel

blackboard chalk – Tafelkreide

blank cap – Blindstutzen, Blindkupplung

blanket – Decke (zum Wärmen)

blanketing ability – Deckvermögen (Löschmittel Schaum)

blast wave – Druckwelle (einer Explosion)

blaze – Feuer, Brand; auch: Blesse, Fleck

blend – Mischung, Verschnitt

blind shell – Blindgänger

blister agent – Hautkampfstoff

blocking – abriegeln

blocking position – Riegelstellung

blood pressure – Blutdruck

blood warfare agents – Blutkampfstoffe

blow up – in die Luft fliegen, sprengen, Durchzündung

blowlamp – Lötlampe

blue flashing light and siren – Sondersignal

blue rotating light – Blaulicht (als RKL)

blue strobe light – Blaulicht (als RKL)

blunt cannula – stumpfe Kanüle

board of health – Gesundheitsbehörde

body fluids – Körperflüssigkeiten

body surface – Körperoberfläche

body temperature – Körpertemperatur

bogus emergency call – falscher Notruf

Bohr atom model – Bohrsches Atommodell

boil over – überkochen

boiler room – Heizungsraum

boiling – kochen

boiling point (bp) – Siedepunkt (Kp)

bolt – Bolzen, Riegel

bolt clipper – Bolzenschneider

bolting – Verriegeln

bomb attack – Bombenanschlag

bond – Bindung, Verbund (Bauelemente)

booster pump – Verstärkerpumpe, Druckerhöhungspumpe

booster-line – Schnellangriffseinrichtung

boot tree – Stiefelspanner

boots – Stiefel

bottle of eyewash solution – Augenspülflasche, Augenwaschflasche

bow saw – Bügelsäge

bowline – Brustbund, Pahlstek, Pfahlstich, Rettungsknoten, Rettungsschlinge (Knoten)

bowline on a bight – doppelter Pfahlstich

brace – Strebe, Stütze, Klammer

brake – Bremse

brake fluid – Bremsflüssigkeit

brake system – Bremsanlage

branch pipe – Strahlrohr, Mehrzweckstrahlrohr

branch saw – Astsäge

branched chain reaction – verzweigte Kettenreaktion

branching-off (pipelines) – Abzweigung

brass – Messing

brass wire brush – Messingdrahtbürste

breach of duty – Pflichtverletzung

break – brechen (Zerbrechen)

breakable container – zerbrechlicher Behälter

breakage – Bruch (mechanisch)

breakdown – Zusammenbruch, Betriebsstörung, Versagen (Bauteile)

breaking tool – Brechwerkzeug

breast rail – Brüstung, Fensterbrüstung

breast wall – Brüstung, Fensterbrüstung

breastwork – Brüstung, Fensterbrüstung

breath – Atem, Atemzug, atmen

breathe in – einatmen

breathe out – ausatmen

breathing – Atmung, Beatmung

breathing air – Atemluft

breathing air cylinder – Atemluftflasche

breathing air filter – Atemschutzfilter

breathing air reserve – Atemluftreserve, Reserveluft

breathing air supply – Atemluftvorrat

breathing apparatus – Atemgerät

breathing apparatus appliance (BA-appliance) – Atemschutzgerätewagen (GW-A)

breathing apparatus fitness – Atemschutztauglichkeit

breathing apparatus instructor – Atemschutzausbilder

breathing apparatus maintainer – Atemschutzgerätewart (AS-Gw)

breathing apparatus tender (BA-tender), (BAT) – Atemschutzgerätewagen (GW-A)

breathing apparatus training facility – Atemschutzübungsanlage

breathing apparatus unit (BA unit) – Gerätewagen Atemschutz (GW-A)

breathing arrest – Atemstillstand

breathing bag – Atembeutel

breathing equipment workshop – Atemschutzgerätewerkstatt

breathing exercise facility – Atemschutzübungsanlage

breathing mask – Atemschutzmaske, Atemmaske

breathing poison – Atemgift

breathing protection – Atemschutz

breathing protection control – Atemschutzüberwachung (ASÜ)

breathing protection equipment – Atemschutzausrüstung

breathing transmission – Atemspende

breathing tube – Atemschlauch

breeder reactor – Brüter

bremsstrahlung – Bremsstrahlung (Kerntechnik)

brick – Ziegelstein

brick hammer – Maurerhammer

brief information – Kurzinformation

briefing – Unterweisung, Einweisung, Anweisung, Einsatzbesprechung

briefing meeting – Lagebesprechung

brigade (Bde) – Verband (Vb)

brigade commander – Feuerwehrkommandant

brigade leader – Verbandsführer (VbFü)

brisance – Brisanz

British horsepower (BHP) – britische Pferdestärke

brook – Bach

broom – Besen

brought in by ambulance (BIBA) – mit dem Rettungswagen eingeliefert

bubonic plague – Beulenpest

bucket – Eimer

bucket pump – Kübelspritze

buffer – dämpfen, abschwächen; auch: Puffer

buffer solution – Pufferlösung

building – Gebäude

building class – Gebäudeklasse

building collapse – Gebäudeeinsturz

building complex – Gebäudekomplex

building control agency – Baubehörde

building control department – Bauaufsichtsbehörde

building insurance – Gebäudeversicherung

building law – Baugesetz

building management system (BMS) – Gebäudeleittechnik (GLT)

building material – Baustoff

building material class – Baustoffklasse

building products – Bauprodukte

building regulations – Bauordnung

building requirements – bauliche Voraussetzungen

building science – Baukunde

building utilisation – Gebäudenutzung

building, n-storey – n-stöckiges Gebäude

building-block design – Baukastensystem, Modulbauweise

built environment – bauliche Anlage

bulk density – Schüttdichte

bumper knot – Wurfleinenknoten

bung wrench – Fassschlüssel

Bunsen burner – Bunsenbrenner

buoyancy – Auftrieb (in Wasser)

buoyancy force – Auftriebskraft

buried pipework – erdverlegtes Rohr

burn – brennen, auch: Brandverletzung, Verbrennung

burn down behaviour – Abbrandverhalten

burn down loss – Abbrandverlust

burn marks – Brandspuren

burn off pyrotechnics – Abbrennen von Feuerwerkskörpern

burn on – Weiterbrennen

burn wound – Brandverletzung

burning – Brennen

burning behaviour – Abbrandverhalten, Brandverhalten, Brennverhalten

burning combustion – Abbrand

burning dripping – brennendes Abtropfen

burning fat – Fettbrand

burning length – Branddauer

burning phase – Brandphase

burning rate – Abbrandgeschwindigkeit

burning rate factor – Abbrandfaktor

burning stability – Abbrandfestigkeit

burning time – Brenndauer

burning velocity – Abbrandgeschwindigkeit

burnt layer – Brandschicht

burst pressure – Zerplatzdruck

bursting – Abplatzung (z. B. bei Beton), auch: Zerknall

bursting disk – Berstscheibe

bursting pressure – Berstdruck

bursting pressure for fire hoses – Berstdruck für Druckschläuche

bush – Buch, Gebüsch

bush fires – Buschbrände

business requirements – betriebliche Voraussetzungen

butterfly needle – Butterfly-Kanüle

button cell – Knopfzelle (Batterie)

buzzer – Summer

BVM ventilation – Beatmung mit Beatmungsbeutel

by-law – Verordnung

bystander – Zuschauer

C

C-14 method – Radiocarbon-Methode, C-14-Methode (Altersbestimmung)

cable – Kabel, Leitung, Seil, Drahtseil

cable cross section – Leitungsquerschnitt

cable duct – Kabelkanal

cable extension reel – Kabeltrommel mit Verlängerungskabel

cable fire – Kabelbrand

cable reel – Kabeltrommel

cable shaft – Kabelschacht

cable spool – Kabeltrommel

cable ties – Kabelbinder

cadaver dog – Leichenspürhund

calculation – Berechnung

calculation of fire load – Brandlastberechnung

calculation pressure – Berechnungsdruck

calibrate – kalibrieren

calibration – Kalibrierung

calibration gas – Prüfgas

calibration source – Kalibrierstrahler (Radioaktivitätsmessung)

call report – Einsatzbericht

call-up – Einberufung

call-up paper – Einberufungsbescheid

caloric energy – Wärmeenergie

caloric value – Brennwert

calorific value – Verbrennungswärme

calorimeter – Kalorimeter

Can do! – Wird erledigt!

can opener – Dosenöffner, Büchsenöffner

cancer – Krebs

cancer causing – krebserzeugend

cancerogenic – krebserregend

canopy – Überdachung, Vordach

cant hook – Wendehaken

canvas – Segeltuch

canvas bucket – Segeltucheimer

cap badge – Mützenabzeichen

cap coupling – Blindkupplung

capacity – Kapazität, Leistung, Leistungsfähigkeit

capillary refill time (CRT) – Nagelbettprobe (NBP)

capture cross-section – Einfangsquerschnitt

car crash – Verkehrsunfall, schwerer Verkehrsunfall

car driver – Fahrer, Autofahrer

carabine – Karabinerhaken, Karabiner

carabine hook – Karabinerhaken, Karabiner

carbohydrates – Kohlehydrate

carbon – Kohlenstoff (C)

carbon dioxide – Kohlenstoffdioxid, Kohlendioxid, Kohlensäuregas, CO_2

carbon dioxide extinguisher – Kohlendioxidlöscher, CO_2-Löscher

carbon filter – Aktivkohlefilter

carbon monoxide – Kohlenstoffmonoxid, Kohlenmonoxid, CO

carbon-dioxide snow – Kohlensäureschnee

carbonic acid – Kohlensäure (H_2CO_3)

T. Schmiermund, *Fachwörterbuch Feuerwehr und Brandschutz*, https://doi.org/10.1007/978-3-662-64120-0_25

carbonisation – Verkohlung

carbonize – verkohlen

carboxylic acid – Carbonsäure

carbs – Kohlehydrate

cardiac arrest (CA) – Herzstillstand, Herz-Kreislauf-Stillstand

cardiac compression – Herzdruckmassage

cardiac infarction – Herzinfarkt

cardiac massage – Herzdruckmassage

cardiac output (per minute) (CO) – Herzminutenvolumen (HMV)

cardiogenic shock – kardiogener Schock

cardiopulmonary arrest – Atem- und Herzstillstand

cardiopulmonary resuscitation (CPR) – Herz-Lungen-Wiederbelebung (HLW)

cardiopulmonary resuscitation training (CPR-training) – Herz-Lungen-Wiederbelebungstraining (HLW-Training)

cardiorespiratory arrest – Atem- und Herzstillstand

careless – leichtsinnig, nachlässig, unachtsam, fahrlässig

careless arson – fahrlässige Brandstiftung

cargo – Fracht, Ladung

carpenter's hammer – Zimmermannshammer

carpenter's hatched – Zimmermannsbeil

Carrick bend – Kreuzknoten

carrier strap – Trageband

carry on a stretcher – auf einer Trage befördern

carry out – ausführen, durchführen

carryover – Übertragung, auch: Verschleppung von Substanzen

cartridge – Kartusche

case – Fall, Angelegenheit; auch: Etui, Hülle, Gehäuse, Schachtel, Zarge (Tür, Fenster)

case of damage – Schadensfall, Schadenslage

case of danger – Gefahrenlage, Gefahrenfall

case red – Fall Rot

case yellow – Fall Gelb

casing – Verschalung, Verkleidung, Gehäuse

cast iron chippings – Gusseisenspäne

castor container – Castor-Behälter

casualties – Verletzte

casualty department (casualty) – Notaufnahme, Unfallambulanz

casualty report – Verlustmeldung

cat's paw (knot) – Katzenpfote (Knoten)

catalysis – Katalyse

catalyst – Katalysator (Kat)

catalyst poison – Katalysatorgift

catalytic – katalytisch

catalytic converter – Katalysator (Fahrzeuge)

catastrophe – Katastrophe (Kat)

catch alight – Feuer fangen

catch eye – Fangöse

catch pot – Auffangwanne

catcher – Fänger; auch: Auffangwanne

catheter – Katheter

cathodic corrosion protection (CCP) – kathodischer Korrosionsschutz (KKS)

cation – Kation

cationic – kationisch

cat-shank – Trompetenknoten

cause of a fire – Brandursache

cause of a malfunction – Störungsursache

cause of conflagration – Brandursache

cause of explosion – Explosionsursache

cause of ignition – Zündursache

causing damage – schädlich, Schaden verursachend

caustic – ätzend (gegen Haut und Gewebe)

caustic effect – Ätzwirkung

caustic gases – ätzende Gase

caustic potash lye – Kalilauge

caustic soda lye – Natronlauge

cauterization – Verätzung

caution – Vorsicht

cautionary measure – Vorsichtsmaßnahme

cave – Höhle

cavitation – Kavitation

CBRN (chemical, biological, radioactive, nuclear) – CBRN (chemisch, biologisch, radioaktiv, nuklear)

CBRNE (chemical, biological, radioactive, nuclear, explosive) – CBRNE (chemisch, biologisch, radioaktiv, nuklear, explosiv)

ceiling – Decke (im Gebäude)

ceiling hook – Einreißhaken (Deckenhaken)

cellar – Keller

cellar fire – Kellerbrand

cement – Zement

Center for Disease Control (CDC) – Seuchenschutzbehörde

Central European Summer Time (CEST) – mitteleuropäische Sommerzeit (MESZ)

Central European Time (CET) – mitteleuropäische Zeit (MEZ)

central nervous system (CNS) – Zentralnervensystem (ZNS)

central water supply – zentrale Wasserversorgung

centrifugal fire pump – Feuerlöschkreiselpumpe (FP)

centrifugal force – Fliehkraft, Zentrifugalkraft

centrifugal force bursting – Fliehkraftzerknall

centrifugal force decay – Fliehkraftzerfall

centrifugal pump – Kreiselpumpe

cerebrovascular accident (CVA) – Schlaganfall

certificate of approval – Abnahmebescheinigung

certificate of dismissal – Entlassungsurkunde

certification – Zertifizierung

cervical collar – Halskrause, HWS-Halskrause

cervical spine – Halswirbelsäule (HWS)

chain – Kette

chain branching – Kettenverzweigung

chain breaking – Kettenabbruch

chain growth – Kettenwachstum

chain initiation – Kettenstart

chain of command – Befehlskette, Befehlsweg

chain of reporting – Meldekette, Meldeweg

chain propagation – Kettenfortpflanzung (Kettenwachstum)

chain reaction – Kettenreaktion

chain saw – Motorsäge

chain sinnet – Kettenknoten, Affenkette

chain termination – Kettenabbruch

chain-reacting mass – kritische Masse

chainsaw chain – Motorkettensägenkette

chalk – Kreide

chamber – Kammer, Raum, Zimmer

change of state – Aggregatzustandsänderung

change-over – Umstellung

changes – Änderungen

channel – Kanal (z. B. für Wasser); auch Funkkanal

character – Zeichen, Buchstabe, Wesen, Charakter, Eigenschaft, Merkmal

characteristic value – Kenngröße, Kennwert

characteristics – Eigenschaften

charcoal – Holzkohle

charcoal layer – Holzkohleschicht

charge – Ladung, Füllmenge; auch: Preis, Gebühr

chargeable fire extinguisher – Aufladelöscher, Aufladefeuerlöscher

charger – Ladegerät

charging current – Ladestrom

charging set – Ladegerät

charging voltage – Ladespannung

charring – Verkohlung, Ankohlung

chart of nuclides – Nuklidkarte

chassis – Fahrgestell (Kfz)

check – überprüfen

check-up – Untersuchung, Überprüfung, ärztliche Kontrolle, Nachuntersuchung

chelating agent – Komplexbildner

chemical (safety) cabinet – Chemikalienschrank

chemical absorbent (agent) – Chemikalienbindemittel, Chemikalienbinder

chemical accident – Chemieunfall

chemical assignment – C-Einsatz, Einsatz mit Chemikalien

chemical bond – chemische Bindung

chemical burn – Verätzung

chemical equation – Reaktionsgleichung, chemische Gleichung

chemical equilibrium – chemisches Gleichgewicht

chemical incident unit (CIU) – ABC-Zug, G-ABC-Zug, Gefahrgutzug, Gefahrstoffzug, Umweltschutzzug, Chemiewehr

chemical law – chemisches Gesetz

chemical protection – Chemikalienschutz

chemical protection suit – Chemikalienschutzanzug (CSA)

chemical protective suit – Chemikalienschutzanzug (CSA)

chemical reaction – chemische Reaktion

chemical splash suit – Flüssigkeitsschutzanzug

chemical warfare – chemischer Krieg, Gaskrieg

chemical warfare agents (CWA) – chemische Kampfmittel, chemische Kampfstoffe

chemical weapon – Chemiewaffe, chemische Waffe

chemical weapons convention – Chemiewaffenübereinkommen (CWÜ), Chemiewaffenkonvention

chemical(s) – Chemikalie(n)

chequered – schachbrettartig

Cherenkov radiation – Tscherenkow-Strahlung

cherry picker – Hubsteiger

chest compression – Herzdruckmassage

chest harness – Brustgurt

chest wader – Wathose

chief – Leiter, Oberhaupt, Anführer

chief ambulance officer – Organisatorischer Leiter Rettungsdienst (OrgL, OLRD), Einsatzleiter Rettungsdienst

chief fire prevention officer – Leiter des Brandschutzes

chief of the fire department – Amtsleiter (AL)

children's fire brigade – Kinderfeuerwehr

chimney – Schornstein, Kamin, Rauchfang

chimney effect – Kamineffekt, Kaminwirkung

chimney fire – Schornsteinbrand, Kaminbrand

chimney sweeper – Schornsteinfeger, Rauchfangkehrer

chimney sweeper set – Schornsteinfegerwerkzeug

chimney sweeping tools – Schornsteinfegerwerkzeug

chin strap – Kinnriemen

chip of wood – Holzspan

chipboard – Spanplatte

chlorine – Chlor

chlorine poisoning – Chlorgasvergiftung

chlorocarbons – Chlorkohlenwasserstoffe (CKW)

chlorofluorocarbons (CFC) – Fluorchlorkohlenwasserstoffe (FCKW)

choking agents – Lungenkampfstoffe

chronic ailment – chronische Erkrankung

chronic disease – chronische Erkrankung

chronic toxic effect – chronische Giftwirkung

church – Kirche

church fire – Kirchenbrand

circuit breaker panel – Stromverteilerkasten

circulatory arrest – Kreislaufstillstand

circulatory system – Blutkreislauf

circumstances – Umstände

cistern – Zisterne

city gas – Stadtgas

city map – Stadtplan

civil defense – Zivilverteidigung, Zivilschutz (ZS)

civil defense siren – Zivilschutzsirene

civil engineering – Bauwesen

civil protection – Bevölkerungsschutz, Zivilschutz (ZS)

civil-military co-operation (CIMIC) – Zivil-militärische Zusammenarbeit (ZMZ)

cladding – Verkleidung (Gebäude)

clamp – Zwinge, klammern, einspannen

clamp shell – Schaufeltrage

clamping jaws – Klemmbacken

clapper valve – Rückschlagklappe

class of danger – Gefahrklassen, Gefahrgutklassen, Klassen gefährlicher Güter

class of fire danger – Brandgefahrenklasse

classes of dangerous goods – Gefahrklassen, Gefahrgutklassen, Klassen gefährlicher Güter

classification – Einstufung

classification system – Klassifizierungssystem

classified document – Verschlusssache (VS)

classified material – Verschlusssache (VS)

clean – sauber, rein

cleaner – Reiniger, Reinigungsmittel

cleaner's naphtha – Waschbenzin

cleaner's solvent – Waschbenzin

cleaning – Reinigung

cleaning agent – Reinigungsmittel, Detergentien

cleaning cloth – Putztuch, Putzlappen

cleaning compound – Reinigungsmittel, Detergentien

cleaning method – Reinigungsmethode

cleaning rag – Putztuch, Putzlappen

cleaning utensils – Putzzeug, Reinigungsutensilien

cleaning work – Reinigungsarbeiten

clean-up – Sanierung (Grund, Boden, Umwelt)

clear up – abräumen, aufräumen

clearing of roofs – Dachräumung (z. B. von Schnee)

clearing-up operations – Aufräumarbeiten

cleavage – Spaltung (mechanisch)

climber's helmet – Bergsteigerhelm

climbing gear – Kletterausrüstung

climbing harness – Klettergurt

clinical waste – klinischer Abfall

clip-on helmet lamp – Helmlampe, ansteckbare

closed-circuit breathing apparatus – Kreislaufgerät (Atemschutz)

closed-circuit oxygen breathing apparatus – Sauerstoffschutzgerät (SSG)

close-meshed – engmaschig

close-up view – Nahaufnahme

closing – Prozessabschluss

closing pressure – Schließdruck (Pumpe)

cloth face mask – Mund-Nase-Bedeckung (MNB)

clothing procedure – Einkleidung

cloud of/from fire – Brandwolke

cloud of smoke – Rauchwolke, Rauchschwaden

clove hitch – Mastwurf, Webleinenstek

club hammer – Fäustel

clump – Klumpen

cluttered – vollgestellt

coarse purification – Grobreinigung

coat – Anstrich

coating – Beschichtung, Anstrich

code of law – Gesetzbuch

coefficient of expansion – Ausdehnungskoeffizient

coefficient of thermal expansion (CTE) – Wärmeausdehnungskoeffizient (WAK)

coefficient of volume expansion – Volumenausdehnungskoeffizient

cohesion – Kohäsion

coin cell – Knopfzelle (Batterie)

coin metals – Münzmetalle

coking – Verkokung

cold protection – Kälteschutz

cold protection clothes/suit – Kälteschutzkleidung

cold store – Kühlraum

collaboration – Zusammenarbeit, Mitwirkung, Mitarbeit

collapse – Einsturz, Zusammenbruch, zusammenbrechen, kollabieren

collapse danger – Einsturzgefahr

collapse zone – Trümmerschatten

collapsible container – Faltbehälter

collapsible spade – Klappspaten

collar band – Ordensband (‚Kragenband‘)

collateral damage – Begleitschaden

collecting head – Sammelstück

collective call – Sammelruf

column – Säule

column of smoke – Rauchsäule

combat – Kampf, bekämpfen, angehen

combination pliers – Kombizange

combined filter – Kombinationsfilter

combined heat and power station (CHP) – Heizkraftwerk (HKW)

combustibility – Brennbarkeit, Zündwilligkeit

combustible – brennbar

combustible dust – brennbarer Staub

combustible gaseous substances – brennbare gasförmige Stoffe

combustible gases – brennbare Gase

combustible liquid substances – brennbare flüssige Stoffe

combustible liquids – brennbare Flüssigkeiten

combustible load – brennbare Masse

combustible material – Brennstoff, Brennmaterial, brennbares Material

combustible mixtures – brennbare Gemische

combustible solid substances – brennbare feste Stoffe

combustible substances – brennbare Stoffe

combustion – Verbrennung

combustion energy – Verbrennungsenergie

combustion equation – Verbrennungsgleichung

combustion gases – Brandgase, Verbrennungsgase

combustion mechanism – Verbrennungsmechanismus

combustion process – Verbrennungsprozess, Verbrennungsvorgang

combustion products – Verbrennungsprodukte

combustion property – Abbrandeigenschaft

combustion rate – Verbrennungsgeschwindigkeit

combustion temperature – Verbrennungstemperatur

combustive agent – Brandbeschleuniger

come along – Mehrzweckzug (Greifzug®)

command – Kommando, Leitung, Führung

command and control process – Führungsvorgang

command and control system – Einsatzleitungssystem, Führungssystem

command and control unit – Abrollbehälter Einsatzleitung (AB-ELW)

command assistant – Führungsassistent

command attitude – Führungsverhalten

command car – Kommandowagen (KdoW)

command facility – Führungseinrichtung

command organisation – Führungsorganisation

command personnel – Führungspersonal

command pod – Abrollbehälter Einsatzleitung (AB-ELW)

command post – Befehlsstelle

command staff, ‚crisis cell' – Stab, Einsatzstab, Krisenstab

command support unit – Einsatzleitwagen (ELW)

command system – Einsatzleitungssystem, Führungssystem

command vehicle – Einsatzleitwagen (ELW)

commander's manuals – Einsatzleiterhandbuch

command-group (command-team of 9) – Führungsgruppe (FüGr)

commanding officer – Einsatzleiter (ELtr), kommandierende Führungskraft

command-process – Führungsvorgang

command-squad (command-team of 5) – Führungsstaffel (FüSt)

command-squadron (command-team of 3) – Führungstrupp (FüTr)

command-staff – Führungsstab, Kommandostab

command-team of 3 (command-squadron) – Führungstrupp (FüTr)

command-team of 5 (command-squad) – Führungsstaffel (FüSt)

command-team of 9 (command-group) – Führungsgruppe (FüGr)

committee of experts – Fachkommission

communication assistant – Melder (Funktion innerhalb der Löschgruppe)

communication network – Fernmeldenetz

communication system – Fernmeldesystem

communications service – Fernmeldedienst

communications-structure – Kommunikationsstruktur

company – Zug, Löschzug

company command crew – Zugtrupp (ZTr)

company leader – Zugführer (ZFü)

company's fire brigade – Betriebsfeuerwehr (BtF)

compartment fires – Brände in geschlossenen Räumen

compatibility class – Verträglichkeitsgruppe

compatibility group – Verträglichkeitsgruppe

compelling necessity – zwingende Notwendigkeit

compensating measures – Ersatzmaßnahmen

competences – Befugnisse, Zuständigkeiten, Fähigkeiten

complete – vollständig

complete combustion – vollständige Verbrennung

complete optional equipment – vollständige Sonderausrüstung, vollständige Zusatzausrüstung

complete special equipment – vollständige Sonderausrüstung

component – Bestandteil, Bauteil, Komponente

composite construction – Verbundbauweise

composite material – Verbundwerkstoff

composite method of building – Verbundbauweise

composition of matter – Aufbau der Materie

compound – Zusammensetzung, Verbindung (Material, Substanz)

compressed air – Druckluft, Pressluft

compressed air breathing apparatus, self-contained breathing apparatus (SCBA) – Pressluftatmer (PA)

compressed air line/pipe – Druckluftleitung

compressed air supply – Druckluftversorgung

compressed gas – Druckgas

compressed gas cylinder – Druckgasflasche

compressed gas cylinder set – Druckgasflaschenbatterie

compressed gas tank – Druckgasbehälter

compressed-air breathing apparatus – Druckluftschlauchgeräte

compressed-air foam (CAF) – Druckluftschaum (DLS)

compressed-air foam system (CAFS) – Druckluftschaumanlage

compressibility – Komprimierbarkeit

compression – Kompression, Verdichtung (von Gasen)

compression heat – Kompressionswärme

compressor – Kompressor, Verdichter

Compton effect – Compton-Effekt

Compton scattering – Compton-Streuung

concentrated – konzentriert

concentration – Konzentration

concentration limits – Konzentrationsgrenzen

concentration-time product – Konzentrations-Zeit-Produkt

concept – Konzept, Vorstellung, Idee

concerned unit – betroffene Einheit

concrete – Beton

concrete reinforcement – Armierung, Bewehrung (Beton)

condensate – Kondensat

condensate separator – Kondensatabscheider

condensation – Kondensation

condensation temperature – Kondensationspunkt, -temperatur

conditioning – Aufbereitung

conductivity – elektrische Leitfähigkeit

conductor – Leiter (für elektr. Strom)

conduit – Leitung, Rohrleitung

confidential – vertraulich

confidentiality – Vertraulichkeit

confirm – bestätigen, zusagen, überprüfen

conflagrate – Feuer fangen, Feuersbrunst

conflagration – Flächenbrand, Großbrand, Feuersbrunst, Großfeuer

confounding effect – Störeinfluss

coniferous wood – Nadelwald

connecting piece for breathing – Atemanschluss

connection – Anschluss

connection line – Anschluss (Gas, Strom, Wasser)

connector – Übergangsstück, Verbindungselement

conscious mind – Bewusstsein

consciousness – Bewusstsein

conscription – Wehrpflicht

conscripts – Wehrpflichtige

consequential damage – Folgeschaden

conservation law – Erhaltungssatz, z. B. Energieerhaltungssatz

considerations – Abwägungen, Erwägungen

constantly – ständig, dauernd, andauernd

construction site – Baustelle

construction time – Aufbauzeit

consumable goods – Verbrauchsmaterial, Betriebsstoffe, Verbrauchsgüter, Konsumgüter

consumables – Verbrauchsmaterial, Betriebsstoffe, Verbrauchsgüter, Konsumgüter

contact – berühren, erreichen, Kontakt, Berührung, Verbindung

contact hazard – Gefahr beim Berühren

contact infection – Kontaktinfektion

contact point – Meldekopf

contact poisons – Kontaktgifte

contain – enthalten, beinhalten; auch: eindämmen

contain spillages – ausgetretene Stoffe eindämmen

container apparatus – Behältergerät (BG)

container explosion – Behälterexplosion

container for extinguishing agents – Löschmittelbehälter

container of lead for radioactive materials – Bleiblock (Radioaktivität)

containers – Behälter, Gefäß, Container

containment – Eindämmung, Sicherheitsbehälter

containment shell – Reaktorhülle, Sicherheitshülle

contaminate – verunreinigen, verschmutzen, verpesten, vergiften

contaminated casualties – kontaminierte Verletzte

contamination – Kontamination, Verunreinigung, Verseuchung, Verpestung

contamination carry-over – Kontaminationsverschleppung

contamination detection area – Kontaminationsnachweisplatz

contamination detector – Kontaminationsnachweisgerät (KNG), Kontaminationsmessgerät

contamination meter – Kontaminationsnachweisgerät (KNG), Kontaminationsmessgerät

contamination monitor (CoMo) – Kontaminationsmonitor

contamination protection hood – Kontaminationsschutzhaube

contamination protective clothing – Kontaminationsschutzkleidung

contamination protective suit – Kontaminationsschutzanzug

contamination spread – Kontaminationsverschleppung

contamination suspected – Kontaminationsverdacht

continuing education – Weiterbildung

continuous tone – Dauerton

contours – Umrisse, Konturen; auch: Höhenlinie (Kartenkunde)

contract – zusammenziehen

contracted pupils – verengte Pupillen

contradictory – widersprüchlich

control – Kontrolle, Regelung, Steuerung

control centre – Leitstelle, Einsatzzentrale, Kontrollzentrum

control device – Kontroll-, Steuergerät; auch: Bedieneinrichtung (BE)

control equipment – Kontrolleinrichtung, Regelanlage

control lamp – Kontrollleuchte

control line – Steuerleitung

control measures – Kontrollmaßnahmen

control panel – Schalttafel, Bedientafel, Bedientableau

control valve – Regelventil, Steuerventil

control zone – Kontrollbereich

controlled dividing breeching – Verteiler mit Überdruckventil

controlled evacuation – kontrollierte Evakuierung

convection – Wärmeauftrieb, Konvektion

convective heat flux – konvektiver Wärmestrom

conversion – Umwandlung, Umrechnung, Umformung

convert – umrechnen, umwandeln, konvertieren

conveying capacity – Fördermenge

convoy – Kolonne, Konvoi

convoy of fire vehicles – Kolonnenfahrt von Feuerwehrfahrzeugen

cool – kühlen, abkühlen

cool down – abkühlen, herunterkühlen

cooling – Abkühlen, Abkühlung

cooling down period – Abkühlzeit; auch: Abklingzeit eines Brands

cooling effect – Kühleffekt

cooling treatment – Kühlverfahren

cooperation – Zusammenarbeit, Mitwirkung, Kooperation

cooperative style – kooperativer (Führungs)Stil

cooperative style of leadership – kooperativer Führungsstil

copy-machine – Kopierer

cord – Kabel, Schnur, Kordel, Leine, Seil

corded ladder – Strickleiter

cordless drill – Akkubohrer

cordless screwdriver – Akkuschrauber

core melt accident – Unfall mit Kernschmelze

core meltdown – Kernschmelze

core of an atom – Atomkern, Kern eines Atoms

corona discharge – Koronaentladung

corona vaccination – Impfung gegen Corona/COVID-19

corona virus disease 2019 (COVID-19) – Coronavirus-Erkrankung 2019

corpuscular radiation – Teilchenstrahlung, Korpuskularstrahlung

correction factor – Korrekturfaktor

corrective factor – Korrekturfaktor

corridor – Flur

corrosion – Korrosion

corrosion damage – Korrosionsschaden

corrosion prevention – Korrosionsschutz

corrosion protection – Korrosionsschutz

corrosive – ätzend (gegen Materialien)

corrosive effect – Korrosionswirkung

corrosive substances – ätzende Stoffe

corrosive vapours – ätzende Dämpfe

corrugated board – Wellpappe
corrugated sheet iron – Wellblech
cortile – Lichthof
cosmic radiation/rays (CRs) – kosmische Strahlung
counter-fire – Gegenfeuer
counterpressure – Gegendruck
counting rate – Zählrate, Impulsrate
counting tube – Zählrohr
countryman's knot – Zimmermannsstich (Knoten)
coupling spanner – Kupplungsschlüssel
course of action – Handlungsablauf, Vorgehensweise
court – Gericht (Justiz)
covalence – Atombindungszahl
covalency – Atombindungszahl
covalent bond – kovalente Bindung, Atombindung, Elektronenpaarbindung
cover – abdecken, Abdeckung, Decke, Umschlag, Hülle, Deckung
cover effect – Deckeffekt (ein Löscheffekt)
coverage – Abdeckung
covering capacity – Deckvermögen (Löschmittel Schaum)
cow hitch – doppelter Ankerstich
craftsman – Handwerker
crane driver – Kranführer
crane driver's cabin – Kranführerkabine
crane hook – Kranhaken, Traghaken
crane truck – Kranwagen (KW)
crash barriers – Leitplanken
crazy glue – Sekundenkleber
creation – Erzeugung, Erschaffung, Entstehung
creek – Bach
crescent wrench – Rollgabelschlüssel
crew – Besatzung
crew cab – Mannschaftsraum (Fahrzeug)
Crimean-Congo fever (CCF) – Krim-Kongo-Fieber
Crimean-Congo haemorrhagic fever (CCHF) – hämorrhagisches Krim-Kongo-Fieber
Criminal Investigation Department – Kriminalpolizei
crisis intervention team (CIT) – Kriseninterventionsteam (KIT)
crisis unit – Krisenstab
critical condition – lebensbedrohender Zustand
critical diameter – kritischer Durchmesser
critical gas – kritisches Gas
critical infrastructures – kritische Infrastrukturen (KRITIS)
critical mass – Masse, kritische
critical point – kritischer Punkt
critical pressure – kritischer Druck
critical temperature – kritische Temperatur
criticality safety – Kritikalitätssicherheit
criticality safety index (CSI) – Kritikalitätssicherheitskennzahl
cross traffic – Querverkehr
crossbar – Querbalken
crosscut saw – Zugsäge
crossing hitch – Halbmastwurfsicherung (HMS)
cross-sensitivity – Querempfindlichkeit
cross-tip screwdriver – Kreuzschraubendreher
crow bar – Krähenfuß, Brechstange
crowding-out effect – Verdrängungseffekt (Löscheffekt)
crown cap – Kronenkorken
crown knot – Kreuzknoten
crucible – Tiegel
crude petroleum – Erdöl
crush-collapsible zone – Knautschzone
cryogenic substances – tiefkalte Stoffe
cryoprotection – Kälteschutz
crystal water – Kristallwasser

cubature – umbauter Raum

cubic law – kubisches Gesetz

curd soap – Kernseife

current distributor – Stromverteiler, Stromverteilung

current strength – Stromstärke

curriculum – Lehrplan

cut – Schnitt; auch: Schnittwunde

cut off by grinding – Trennschleifen

cut protection – Schnittschutz

cut protection gloves – Schnittschutzhandschuhe

cut protection jacket – Schnittschutzjacke

cut protection trouser – Schnittschutzhose

cutaneous anthrax – Hautmilzbrand

cut-off device – Trennvorrichtung, Absperrarmatur

cut-off flap – Absperrklappe

cut-off gate – Absperrschieber

cut-off valve – Absperrventil

cutting – Schneiden; auch: Trennschneiden

cutting disk – Trennscheibe

cutting pliers – Drahtschere

CWA detection-paper – Kampfstoffspürpapier

C-wrench – Hakenschlüssel

cylinder – Zylinder

cylinder lock – Zylinderschloss

cylinder projection – Zylinderprojektion (Kartenkunde)

cylinder valve – Druckgasflaschenventil

cytocidal – zelltötend

cytopathic – zellschädigend

cytotoxic – zellschädigend

D

dairy coupling – Milchrohrverschraubung

Dalton's (gas) laws – Dalton'sche (Gas)Gesetze

dam – Staudamm

dam point – Staustelle

dam up – Anstauen

damage – Beschädigung, Schaden

damage analysis – Schadensanalyse (z. B. nach Sturm)

damage area – Schadenstelle

damage caused by extinguishing agent – Löschmittelschaden

damage caused by fire – Brandschaden

damage class – Schadensklasse

damage control – Schadensabwehr, Schadensbekämpfung

damage repair – Schadensbehebung

damage territory – Schadensgebiet

damage to property – Sachschaden

damaged area – Fläche, beschädigte

damage-impact – Schadenswirkung

damaging – schädlich

damaging event – Schadensereignis

danger – Gefahr

danger allowance – Gefahrenzulage

danger area – Gefahrenbereich

danger point – Gefahrenstelle

danger sign – Gefahrschild

danger spot – Gefahrenstelle

danger to life and health – Gefahr für Leben und Gesundheit

danger zone – Gefahrenzone, Gefahrenbereich

dangerous – gefährlich, riskant, unsicher

dangerous area – Gefahrenzone, Gefahrenbereich

dangerous articles – gefährliche Gegenstände

dangerous goods (DG) – Gefahrgut, gefährliche Güter

dangerous goods emergency action code (DG-EA-Code) – Gefährliche Stoffe und Güter: Kodierung der Notfallmaßnahmen {GB}

dangerous substances – gefährliche Stoffe

dangers – Gefahren

darting flame – Stichflamme

data sheet – Datenblatt

data transfer – Datenübertragung

daughter nuclide – Tochternuklid

day(time) shift – Tagschicht

day-to-day operations – tägliche Einsätze

dead-ender – Sackgasse; auch: Fanatiker

deaeration – Entlüftung (Vorgang)

deaf-aid – Hörgerät

debriefing session – Einsatzabschlussbesprechung

debris – Trümmer, Schutt; auch: Brandrückstände, Brandschutt

debris trough – Schuttmulde

T. Schmiermund, *Fachwörterbuch Feuerwehr und Brandschutz*, https://doi.org/10.1007/978-3-662-64120-0_26

deburring – entgraten

decay – zerfallen, verfallen, verfaulen, Zerfall, Verfall, Fäulnis

decay chain – Zerfallsreihe, Zerfallskette

decay constant – Zerfallskonstante

decay energy – Zerfallsenergie

decay period – Zerfallszeit

decay series – Zerfallsreihen

decay time – Zerfallszeit, Abklingzeit (Kerntechnik)

decaying fire – abklingender Brand

decentralized water supply – dezentrale Wasserversorgung

deciduous wood – Laubwald

decision – Entscheidung, Entschluss

decisiveness – Entschlossenheit, Zielbewusstsein

deck gun – Dachmonitor

decompose – zersetzen, zerfallen, verrotten, verwesen

decomposition – Zersetzung

decomposition effect – Zersetzungseffekt

decomposition products – Zersetzungsprodukte

decomposition rate – Zersetzungsgeschwindigkeit

decomposition temperature – Zersetzungstemperatur

decompression chamber – Dekompressionskammer

decompression stop – Haltezeit (Tauchen)

decompression stop depth – Austauchstufe, Austauchtiefe (Tauchen)

decontamination – Dekontamination

decontamination area (decon area) – Dekontaminationsplatz (Dekon-Platz)

decontamination of equipment – Dekon G (G = Gerät)

decontamination of persons/people – Dekon P (P = Personen)

decontamination shower – Dekontaminationsdusche

decontamination station (decon station) – Dekontaminationsplatz (Dekon-Platz)

decoration – Dekoration, Verzierung; auch: Auszeichnung, Orden

decrease – Abnahme, Minderung, Verkleinerung, Verringerung

dedicated line – Festverbindung (Standleitung)

dee – Öse

deep well – Tiefbrunnen

defence – Verteidigung

defervescence – Siedeverzug

deficiencies – Mängel

deficiency report – Mängelbericht

deflagration – Verpuffung, Deflagration, Aufflammung

deflagration limit pressure – Deflagrationsgrenzdruck

deflagration pressure – Deflagrationsdruck

deflection – Umlenkung

deform – Verformen

deformation – Verformung

deformation energy – Verformungsenergie

deformation resistance – Verformungswiderstand

degas – ausgasen

degrade – zersetzen; auch: degradieren, erniedrigen

degrease – entfetten

degree – Grad

degree of danger – Gefährdungsgrad

degree of dispersion – Dispersionsgrad

degree of dissociation – Protolysegrad

delay – Verzögerung

delay in boiling – Siedeverzug

delay period – Verzögerungsdauer, Verzögerungszeit

delay treatment (DT) – verzögerte Behandlung

delayed alarm – verzögerter Alarm

delayed damages – Spätschäden

delayed effect – Verzögerungseffekt

deliberate – vorsätzlich, bewusst, absichtlich

deliberate arson – vorsätzliche Brandstiftung

delivery – Versand

delivery distance – Förderstrecke

delivery height – Förderhöhe

delivery hose – Druckschlauch, Zuleitung

delivery hose adaptor – Übergangsstück für Feuerwehrschläuche

delivery hose coupling – Druckkupplung

delivery line – Förderleitung

delivery of water – Wasserförderung, Löschwasserförderung

delivery pressure – Förderdruck

delivery volume – Förderleistung

delusive alarm – Täuschungsalarm

demarcation – Abgrenzung, Begrenzung

demolition – Abriss, Abbruch (Gebäude, Anlagen)

demountable – zerlegbar, demontierbar

demountable pod – Abrollbehälter (AB)

denial of access – Zutrittsverweigerung

denseness of smoke – Rauchdichtheit

densimeter – Dichtemessgerät

density – Dichte

density anomaly – Dichteanomalie

density of discharge – Entladungsdichte; auch für: Wasserbeaufschlagung

density value – Dichtewert

departure – Abfahrt, Abmarsch

dependent water supply – abhängige Löschwasserversorgung

depleted – abgereichert (Kernphysik)

deployment of chemical weapons – Chemiewaffeneinsatz

deployment of drones – Drohneneinsatz

depress – neigen, dämpfen, schwächen

depression – Senke, Bodensenke, Niederung

depressurized – drucklos

depth of penetration – Eindringtiefe (von z. B. Löschwasser)

deregister – abmelden

desensitization – Phlegmatisierung, Desensibilisierung

desensitization agent – Phlegmatisierungsmittel, Desensibilisierungsmittel

desensitized – phlegmatisiert, desensibilisiert

desensitizer – Phlegmatisierungsmittel, Desensibilisierungsmittel

desiccation – Trocknung, Entwässerung, Austrocknung

designation – Bezeichnung, Benennung

designation of responsibilities – Aufgabenverteilung, Benennung der Verantwortlichkeiten

desorption – Desorption

destruction – Zerstörung, Vernichtung

destructive fire – Schadensfeuer

desublimation – Resublimation

detailed information – detaillierte Information

detect – nachweisen, ermitteln, feststellen, identifizieren

detection devices – Nachweisgeräte

detection of contamination – Kontaminationsnachweis

detection paper – Spürpapier

detection powder – Spürpulver

detection time – Entdeckungszeit, Brandentdeckungszeit

detector – Detektor, Nachweisgerät

detector housing – Detektorgehäuse, Meldergehäuse

detector tube – Prüfröhrchen

detergent – Reinigungsmittel, Detergentien

determination – Bestimmung, Ermittlung (einer Substanz oder Größe)

determination limit – Bestimmungsgrenze

determination procedure – Bestimmungsverfahren

deterministic radiation effect – deterministische Strahlenwirkung

detonation – Detonation

detonation limits – Detonationsgrenzen

detonation pressure – Detonationsdruck

detonation range – Detonationsbereich

detonation velocity – Detonationsgeschwindigkeit

detonation wave – Detonationswelle

detoxification – Entgiftung

detoxify – entgiften

deuterium – Deuterium (^2H, D), schwerer Wasserstoff

deuterium oxide – Deuteriumoxid (D_2O), schweres Wasser

deuterium-depleted water – Leichtwasser

development – Entwicklung; auch: Ablauf

development phase – Entwicklungsphase

development stage – Entwicklungsphase, Entwicklungsstadium

development time – Entwicklungszeit

deviation – Abweichung

device group – Gerätegruppe

dew point – Taupunkt

diagonal cutter – Seitenschneider

diameter – Durchmesser

diarrhoea – Durchfall (Erkrankung)

dictaphone – Diktiergerät

dielectric gloves – Elektrohandschuhe, Elektroschutzhandschuhe

diesel emergency power aggregate – Notstromaggregat, mit Diesel betrieben

different – verschieden, unterschiedlich, andere

differential pressure – Differenzdruck

diffuser – Zerstäuber; auch: Pumpenleitrad

diffuser nozzle – Sprühstrahlrohr

diffusibility – Diffusionsvermögen

diffusion – Ausbreitung, Diffusion, Zerstreuung

diffusion coefficient – Diffusionskoeffizient

diffusion flame – Diffusionsflamme

diffusion rate – Diffusionsgeschwindigkeit

diffusion velocity – Diffusionsgeschwindigkeit

diffusion-impending – diffusionshemmend

diffusion-resistant – diffusionsdicht

diffusivity – Diffusionsvermögen

digital enhanced cordless telecommunication (DECT) – Telekommunikation, digital verstärkte schnurlose

digital paging device – digitaler (Funk-) Meldeempfänger (DME)

digital radio – Digitalfunk (Begriff)

digital radio network – Digitalfunknetz

digital radio transmitter – Digitalfunkgerät

dike defence – Deichverteidigung

dilatability – Ausdehnungsvermögen

dilatation joint – Dehnfuge

dilated pupils – Pupillen, erweiterte

diluent – Verdünnungsmittel

diluting effect – Verdünnungseffekt (Löscheffekt)

dilution – Verdünnung

dilution effect – Verdünnungseffekt (Löscheffekt)

dilution rule – Mischungsregel

dimensions (height, width, depth) – Abmessungen (Höhe, Breite, Tiefe)

diminution – Reduktion

dioxins – Dioxine

dipping – Eintauchung, Tränkung

direct attack – direkter Angriff/Löschangriff

direct cooling – direkte Kühlung

direct current (dc) – Gleichstrom

direct fire attack – direkter Löschangriff

direct mode operation (DMO) – Direktmodus (Digitalfunk)

direction arrow – Richtungspfeil

direction of attack – Angriffsrichtung

direction of flow – Fließrichtung

direction sign – Hinweisschild (z. B. für Hydrant)

directional radio – Richtfunk

directive – Richtlinie (RL), Weisung, Direktive

dirty bomb – schmutzige Bombe

disadvantage – Nachteil

disaster – Katastrophe (Kat)

disaster alert – Katastrophenalarm

disaster control – Katastrophenschutz (KatS)

disaster control exercise – Katastrophenschutzübung

disaster control regulation – Katastrophenschutz-Dienstvorschrift (KatS-DV)

disaster management – Krisenstab

disaster relief – Katastrophenschutz (Organisation), Katastrophenhilfe

disaster relief crew – Katastrophenschutzhelfer

disaster relief organisations – Katastrophenschutz-Organisationen

disaster relief worker – Katastrophenschutzhelfer

disaster response exercise – Katastrophenschutzübung

disaster situation – Katastrophenfall

discharge – Entlassung, Abschied aus dem Dienst; auch: Entladung

discharge cock – Ablasshahn

discharge event – Entladungsvorgang (Elektrizität)

discharge outlet – Ausfluss (aus einem Strahlrohr)

discharge pressure – Enddruck, Abgabedruck

discharge spark – Entladungsfunke

disciplinary action – Disziplinarmaßnahme, Disziplinarstrafe

disconnect from the mains – Freischalten

disease (Dz.) – Erkrankung, Krankheit

disease control – Seuchenschutz

disease control centre – Seuchenschutzbehörde, Seuchenschutzzentrum

disease-causing agent – Krankheitserreger

disinfectants – Desinfektionsmittel

disinfection – Entseuchung, Desinfektion

disintegrate – zerfallen, zersetzen, auseinander fallen

disintegration – Zerfall, Zersetzung, Auflösung

dislodge – verschieben, verdrängen

dismutation – Disproportionierung, Dismutation

dispatch – Depesche, Nachricht; auch: Versand, Absendung

dispatch centre – Leitstelle

dispatcher – Disponent, Leitstellendisponent, Leistellenmitarbeiter

dispersed substances – dispergierte Stoffe

dispersing agent – Dispergiermittel

dispersion – Dispersion

displace – verdrängen, verschieben

displacing effect – Verdrängungseffekt (Löscheffekt)

disposable gloves – Einmalhandschuhe, Wegwerfhandschuhe

disposable syringe – Einwegspritze

disposable use – Einmalgebrauch

disposal – Entsorgung

disposal of chemical weapons – Chemiewaffenentsorgung

disproportionation – Disproportionierung

dissociation – Dissoziation

dissolving behaviour – Löseverhalten

dissolving capacity – Lösevermögen

dissous gas – Dissousgas

distance – Entfernung, Abstand, Distanz

distance law – Abstandsgesetz

distress at sea – Seenot

distress in mountains – Bergnot

distress signal – Notsignal, Notzeichen

distress signal unit (DSU) – Notsignalgeber, Totmanneinrichtung (Atemschutzeinsatz)

distress-signal devices – Notsignalgeräte

distribution – Verteilung

distribution list – Verteiler (Empfänger-Liste)

distributor – Verteiler; auch Zündverteiler oder Verleiher

distributor of electricity – Elektroverteiler

district master chimney sweeper – Bezirksschornsteinfegermeister

disturbances – Störungen

dive – tauchen

diver's decompression chamber – Taucherdruckkammer

diver's disease – Taucherkrankheit

diversity – Vielfalt, Verschiedenartigkeit, Vielfältigkeit

dividing breeching – Verteiler (Feuerwehr-Gerät)

diving – Tauchen

diving apparatuses – Tauchgeräte

diving equipment – Tauchausrüstung

diving knife – Tauchermesser

Do not breathe dust. – Staub nicht einatmen.

doctor's emergency ambulance – Notarztwagen (NAW)

doctor's incident vehicle – Notarztwagen (NAW)

dog of a hose coupling – Knaggenteil

donor – Donator

door closer – Türschließer

door hinge – Türband

door leaf – Türblatt

doped – dotiert

dormer – Dachgaube

dormer window – Dachfenster

dose – Dosis

dose control – Dosiskontrolle

dose equivalent – dosisäquivalent

dose rate – Dosisleistung

dose rate change – Dosisleistungsänderung

dose rate constant – Dosisleistungskonstante

dose rate meter – Dosisleistungsmesser (DLM), Dosisleistungsmessgerät (DLM), Dosimeter

dose rate warning device – Dosisleistungswarner (DLW), Dosisleistungswarngerät (DLW)

dose reduction – Dosisminderung, Dosisverringerung

dose warning device – Dosiswarngerät (DWG)

dosimeter – Dosisleistungsmesser (DLM), Dosisleistungsmessgerät (DLM)

dosimeter badge – Dosimeterplakette

dosimetry – Dosimetrie, Dosismessung

dosing device – Dosiervorrichtung

dosing range – Dosierbereich

DOT classes {US} – Gefahrklassen, Gefahrgutklassen, Klassen gefährlicher Güter

double clove hitch – doppelter Mastwurf

double door system – Schleuse (im Gebäude)

double knot – Doppelknoten

double ladder – Bockleiter

double meter stick – Gliedermaßstab, „Zollstock“

double sheet bend – doppelter Schotenstich

double-sided adhesive tape – Doppelklebeband

doubt – Zweifel

dowel – Dübel

down-blended – abgereichert (Chemie)

downwind – windabgewandt

downwind side – windabgewandte Seite

draegerman {US} – Atemschutzgeräteträger (AGT)

draft – Zugluft, Luftzug; auch: Entwurf, Vorlage, Konzept

drag shoe – Unterlegkeil

dragging basket – Schleifkorbtrage

drain – Abfluss, Ablauf, ablasse, leeren, entwässern

drain seal – Kanalabdeckung

drain tube – Abflussrohr; auch: Kanüle

draught – Zugluft, Luftzug; auch: Entwurf, Vorlage, Konzept

draw up into a syringe – Spritze aufziehen

dress uniform – Ausgehuniform

dressing roll – Verbandpäckchen

dressing station – Verbandplatz

dressings – Verbände (bei Verletzungen)

drift distance – Driftweite

drill – bohren, Bohrer, Bohrmaschine; auch: Drill, Übung

drill chuck key – Bohrfutterschlüssel

drill hammer – Bohrhammer

drill tower – Übungsturm; auch: Bohrturm

drilling machine – Bohrmaschine

drinking water – Trinkwasser

drinking water pipe – Trinkwasserleitung

drinking water pipework system – Trinkwasserleitungsanlage

drip – tropfen, abtropfen, tröpfeln

drip bottle – Infusionsflasche, Infusionstropfflasche

drip stand – Infusionsständer

dripping – tropfen, abtropfen

driver – Fahrer, Kraftfahrer

driver airbag – Fahrerairbag

driver's cab – Führerhaus

driver's license revocation – Entziehung der Fahrerlaubnis

driveways for the fire brigade – Zufahrten für die Feuerwehr

driving ban – Fahrverbot

driving licence – Fahrerlaubnis

drizzling fog – Sprühnebel

drone – Drohne

drop – Tropfen; auch: Fall, Sturz

drop line – Arbeitsleine (insbes. bei Verwendung a. d. Saugleitung)

drop shaft – Fallschacht

droplet infection – Tröpfcheninfektion

droplets – Tröpfchen

drug poisoning – Arzneimittelvergiftung

drugs – Medikament, Arzneimittel; auch: Rauschgift

drum – Trommel, Fass, Tonne

drum pump – Fasspumpe

drum wrench – Fassschlüssel

dry – trocken

dry cell – Trockenbatterie

dry cell battery – Trockenbatterie

dry extinguishing process – Trockenlöschverfahren, mit Pulver löschen

dry powder appliance – Pulverlöschfahrzeug

dry powder extinguisher – Pulverlöscher

dry powder extinguishing method – Pulverlöschverfahren

dry powder nozzles – Pulverdüse

dry riser – Trockensteigleitung

dry rising main – Trockensteigleitung

dry suction test – Trockensaugprüfung

dry-ice – Trockeneis (CO_2)

dual-circuit brake – Zweikreisbremse

dual-use goods – Dual-Use-Güter

Duck Tape™ – Universalklebeband, Industrieklebeband, Panzerband

duct – Leitung, Rohr, Kabelführung; auch: Duktus

duct tape – Universalklebeband, Industrieklebeband, Panzerband

dues – Gebühren, Beiträge

dummy – Übungspuppe, Rettungspuppe, Attrappe, Puppe

dump – Müllhalde, Müllkippe, Schutthaufen

dump fire – Müllhaldenbrand

dunk tank – Auffangbehälter (für Chemikalien)

durable – strapazierfähig

duration of action – Einsatzzeit, Einsatzdauer, Aktionsdauer

duration of inflammation – Entflammungsdauer, Entflammungszeit

duration of operation – Einsatzdauer, Operationsdauer

duration of the mission – Einsatzdauer, Missionsdauer

duroplasts – Duroplaste

dust – Staub

dust explosion – Staubexplosion

dust fire – Staubbrand

dust ignition proof – staubexplosionsgeschützt

dust mask – (Grob-)Staubmaske

dust respirator – (Grob-)Staubmaske

dust-air-mixtures – Staub-Luft-Gemische

dust-explosion hazard – Staubexplosionsgefährdung

dustpan – Kehrblech

duty instruction – Dienstvorschrift

duty of control – Kontrollpflicht

duty regulation – Dienstvorschrift

duty station – Dienststelle

duty to inform – Informationspflicht

duty-dress – Dienstkleidung

dynamic kernmantle rope – Kernmanteldynamikseil, Dynamik-Kernmantelseil

dynamic pressure – dynamischer Druck

E

early detection – Früherkennung

early fire detection – Brandfrüherkennung

early fire detection system – Brandfrüherkennungssystem

early shift – Frühschicht

early warning – Frühwarnung

earmuff – Kapselgehörschutz

earplugs – Gehörschutzstöpsel

earth pressure – Erddruck

earth rod – Erdungsstange

earth wire – Erdungskabel

earthing – Erdung (Elektrizität)

earthing accessories – Erdungszubehör

ease of ignition – Zündfähigkeit, Entzündlichkeit

easily combustible – leicht brennbar

easily ignitable – leichtentzündlich

eave – Traufe

Ebola virus – Ebola-Virus

Ebola virus disease (EVD) – Ebola-Fieber (EF)

eCall (automated emergency call system for motor vehicles) – eCall (Kfz)

eccentric screw pump – Exzenterschneckenpumpe

echelon – Rang, Staffelung

echelons of command – Führungsstufen

eddy (current) brake – Wirbelstrombremse

effective dose (ED) – effektive Dosis (ED)

effective power – Nutzleistung, effektive Leistung

effectiveness – Wirksamkeit, Effektivität

efficiency – Leistungsfähigkeit, Nutzeffekt, Wirksamkeit; auch: Wirkungsgrad

effluent water – Abwasser

effort – Anstrengung, Bemühung

effusion – ausströmen (Gas); auch: Erguss

ejection of combustible liquids – Auswerfen brennbarer Flüssigkeiten

elasticity – Elastizität, Dehnbarkeit

elastomers – Elastomere

elbow – Ellbogen; auch: Krümmer, Knie, Rohrbogen

electric accident – Elektrounfall, Stromunfall

electric arc – Lichtbogen

electric burn – Strommarke (nach Elektrounfall)

electric conductivity – elektrische Leitfähigkeit

electric current – elektrischer Strom

electric equipment – elektrische Betriebsmittel

electric field strength – elektrische Feldstärke

electric resistance – elektrischer Widerstand

electric sparks – elektrische Funken

electric tools – Elektrowerkzeug

electric tools box – Elektrowerkzeugkasten

electric welding – Elektroschweißen

electrical appliances – Elektrogeräte

electrical charge – elektrische Ladung

electrical circuit – Stromkreis

electrical devices – Elektrogeräte

T. Schmiermund, *Fachwörterbuch Feuerwehr und Brandschutz*, https://doi.org/10.1007/978-3-662-64120-0_27

electrical discharge – elektrische Entladung

electrical discharge spark – Entladungsfunke

electrical fire – Elektrobrand

electrical leakage – Ableitung (von elektr. Strom)

electrical shock – Elektroschock

electrical short – Kurzschluss (elektr.)

electrical voltage – elektrische Spannung

electrically qualified person – Elektrofachkraft (EFK)

electric-arc welding – Lichtbogenschweißen

electrician – Elektriker

electricity – Elektrizität

electrocardiogram (ECG) – Elektrokardiogramm (EKG)

electrochemical series – Spannungsreihe

electroencephalogram (EEG) – Elektroenzephalogramm (EEG)

electromagnetic compatibility (EMC) – elektromagnetische Verträglichkeit (EMV)

electromagnetic compliance (EMC) – elektromagnetische Verträglichkeit (EMV)

electromagnetic environmental compatibility (EMEC) – elektromagnetische Umweltverträglichkeit (EMUV)

electromagnetic radiation – elektromagnetische Strahlung

electromagnetic spectrum – elektromagnetisches Spektrum

electron – Elektron

electron detachment – Abtrennung eines Elektrons

electron flux – Elektronenfluss

electron gas – Elektronengas

electron pair – Elektronenpaar

electron radiation – Elektronenstrahlung

electron rays – Elektronenstrahlen

electron shell – Elektronenschale

electron stream – Elektronenstrahl (ES)

electronegativity – Elektronegativität

electronic beam (EB) – Elektronenstrahl (ES)

electronic personal dosimeter (EPD) – elektronisches Personendosimeter (EPD)

electronic region – Atomhülle (Elektronenbereich)

electron-nucleus attraction – Anziehung (der Elektronen durch den Atomkern)

electrostatic charge/charging – elektrostatische Aufladung

electrostatic discharge (ESD) – elektrostatische Entladung (ESD, ESE)

element – Element

elemental particle – Elementarteilchen

elementary quantum of action – Planck'sches Wirkungsquantum

elevated reservoir – Hochbehälter

elevated tank – Hochbehälter

elevated temperature substances – erwärmte Stoffe (Gefahrgut)

elevator – Aufzug, Fahrstuhl; auch: Winde, Silo

embers – Glut, glühende Asche/Kohle

embolism – Embolie

emergence – Entstehung, Aufkommen

emergency action – Notfallmaßnahmen

emergency action code (EAC) – Notfallmaßnahmen-Code

emergency admission – Notaufnahme eines Patienten

emergency ambulance – Rettungswagen (RTW)

emergency backpack – Notfallrucksack

emergency bridge – Behelfsbrücke, Notbrücke

emergency call – Notruf

emergency call centre – Notrufzentrale, Rettungsleitstelle

emergency call service – Notrufdienst

emergency case – Notfallkoffer

emergency channel – Notrufkanal (Funk)

emergency chute – Notrutsche

emergency cooling – Notkühlung

emergency decontamination – Not-Dekontamination

emergency department (ED) – Notaufnahme (Einrichtung)

emergency dispatcher – Leitstellendisponent

emergency diving squad – Einsatztauchtrupp

emergency doctor – Notarzt

emergency doctor in charge – leitender Notarzt (LNA)

emergency doctor mission – Notarzteinsatz

emergency doctor's car – Notarzteinsatzfahrzeug (NEF)

emergency drill – Notfallübung

emergency entrance – Notfalleingang

emergency escape breathing device (EEBD) – Druckluft-Fluchtgerät

emergency escape mask – Fluchtmaske, Notfallfluchtmaske

emergency escape route – Rettungsweg

emergency evacuation – Notfallevakuierung

emergency exit – Notausgang, Notausstieg

emergency filtering apparatus – Fluchtfiltergerät

emergency fire door – Brandschutztür

emergency generator – Notstromaggregat

emergency hammer – Nothammer, Rettungshammer

emergency hospitalization – Notaufnahme eines Patienten

emergency lane – Rettungsgasse

emergency light – Notbeleuchtung

emergency lighting – Notbeleuchtung

emergency lights – Notfalllichter; auch: Blaulicht

emergency line – Notrufkanal (Funk)

emergency medical services (EMS) – Rettungsdienst (RD)

emergency medical technician – i. d. R.: Rettungsassistent (RettAss, RA), Notfallsanitäter (NotSan)

emergency medical technician-basic (EMT-B) – Rettungshelfer (RH), Sanitäter (Sani), Sanitätshelfer (Sanhelfer)

emergency medical technician-intermediate (EMT-I) – Rettungssanitäter (RettSan, RS)

emergency medical technician-paramedic (EMT-P) – Rettungsassistent (RettAss, RA), Notfallsanitäter (NotSan)

emergency mode – Notbetrieb

emergency nail – Notnagel

emergency number – Notrufnummer

emergency operation – Einsatz; auch: Notoperation

emergency operator – Leitstellendisponent (Rettungsdienst)

emergency patient – Notfallpatient

emergency power – Notstrom

emergency power aggregate – Notstromaggregat

emergency power battery – Notstromakku

emergency power generator – Notstromerzeuger

emergency power source – Notstromanlage

emergency power supply – Notstromversorgung

emergency power supply unit (EPSU) – Notstromaggregat

emergency rescue helicopter – Rettungshubschrauber (RTH)

emergency rescue knife – Rettungsmesser

emergency responder – Ersthelfer (professionell)

emergency response operations – Gefahrenabwehrmaßnahmen

emergency room (ER) – Notaufnahme (Einrichtung)

emergency service – Notdienst

emergency services – Rettungswesen

emergency shelter – Notunterkunft

emergency shower – Notdusche

emergency sign – Rettungszeichen

emergency staircase – Nottreppenhaus, Feuertreppe

emergency standby power system (ESPS) – Netzersatzanlage (NEA)

emergency vehicle – Einsatzfahrzeug

emergency vehicles only! (sign) – Einsatzfahrzeuge frei! (Schild)

emergency ward (EW) – Notaufnahme, Notaufnahmeabteilung, Notfallstation (NFS)

emery paper – Schmirgelpapier, Sandpapier

emetic – Brechmittel

emission – Emission

emit – aussenden, ausstrahlen, abstrahlen, emittieren

emitted radiation – Abstrahlung, ausgesendete Strahlung

empiric formula – Summenformel

empowered – ermächtigt, berechtigt

emulsifier – Emulgator

emulsifying agent – Emulgator

emulsion – Emulsion

enabled – betriebsbereit, aktiviert

encircle – Umfassen, Einkreisen, Einkesseln, Abriegeln

enclosed person – eingeschlossene Person

enclosure – Gehäuse; auch: geschlossener Raum, Gehege

encryption – Verschlüsselung

end of duty – Wachablösung

endanger – gefährden, bedrohen

endangerment – Gefährdung, Bedrohung

endothermal – wärmeaufnehmend, endotherm

endothermic – wärmeaufnehmend, endotherm

endothermic reaction – endotherme Reaktion

endotracheal tube – Endotrachealtubus

end-to-end encryption (E2EE) – Ende-zu-Ende-Verschlüsselung

end-user certificate (EUC) – Endverbleibserklärung (EVE)

energy – Energie; auch: Tatkraft, Schwung

energy absorber – Falldämpfer

energy level – Energieniveau

energy of activation – Aktivierungsenergie

energy of combustion – Verbrennungsenergie

energy release – Energiefreisetzung

energy release rate – Energiefreisetzungsrate

energy source – Energiequelle

energy theorem – Energieerhaltungssatz, Satz von der Erhaltung der Energie

engage – engagieren, beteiligen; auch: einrasten

engine driver – Fahrer, Maschinist

engine fire – Triebwerksbrand (Flugzeug)

engine mount damper – Schwingungsdämpfer (im Kfz)

engine oil – Motoröl

engine power – Motorleistung, Leistung

engine revolution – Motordrehzahl

English knot – Spierenstich, Anglerknoten

enhanced decontamination – erweiterte Dekontamination

enhancement forces – Verstärkung

enriched – angereichert (Kernphysik)

enriched uranium – angereichertes Uran

enrichment facility – Anreicherungsanlage

enrichment plant – Anreicherungsanlage

enrolled nurse – Krankenschwester, examinierte

ensuring – Gewährleistung, Sicherstellung

enteritis – Durchfall, Dünndarmentzündung

enthalpy – Enthalpie

enthalpy of sublimation – Sublimationsenthalpie, Sublimationswärme

enthalpy of vaporization – Verdampfungswärme, Verdampfungsenthalpie

entity – Einheit (Messgröße)

entrances for the fire brigade – Zugänge für die Feuerwehr

environment – Umwelt

environment conditions – Umweltbedingungen

environment law – Umweltrecht

environmental authority – Umweltbehörde

environmental burden – Umweltbelastung

environmental degradation – Umweltzerstörung

environmental impact assessment (EIA) – Umweltverträglichkeitsprüfung (UVP)

environmental load – Umweltbelastung

environmental pollution – Umweltverschmutzung

environmental protection – Umweltschutz

environmental radioactivity – Umweltradioaktivität

environmentally compatible – umweltverträglich

environmentally friendly – umweltverträglich

environmentally hazardous substances – umweltgefährdende Stoffe

equipment – Ausrüstung

equipment group (EG) – Gerätegruppe

equipment manager – Gerätewart (Gw)

equipment tender – Rüstwagen (RW)

equivalent dose – Äquivalentdosis

equivalent dose rate – Äquivalentdosisleistung

equivalent sound pressure level (ESPL) – äquivalenter Schalldruckpegel

erecting angle – Aufrichtwinkel

escape – Flucht

escape hatch – Ausstieg

escape hatch – Notausstieg

escape smoke hood – Brandfluchthaube, Fluchthaube

escape symbol – Fluchtweg-Beschilderung

escape tunnel – Fluchttunnel

escaping gas – austretendes Gas, ausströmendes Gas

escutcheon – Türschild; auch: Wappenschild

Esmarch mitella – Dreieckstuch (Erste Hilfe)

establishing the situation – Lagefeststellung

estimate of the situation – Lagebeurteilung, Situationseinschätzung

estimation – Schätzung, Einschätzung

etch-proof – ätzfest

ethine (acetylene) – Ethin (Acetylen)

European Atomic Energy Community – Europäische Atomgemeinschaft (EURATOM)

European death knot (EDK) – Sackstich, Schlaufenknoten, Bandschlingenknoten

European Standard, EU standard – Europäische Norm (EN)

evacuated persons – evakuierte Personen

evacuation – Evakuierung

evacuation chut – Rettungsrutsche

evacuation exercise – Räumungsübung, Evakuierungsübung

evacuation hood – Fluchthaube, Brandfluchthaube

evacuation order – Evakuierungsbefehl

evacuation practice – Evakuierungsübung, Räumungsübung

evacuation route – Rettungsweg

evacuation signal – Räumungssignal

evacuation strategy – Evakuierungsstrategie

evacuation test – Evakuierungsübung

evacuation time – Räumungszeit, Evakuierungszeit

evacuation tunnel – Evakuierungstunnel

evaluation – Beurteilung, Einschätzung, Auswertung, Bewertung

evaluation of the situation – Bewertung der Lage

evaporation – Verdampfung, Verdunstung

evaporation chill – Verdunstungskälte

evaporation number – Verdunstungszahl

evaporation rate – Verdunstungszahl, Verdunstungsrate

evaporation surface – Verdunstungsfläche

evaporative cooling – Verdunstungskälte

evaporize – verdampfen, verdunsten

evening shift – Spätschicht

evidence of contamination – Kontaminationsnachweis

examination (exam) – Untersuchung, Überprüfung, Prüfung, Abschlusstest

exceed – überschreiten

exception – Ausnahme

exceptional permission – Ausnahmegenehmigung

excess of air – Luftüberschuss

exchange – Austausch, Umtausch

exchange of air – Luftwechsel

exchangeable – austauschbar

excited atom – angeregtes Atom

exclusion – Ausschluss

exclusion area – Sperrbereich, Sperrgebiet

executable – ausführbar, durchführbar

execute a command – Befehl ausführen

executive fiat – Ersatzvornahme; auch: Direktionsrecht

exemption – Freistellung

exemption limit – Freigrenze

exercise – Übung, Training, Aufgabe

exhalation – Ausatmung

exhalation resistance – Ausatemwiderstand

exhalation valve – Ausatemventil

exhalation valve disk – Ausatemventilscheibe

exhale – ausatmen

exhaled air – Ausatemluft

exhaust – Abgase; auch: Lüftung

exhaust air – Abluft

exhaust air opening – Abluftöffnung

exhaust brake – Auspuff(klappen)bremse, Motorbremse

exhaust duct – Abluftschacht

exhaust ejector primer – Gasstrahler

exhaust equipment – Absauganlage

exhaust fan – Abluftventilator

exhaust gases – Abgase

exhaust gases hose – Abgasschlauch

exhaust hood – Abzugshaube, Abzug, Absaughaube

exhauster – Absaugung, Absauganlage

exhaustible water supply for fire-fighting – erschöpfliche Löschwasserversorgung

existing dangers – bestehende Gefahren

exited state – angeregter Zustand

exothermal – exotherm, wärmeabgebend, wärmefreisetzend

exothermic – exotherm, wärmeabgebend, wärmefreisetzend

exothermic decomposition – exotherme Zersetzung

exothermic reaction – exotherme Reaktion

expansion – Ausdehnung, Expansion, Erweiterung, Vergrößerung

expansion adapter – Expansionsstück, Erweiterungsadapter

expansion behaviour – Ausdehnungsverhalten

expansion ratio – Verschäumungszahl (VZ)

expectant treatment – abwartende Behandlung

expellant – Treibmittel

expendable parts – Verschleißteile, Verbrauchsteile

experience – Erfahrung; auch: Erlebnis

expert – Experte, Sachverständiger, Gutachter, fachkundige Person, Sachkundiger

expert assessment – Gutachten

expert evidence – Sachverständigengutachten

expert knowledge – Expertenwissen

expert witness – Gutachter, Sachverständiger

expert's information – Experteninformation

expertise – Expertise, Sachverstand, Expertenwissen

expiration – Ausatmung; auch: Ablauf

explanation of the map – Legende (Kartenkunde)

explanatory leaflets – Merkblätter, erklärende Broschüren

explode – explodieren, sprengen, platzen

exploder – Sprengmittel

exploring time – Erkundungszeit

explosible – explosionsfähig (= explosibel)

explosible atmosphere – explosionsfähige Atmosphäre

explosible mixture – explosionsfähiges Gemisch (= explosibles Gemisch)

explosion – Explosion

explosion action – Explosionswirkung, explosive Wirkung

explosion hazard – Explosionsgefährdung (Explosionsgefahr)

explosion hazard concentration – explosionsgefährliche Konzentration

explosion hazard indexes – Explosionskennzahlen

explosion hazardous room – explosionsgefährdeter Raum

explosion heat – Explosionswärme

explosion indexes – Explosionskennzahlen

explosion limits – Explosionsgrenzen

explosion points – Explosionspunkte

explosion pressure – Explosionsdruck

explosion propagation – Explosionsfortpflanzung

explosion protected – explosionsgeschützt

explosion protection – Explosionsschutz

explosion range – Explosionsbereich

explosion spread – Explosionsausbreitung

explosion strength – Explosionsstärke

explosion suppression – Explosionsunterdrückung

explosion suppression system – Explosionsunterdrückungsanlage

explosion volume – Explosionsvolumen

explosion wave – Explosionswelle

explosion zone – Explosionszone

explosion-hazardous area – explosionsgefährdeter Bereich

explosion-limits warning device – Explosionsgrenzenwarngerät

explosion-proof electric equipment – explosionsgeschützte elektrische Betriebsmittel

explosive – explosiv, auch: Sprengstoff

explosive articles – Gegenstände mit Explosivstoff

explosive combustion – explosionsartige Verbrennung

explosive effect – Explosionswirkung, explosive Wirkung

explosive flame – Stichflamme

explosive force – Sprengkraft

explosive ordnance disposal service – Kampfmittelräumdienst

explosive power – Sprengkraft

explosive strike – Zünddurchschlag

explosives – Explosivstoffe

exposure – Exposition, ausgesetzt sein; auch für: Strahlenbelastung

exposure factor (EF) – Expositionsfaktor (EF)

exposure time – Einwirkzeit

expressway – Schnellstraße

EX-protection – EX-Schutz

extend – verlängern, erweitern

extender cable – Verlängerungskabel

extending jib crane – Teleskopkran

extending turntable ladder – Teleskopdrehleiter

extension – Verlängerung, Erweiterung, Ausdehnung; auch: Anbau, Erweiterungsbau

extension cord – Verlängerungskabel

extension ladder – Schiebleiter

extension lead – Verlängerungskabel

extension length of ladder – Auszugslänge einer Leiter

exterior temperature – Außentemperatur

exterior wall – Außenwand

external danger – gefährliche Einwirkung von außen, Gefahr von außen

external dangerous influence – gefährliche Einwirkung von außen

external heating – Fremderwärmung

external irradiation – äußere Bestrahlung

external pressure – Außendruck

external staircase – Außentreppenhaus

external thread – Außengewinde

externally supplied ignition – Fremdzündung

extinction by dry powder – Trockenlöschverfahren, mit Pulver löschen

extinction by water – Nasslöschverfahren, mit Wasser löschen

extinguish – löschen

extinguishing agent – Löschmittel

extinguishing agent additives – Löschmittelzusätze

extinguishing agent jet – Löschmittelstrahl

extinguishing agent unit – Löschmitteleinheit (LE)

extinguishing boom – Gelenklöscharm (GLA)

extinguishing capacity – Löschvermögen

extinguishing concentration – löschwirksame Konzentration

extinguishing crew – Löschtrupp

extinguishing device – Löscheinrichtung

extinguishing dry powder – Löschpulver

extinguishing effect – Löschwirkung, Löscheffekt

extinguishing efficiency – Löschwirksamkeit, Löscheffektivität

extinguishing equipment – Löschgeräte

extinguishing exercise – Löschübung (Unterricht)

extinguishing foam – Löschschaum

extinguishing gas – Löschgas

extinguishing mechanism – Löschmechanismus

extinguishing medium – Löschmittel

extinguishing method – Löschmethode, Löschverfahren

extinguishing nozzle – Löschdüse (Sprinkler)

extinguishing powder – Löschpulver

extinguishing powder rate – Löschpulverrate

extinguishing powder type ABC – ABC-Löschpulver, Glutbrandpulver

extinguishing powder type BC – BC-Löschpulver

extinguishing powder type D – D-Löschpulver, Metallbrandpulver

extinguishing rate – Löschmittelrate

extinguishing substance – Löschmittel

extinguishing time – Löschzeit

extinguishing unit – Löschmitteleinheit (LE)

extinguishing vapour – Löschdampf

extinguishing water – Löschwasser

extinguishing water demand – Löschwasserbedarf

extinguishing water pipe – Löschwasserleitung

extinguishing water reservoir – Löschwasserbehälter

extinguishing water supply – Löschwassereinspeisung

extinguishing water supply pipe – Löschwasserzuleitung (Schlauch)

extinguishing water tank – Löschwasserbehälter

extinguishing-water retention – Löschwasserrückhaltung

extinguishment – Löschvorgang

extraneous ignition – Fremdzündung

extranuclear region – Atomhülle („nicht-Kern-Bereich“)

extreme fire behaviour – extremes Brandverhalten

extremely flammable – hochentzündlich

extremely flammable substance – hochentzündlicher Stoff

extremely toxic – hochgiftig

extremely toxic substance – hochgiftiger Stoff

extrication – Befreiung; auch für: technische Rettung

eye dropper bottle – Augenspülflasche, Augenwaschflasche

eye irritant – Augenreizstoff

eye irritation – Augenreizung

eye protection – Augenschutz

eye rinsing liquid – Augenspülflüssigkeit

eye-wash station – Augendusche

F

fabric – Stoff, Gewebe, Tuch

facade – Fassade

face covering – Mund-Nase-Bedeckung (MNB)

face guard – Gesichtsschutz

face piece – Atemanschluss

face shield – Gesichtsschutz, Gesichtsschutzschirm

factor of distribution – Verteilungsfaktor

factory fire brigade – Betriebsfeuerwehr (BtF)

factual – sachlich

fail-safe – eigensicher

failure analysis – Fehlersuche, Fehleranalyse, Schadensanalyse

failure rate – Ausfallquote, Ausfallrate, Fehlerquote, Fehlerrate

fairground rides – fliegende Bauten

fall arrest harness – Absturzsicherungsgeschirr

fall arrester – Absturzsicherung

fall danger – Absturzgefahr

fall factor – Sturzfaktor

fall protection – Absturzsicherung, Höhensicherung

fall protection harness – Absturzsicherungsgeschirr

fallback level – Rückfallebene

fall-impact absorber – Falldämpfer

falling ill – krank werden

fallout – radioaktiver Niederschlag

fallout shelter – Atombunker

false alarm – Fehlalarm, Falschalarm

false alarm with good intent – blinder Alarm

false-negative – falsch-negativ

false-positive – falsch-positiv

fan – Lüfter, Ventilator, Gebläse

fanlight – Oberlicht

Faraday's law – Faraday'sches Gesetz

fastener – Verbindungsmittel (Absturzsicherung)

fastening point – Befestigungspunkt

fastening strap – Befestigungsgurt

fat fire – Fettbrand

fat fire extinguisher – Fettbrandlöscher

fatal injury – tödliche Verletzung

fatigue – Müdigkeit, Ermüdung, Erschöpfung, Übermüdung

fatigue failure – Ermüdungsbruch

faucet – Wasserhahn

fault – Fehler, Störung, Mangel, Schaden

fault current – Fehlerstrom

fault report – Störungsmeldung, Fehlerbericht

fault-clearing service – Störungsdienst

faulty mission – Fehleinsatz

faulty planning – falsche Planung

FCI-steam explosion between oil and water – Fettexplosion

fear reaction – Angstreaktion

feature – Eigenschaft, Merkmal, Charakteristikum

T. Schmiermund, *Fachwörterbuch Feuerwehr und Brandschutz*, https://doi.org/10.1007/978-3-662-64120-0_28

Federal Office for Civil Protection and Disaster Assistance – Bundesamt für Bevölkerungshilfe und Katastrophenschutz (BBK)

Federal Office for Radiation Protection (Germany) – Bundesamt für Strahlenschutz (BfS)

feed – Futter, füttern, versorgen, auch: Zufuhr, Zuführung, zuführen

feeder water conduit – Zubringerleitung, Speisewasserleitung

feeding – Einspeisung

feeding point – Einspeisepunkt

feel – fühlen, spüren, empfinden, Gefühl

fees – Gebühren

felling axe – Holzaxt, Fällaxt

felling wedge – Fällkeil

felt-tipped pen – Filzschreiber

female thread – Innengewinde

fermentation – Gärung, Fermentation

ferry crash – Fährunglück

fertilizer – Düngemittel

fever – Fieber

fever of undetermined origin (FUO) – Fieber unbekannter Ursache (FUU)

fibreboard – Holzfaserplatte

field glass – Fernglas

field kitchen – Feldküche, „Gulaschkanone“

field of application – Anwendungsbereich, Anwendungsgebiet

field strength – Feldstärke

figure eight descender – Abseilachter

figure-eight knot – Achterknoten (als Stopperknoten)

figure-eight knot loop – Achterknoten (mit fester Schlaufe)

fill a syringe – Spritze aufziehen

fill level – Füllstand

filling degree – Füllgrad

filling material – Füllmaterial

filling neck – Einfüllstutzen

filling pressure – Fülldruck

filling quantity – Füllmenge

filling station – Füllstation, Abfüllstation; auch: Tankstelle

film – Folie, Film, Schicht; auch: Film (i. S. v. Kinofilm)

film badge – Filmplakette

film badge meter – Filmdosimeter

film dosimeter – Filmdosimeter

film forming additive – filmbildender Zusatz

film tube – Folienschlauch

filter appliances – Filtergeräte

filter capacity – Filterleistung

filter layer – Filterschicht

filter respirator – Filtergerät, Atemschutzgerät mit Filter

filtering escape device – Fluchtfiltergerät

filtering facepiece (FFP) – Feinstaubmaske

final deposition – Endlagerung

final disposal site – Endlager, Endlagerstätte

final exit – Ausgang als Tür ins Freie, „letzter Ausgang“

final temperature – Endtemperatur

financial penalty – Geldbuße, Geldstrafe, Bußgeld, finanzielle Strafe

financial responsibilities – finanzielle Verantwortlichkeiten

fine – Geldbuße, Geldstrafe, Bußgeld; auch: fein, gut, zart, heiter

fine dust – Feinstaub, feiner Staub

fine particulate matter – Feinstaub

fingerguard – Handschutz

fire – Feuer, Brand, brennen, entzünden, anzünden

fire accelerant – Brandbeschleuniger

fire alarm – Feueralarm; auch: Feuermelder

fire alarm centre – Brandmeldezentrale (BMZ)

fire alarm control panel (FACP) – Brandmeldezentrale (BMZ)

fire alarm display – Feuerwehranzeigetableau (FAT)

fire alarm drill – Feuerwehralarmübung

fire alarm system – Brandmeldeanlage

fire alert – Feueralarm

fire analysis – Brandanalyse

fire and rescue station – Feuer- und Rettungswache (FRW)

fire and rescue training centre (FRTC) – Feuerwehr- und Rettungs-Trainingscenter (FRTC)

fire apparition – Feuererscheinung

fire appliance – Feuerwehrfahrzeug

fire area – Brandumfang, Brandfläche

fire arrester – Flammenrückschlagsicherung

fire atmosphere – Brandatmosphäre

fire attack – Löschangriff; auch: Brandanschlag

fire audit – Feuerbeschau, Brandschau

fire axe – Brandaxt, Feuerwehraxt

fire barrier – Brandschutzwand, Brandschutzbarriere, Feuerschutzabschluss, Brandsperre

fire beater – Feuerpatsche

fire blanket – Löschdecke, Feuerlöschdecke

fire boat – Feuerwehrboot, Feuerlöschboot

fire break – Brandschneise; auch: Abtrennung zwischen zwei Etagen

fire bridge – Feuerbrücke

fire brigade – Feuerwehr

fire brigade access point – Feuerwehrzugang

fire brigade access road – Feuerwehrzufahrt

fire brigade attendance time – Anmarschzeit (der Feuerwehr)

fire brigade diver – Feuerwehrtaucher

fire brigade dress uniform – Feuerwehr-Dienstanzug

fire brigade duty – Feuerwehrdienst

fire brigade federation – Feuerwehrverband

fire brigade headquarters – Hauptfeuerwache

fire brigade medical technician – Feuerwehrsanitäter

fire brigade operation – Feuerwehreinsatz

fire brigade paramedic – Feuerwehrsanitäter

fire brigade radio – Feuerwehrfunk

fire brigade restraint belt – Feuerwehr-Haltegurt

fire brigade service – Feuerwehrdienst

fire brigade shed – Feuerwehrschuppen

fire brigade signal flag – Warnflagge, Signalflagge

fire brigade tactics – Feuerwehrtaktik

fire brigade uniform – Feuerwehruniform

fire brigade's institute – Institut der Feuerwehr (IdF)

fire bucket – Feuerlöscheimer, Feuereimer

fire cabinet – Wandhydrant

fire call – Brandmeldung (telefonisch)

fire casualties – Brandopfer, Brandverletzte, Brandtote

fire cause – Brandursache

fire cause investigator – Brandursachenermittler

fire certificate – Brandschutzbescheinigung

fire characteristics – Brandkenngrößen

fire chief – Amtsleiter (AL)

fire class – Brandklasse

fire class A: solid combustible materials {EU} – Brandklasse A: feste, glutbildende brennbare Stoffe {EU}

fire class B: flammable liquids {EU} – Brandklasse B: brennbare flüssige Stoffe {EU}

fire class B: flammable liquids and gases {US} *– Brandklasse B: brennbare flüssige Stoffe und brennbare Gase {US}*

fire class C: electrical fires {US} *– Brandklasse C: Elektrobrände {US}*

fire class C: flammable gases {EU} – Brandklasse C: brennbare Gase {EU}

fire class D: combustible metals {EU} – Brandklasse D: brennbare Metalle {EU}

fire class E: electrical fires {EU, retracted} – Brandklasse E: Elektrobrände {EU, zurückgezogen}

fire class F: cooking oils/fats {EU} – Brandklasse F: Speiseöl/-fett {EU}

fire class K: cooking oils/fats {US} (K = kitchen) *– Brandklasse K: Speiseöl/-fett {US} (K = Küche)*

fire clinker – Brandschlacke

fire college – Feuerwehrschule

fire compartment – Brandabschnitt, Brandraum

fire compartment door – Brandabschnittstür

fire control and protection – Brandschutz (BS)

fire course – Brandverlauf

fire crackers – Feuerwerkskörper

fire curtain – Rauchschürze, Eiserner Vorhang (Bühne, Theater)

fire cut – Brandabschnitt

fire dam – Feuerschutzdamm

fire damper – Brandschutzklappe

fire debris – Brandschutt

fire decay – Abklingphase (eines Brands)

fire defence – abwehrender Brandschutz

fire department – Feuerwehrwache, Feuerwehr (als kommunales Amt)

fire department supply connection – Einspeisung für die Feuerwehr

fire dept crew car – Mannschaftstransportfahrzeug (MTF)

fire detection – Brandentdeckung, Brandererkennung, Brandmeldung

fire detection and alarm system – Brandmeldeanlage

fire detector – Brandmelder

fire development – Brandentwicklung

fire director – Branddirektor

fire disaster – Brandkatastrophe

fire district – Ausrückebereich

fire diver – Feuerwehrtaucher

fire door – Brandschutztür

fire drencher – Feuerlöscher

fire drill – Alarmübung, Feuerwehrübung; auch: Brandschutzübung

fire drill house – Brandhaus, Übungshaus

fire duct – Brandkanal (z. B. in Heu)

fire duration – Branddauer

fire duty – Feuerwehrdienst

fire effects – Brandeffekte

fire effluent – Rauchgas

fire emergence – Brandentstehung

fire emergency telephone – Brandmeldetelefon

fire engine – Feuerwehrfahrzeug

fire engine (group fire-fighting vehicle) – Einsatzfahrzeug, Löschfahrzeug, Löschgruppenfahrzeug (LF)

fire engine house – Spritzenhaus

fire equipment building – Feuerwehrgerätehaus (FGH)

fire equipment house – Feuerwehrgerätehaus (FGH)

fire equipment storage shed – Feuerwehrgeräteschuppen

fire escape – Feuerleiter

fire escape hood – Brandfluchthaube, Fluchthaube

fire event – Brandereignis

fire exit – Notausgang

fire expert – Brandsachverständiger

fire exposure – Brandeinwirkung

fire extinguisher – Feuerlöscher

fire extinguishing agent for metals – Löschmittel für Metallbrände

fire extinguishing strategy – Löschstrategie

fire extinguishing unit – Feuerlöscheinrichtung

fire fatalities – Brandopfer, Brandtote

fire fighters – Feuerwehrangehörige (FwA), Brandbekämpfer, Feuerwehrleute

fire fighting – Brandbekämpfung

fire flap – Brandschutzklappe

fire flash – Feuerübersprung

fire frequency – Brandhäufigkeit

fire funnel – Brandtrichter

fire gas volume – Brandgasvolumen

fire gases – Brandgase

fire gate – Brandschutztor

fire ground – Brandstelle (BSt)

fire growth – Brandentwicklung

fire hazard – Brandgefährdung, Brandgefahr, Feuergefahr

fire hazard class – Brandgefahrenklasse

fire hose – Feuerwehrschlauch, Druckschlauch

fire hose depot – Schlauchlager

fire hose pipe – Schlauchleitung

fire hose type B, C, D – B-, C-, D-Druckschlauch

fire house – Brandhaus, Übungshaus

fire hydrant – Hydrant (H)

fire hydrant key and bar – Hydrantenschlüssel

fire impact – Brandauswirkung, Brandeinwirkung, Brandbeanspruchung

fire incident – Brandereignis

fire inspection – Feuerbeschau, Brandschau

fire insurance – Brandversicherung, Feuerversicherung

fire insurance police – Feuerversicherungspolice

fire integrity – Feuerwiderstand, Feuerfestigkeit

fire intensity – Brandintensität

fire investigation – Branduntersuchung, Brandermittlung, Brandursachenermittlung

fire investigator – Brandermittler, Brandursachenermittler

fire ladders – Feuerwehrleitern

fire lane – Brandschneise

fire law – Feuerwehrgesetz, Brandschutzgesetz

fire load – Brandlast

fire load density – Brandlastdichte, flächenbezogene Brandlast, Brandintensität, Brandbelastung

fire lookout (a person) – Brandbeobachter

fire loss – Brandschaden

fire manner – Brandart

Fire Marshall – Ermittlungsbeamter (bei Brandstiftung)

fire model – Brandmodell

fire monitor – Monitor, Wasserwerfer, Schaum-Wasser-Werfer

fire nozzle – Strahlrohr

fire object – Brandobjekt

fire officer – Feuerwehrführungskraft

fire operator panel – Feuerwehrbedienfeld (FBF)

fire origin – Brandursprung, Brandentstehung

fire pentacle – Feuerfünfeck

fire performance – Verhalten bei Brandeinwirkung (Bauteile)

fire phase – Brandphase

fire photographer – Einsatzfotograf

fire picket – Brandschutzposten, Brandwache, Brandsicherheitswache

fire piquet – Brandsicherheitswache, Brandschutzposten, Brandwache

fire place – Brandort; auch: Kamin, Feuerstelle

fire plug – Hydrant (H)

fire pockets – Glutnester, Brandnester

fire policy – Brandversicherungspolice

fire pond – Feuerlöschteich, Löschwasserteich

fire practice – Feuerwehrübung

fire prevention (FP) – vorbeugender Brandschutz (VB), Brandschutz, Brandverhütung

fire prevention engineer – Brandschutzingenieur (BSIng)

fire prevention history – Brandschutzgeschichte

fire progress(ion) curve – Brandverlaufskurve

fire propagation – Brandausbreitung

fire properties – Brandverhalten

fire protection – vorbeugender Brandschutz (VB), Brandschutz

fire protection act – Brandschutzgesetz (konkretes, bestimmtes)

fire protection assistant – Brandschutzhelfer

fire protection cladding – Brandschutzverkleidung

fire protection classes – Brandschutzklassen

fire protection curtain – Brandschutzvorhang

fire protection education – Brandschutzerziehung

fire protection enlightenment – Brandschutzaufklärung

fire protection helper – Brandschutzhelfer

fire protection indexes – brandschutztechnische Kennzahlen

fire protection installation – Brandschutzanlage

fire protection law – Brandschutzrecht, Brandschutzgesetz

fire protection measure – Brandschutzmaßnahme (konkret)

fire protection method – Brandschutzmethoden, Brandschutzmaßnahmen (allgemein)

fire protection officer – Brandschutzbeauftragter

fire protection products – Brandschutz-Produkte

fire protection regulations – Brandschutzvorschriften, Brandschutzbestimmungen

fire protection sliding door – Brandschutzschiebetür

fire protection training – Brandschutzausbildung

fire protection valve – Brandschutzventil

fire protective treatment – Flammschutzbehandlung

fire proximity suit – Hitzeschutzanzug

fire pump – Feuerwehrpumpe, Feuerlöschkreiselpumpe (FP)

fire pump for high pressure (FPH) – Feuerlöschkreiselpumpe, Hochdruck-Ausführung

fire pump for normal pressure (FPN) – Feuerlöschkreiselpumpe, Normaldruck-Ausführung

fire quad/quadrangle – Feuerviereck

fire regulations – Brandschutzbestimmungen, Brandschutzverordnung

fire report – Brandbericht

fire rescue path – Feuerwehrzufahrt

fire research – Brandforschung, Brandschutzforschung

fire resistance – Feuerwiderstand, Feuerwiderstandsfähigkeit (FWF), Feuerbeständigkeit

fire resistant closure – feuer-/flammendicher Abschluss

fire resistant damper – feuer-/flammendicher Abschluss

fire resistant glazing – Brandschutzverglasung

fire resisting door – feuerfeste Tür

fire resisting smoke door – rauch- und flammendichte Tür

fire retardant – Brandschutzmittel, Feuerschutzmittel

fire retardant paint – Brandschutzanstrich

fire risk – Brandrisiko

fire road – Feuerschneise

fire room – Brandraum

fire run – Feuerwehreinsatz; auch: Brandserie

fire safety – vorbeugender Brandschutz (VB), Brandschutz, Brandverhütung

fire safety engineer – Brandschutzingenieur (BSIng)

fire safety regulations – Brandschutzordnung (BSO), Brandschutzvorschriften

fire safety signs – Brandschutzzeichen

fire scenario – Brandszenario

fire scene residue – Brandreste

fire seal – Feuerschutzabschluss

fire sector commander – Abschnittsleiter

fire series – Brandserie

fire service – Feuerwehr (Oberbegriff), Feuerwehrwesen

fire service association – Feuerwehrverband

fire service regulation – Feuerwehrdienstvorschrift (FwDV)

fire signal – Brandalarm

fire simulation – Brandsimulation

fire siren – Martinshorn

fire size – Brandumfang, Brandgröße

fire smoke – Brandrauch

fire source – Brandherd, Brandquelle

fire spark over – Feuerüberschlag

fire spread(ing) – Brandausbreitung, Feuerüberschlag

fire sprinkler – Sprinkler, Sprinkleranlage

fire sprinkler system – Sprinkleranlage

fire squad – Löschstaffel

fire stability – Brandstabilität

fire stairs – Feuertreppe

fire standards – Brandschutznormen

fire station – Feuerwehrhaus (FwH), Feuerwache (FW), Feuerwehrhalle

fire statistics – Brandstatistik

fire storm – Feuersturm

fire suit – Einsatzkleidung

fire suppression – Brandunterdrückung

fire suppression system (FSS) – Brandunterdrückungsanlage (BUA)

fire swatter – Feuerpatsche

fire tender – Löschfahrzeug, Feuerwehrfahrzeug

fire test – Brandprüfung

fire theory – Brandlehre

fire tornado – Feuertornado, Feuerhose

fire traces – Brandspuren

fire training centre (FTC) – Feuerwehrtrainingscenter (FTC)

fire triangle – Feuerdreieck

fire truck – Löschfahrzeug, Feuerwehrfahrzeug

fire victim – Brandopfer (Person)

fire wall – Brandwand, Brandschutzwand

fire watch – Brandsicherheitswache, Brandwache, Brandschutzposten

fire water pond – Feuerlöschteich, Löschwasserteich

fire well – Löschwasserbrunnen

fire zone – Brandzone

firebox – Feuermelder

fire-extinguisher cabinet – Feuerlöscherkasten

fire-extinguishing agent – Feuerlöschmittel

fire-extinguishing equipment – Feuerlöschgerät (Ausrüstung)

fire-extinguishing foam – Feuerlöschschaum

fire-extinguishing lance – Löschlanze

fire-extinguishing medium – Feuerlöschmittel

fire-extinguishing powder – Feuerlöschpulver, Löschpulver

fire-extinguishing pump – Feuerwehrpumpe, Feuerlöschkreiselpumpe (FP)

fire-extinguishing substance – Feuerlöschmittel

fire-extinguishing system – Löschanlage, Feuerlöschanlage, Feuerlöschsystem

fire-extinguishing tender – Tanklöschfahrzeug (TLF)

fire-extinguishing water – Löschwasser

fire-fighter (Ff) – Truppmann (TrM)

fire-fighter boots – Feuerwehr-Sicherheitsstiefel, Feuerwehrstiefel

fire-fighter gloves – Feuerwehrhandschuhe, Feuerwehr-Schutzhandschuhe

fire-fighter hood – Flammschutzhaube

firefighter uniform – Feuerwehruniform

fire-fighter's helmet – Feuerwehrhelm

fire-fighter's rope – Feuerwehrleine

fire-fighters – Feuerwehrangehörige (FwA)

fire-fighting – abwehrender Brandschutz

fire-fighting appliance – Feuerlöschgerät

fire-fighting axe – Feuerwehraxt

fire-fighting cage – Arbeitskorb (einer Drehleiter)

fire-fighting clothes – Feuerschutzkleidung (FSK)

fire-fighting company – Löschzug (LZ)

fire-fighting equipment – Löschgeräte, Brandbekämpfungsausrüstung

fire-fighting exercise – Feuerwehrübung

fire-fighting group – Löschgruppe

fire-fighting helicopter – Löschhubschrauber

fire-fighting lift – Feuerwehraufzug

fire-fighting nozzle – Strahlrohr

fire-fighting operation – Löscheinsatz

fire-fighting operations – Löscharbeiten

fire-fighting pond – Löschteich

fire-fighting pump – Feuerwehrpumpe, Feuerlöschkreiselpumpe (FP)

fire-fighting sector – Brandbekämpfungsabschnitt

fire-fighting tactics – Brandbekämpfungstaktik, Löschtaktik

fire-fighting tank – Löschpanzer

fire-fighting team – Löschtrupp

fire-fighting technic – Feuerwehrtechnik

fire-fighting techniques – Löschtechnik

fire-fighting truck – Löschfahrzeug

fire-fighting tunic – Einsatzkleidung zur Brandbekämpfung, Feuerschutzkleidung (FSK)

fire-fighting unit – Löschzug (LZ)

firefighting way – Angriffsweg

fireground – Einsatzstelle (ESt)

firehouse – Feuerwache (FW)

fireman – Feuerwehrmann

fireman's axe – Feuerwehraxt

fireman's axe – Rettungsaxt

fireman's belt with safety rope – Feuerwehrsicherheitsgurt

fireman's boots – Feuerwehr-Sicherheitsstiefel, Feuerwehrstiefel

fireman's hatchet – Feuerwehrbeil

fireman's helmet – Feuerwehrhelm

fireman's hook – Einreißhaken

fireman's life belt – Feuerwehrsicherheitsgurt

fireman's pole – Rutschstange

fireman's rope – Fangleine

fireman's belt – Feuerwehrgurt

fireman's safety belt – Feuerwehrsicherheitsgurt

fireman's waist belt – Feuerwehrsicherheitsgurt

firemanship – Feuerwehr, Feuerwehrwesen

firemen – Feuerwehrangehörige (FwA), Brandbekämpfer, Feuerwehrleute

firemen association – Feuerwehrverein

firemen club – Feuerwehrverein

firemen society – Feuerwehrverein

firenado (short for fire tornado) – Feuertornado, Feuerhose

fire-proof – brandbeständig, feuerbeständig, feuerfest, feuersicher

fireproof suit – Feuerschutzanzug

fire-proofing agent – Brandschutzmittel, Feuerschutzmittel

fire-protection appliance – Brandschutzeinrichtungen

fire-protection facade – Brandschutzfassade

fire-resistance class – Feuerwiderstandsklasse

fire-resistance duration – Feuerwiderstandsdauer (eines Bauteils)

fire-resistance rating – Feuerwiderstandsdauer (eines Bauteils)

fire-resistant – feuerbeständig, brandhemmend, brandbeständig

fire-resistant finish – Flammschutzimprägnierung

fire-resistant rating – Feuerwiderstandsklasse

fire-resisting closure – Brandschutzabschluss

fire-resisting wall – Brandwand, Brandschutzwand

fire-retardant – feuerhemmend, brandverzögernd, abbrandverzögernd

fire-retardant door – feuerhemmende Tür

fire-safety engineering – Brandschutzingenieurwesen

fire-service key safe – Feuerwehrschlüsseldepot (FSD), Feuerwehrschlüsselkasten (FSK)

fire-service school – Feuerwehrschule

firestop seal – Brandschutzschott

firewoman – Feuerwehrfrau

fireworks – Feuerwerk

firing point – → inflammation point

first aid – Erste Hilfe (EH)

first aid kit (FAK) – Sanitätskasten, Erste-Hilfe-Set

first aid training (FAT) – Erste-Hilfe-Training (EHT, EH-Training)

first alarm – erster Abmarsch

first escape route – erster Rettungsweg

first fire escape – erster Rettungsweg

first responder – Ersthelfer (professionell)

first-aid box – Verbandskasten, Erste-Hilfe-Kasten

first-aid case – Notfallkoffer, Erste-Hilfe-Kasten

first-aid scissors – Verbandschere

first-aid station – Sanitätsstation, Erste-Hilfe-Stelle

first-aider (FA) – Ersthelfer (Laie)

first-degree burn – Verbrennung ersten Grades

first-degree frostbite – Erfrierung ersten Grades

first-order conductor – Leiter erster Ordnung

fisherman's bend – Ankertauknoten

fisherman's knot – Spierenstich, Anglerknoten

fissile material – spaltbare Stoffe (Kerntechnik)

fission – Spaltung (Kerntechnik)

fission bomb – Atombombe, Spaltbombe

fission products – Spaltprodukte (Kerntechnik)

fission reaction – Spaltreaktion, Atomspaltungsreaktion

fissionable material – spaltbares Material

fitness – Eignung, Tauglichkeit

fitting – Einbau, Armatur, Beschlag

fix – einspannen, fixieren

fixation – Ruhigstellung

fixed command post – ortsfeste Befehlsstelle

fixed coupling – Festkupplung

fixed installation – feste Installation; auch: ortsfeste Anlage

fixing point – Fixpunkt, Haltepunkt, Befestigungspunkt

fixture – Vorrichtung, Halterung, Befestigung

flag-bearer – Bannerträger

flame – Flamme

flame application time – Beflammungsdauer

flame arrester/arrestor – Flammensperre, Flammendurchschlagsicherung

flame barrier – Flammensperre, Flammendurchschlagsicherung

flame colour – Flammenfärbung

flame cutting – Brennschneiden (Acetylen-Sauerstoff)

flame detector – Flammenmelder

flame front – Flammenfront

flame height – Flammenhöhe

flame ionisation detector (FID) – Flammenionisationsdetektor (FID)

flame of explosion – Explosionsflamme

flame penetration – Flammendurchschlag

flame propagation – Flammenfortpflanzung, Flammenausbreitung

flame propagation rate – Flammenfortpflanzungsgeschwindigkeit

flame resistant – flammfest, flammenfest, feuerfest, feuerresistent

flame retardant agents – Flammschutzmittel, flammhemmendes Mittel

flame retardant product – Flammschutzmittel, flammhemmendes Mittel

flame retardant treatment – Flammschutzbehandlung, Flammschutzausrüstung

flame spread – Flammenausbreitung

flame temperature – Flammentemperatur

flame velocity – Flammengeschwindigkeit

flame zones – Flammenzonen

flameless combustion – flammenlose Verbrennung

flameproof clothes – Feuerschutzkleidung (FSK)

flame-retardant – flammhemmend

flame-retardant clothing – Flammschutzbekleidung, flammhemmende Kleidung

flame-retardant finish – Flammschutzimprägnierung

flame-spread time – Flammenausbreitungszeit

flaming fire – Flammenbrand

flammability – Entflammbarkeit

flammability limits – Entflammbarkeitsgrenzen

flammability test – Brennbarkeitsprüfung

flammable – brennbar, entzündbar, feuergefährlich

flammable dust – entzündbarer Staub, brennbarer Staub

flammable gas – entflammbares Gas, brennbares Gas

flammable limit – Zündgrenze

flammable solids – feste entzündbare Stoffe, brennbare Feststoffe

flammable solvents – brennbare Lösemittel

flammable vapours – brennbare Dämpfe

flange – Flansch

flange connection – Flanschverbindung

flank – Flanke, Flügel

flare back – zurückschlagen; auch: Flammenrückschlag

flare light – Streulicht („Flackerlicht“)

flaring – Abfackeln

flaring up – aufflammen, auflodern

flash – Blitz; auch: Stichflamme

flash back – zurückblitzen; auch: Flammenrückschlag, Rückzündung

flash light – Taschenlampe

flash point – Flammpunkt

flashing – blinken; auch: aufflammen, auflodern

flashing blue light – Blaulicht

flashing light – Blinklicht, Warnblinkleuchte, Blitzleuchte

flash-over – Feuerübersprung

flat fire – Wohnungsbrand

flat gasket – Flachdichtung

flat roof – Flachdach

flat seal – Flachdichtung

flat-nosed plier – Flachzange

flexible barrier – Schlauchsperre

flexible stretcher – Falttrage

flexural strength – Biegefestigkeit

float switch – Schwimmerschalter

float valve – Schwimmerventil

floating roof tank – Schwimmdachtank

flood – Überschwemmung, Flut

flood barrier – Hochwassersperre

flood-gate – Schleuse (im Gewässer)

flooding – Überschwemmung, Hochwasser

flood-light mast – Flutlichtmast

flood-lightning – Flutlichtbeleuchtung

floodlights – Flutlichtscheinwerfer

floods – Hochwasser; auch: Flutlichter

floor – Fußboden, Boden; auch: Stockwerk, Etage, Geschoss

floor area – Grundfläche, Bodenfläche

floor space – Grundfläche

floor-space index (FSI) – Geschossflächenzahl (GFZ)

flotation suit – Rettungsweste, Schwimmweste

flow – fließen, strömen; auch i. S. v. Durchfluss, Fluss

flow chart – Flussdiagramm

flow measurement – Durchflussmessung

flow pressure – Fließdruck, Durchflussdruck

flow rate – Durchfluss, Durchflussmenge, Flussrate, Fließrate, Förderstrom

flow speed – Strömungsgeschwindigkeit, Durchflussgeschwindigkeit

flow velocity – Fließgeschwindigkeit, Durchflussgeschwindigkeit

flu vaccination – Impfung gegen Grippe

flue gas cooling – Rauchgaskühlung

flue gas ignition – Rauchgasentzündung

flue gas layer – Rauchgasschicht

flue gases – Rauchgase

fluence rate – Bestrahlungsintensität

fluid fire – Flüssigkeitsbrand

fluorescent jerkin – Warnweste, Sicherheitsweste

fluorescent tube – Leuchtstoffröhre

fluorinated hydrocarbons – fluorierte Kohlenwasserstoffe (FKW)

fluorine – Fluor

fluorobromocarbons (FBC) – Fluor-Brom-Kohlenwasserstoffe (FBKW)

fluorocarbons – Fluorkohlenwasserstoffe (FKW)

fluoro-protein foam compound – Fluorproteinschaummittel

fluorotelomer alcohols (FTOH) – Fluortelomeralkohole (FTOH)

flux of information – Informationsfluss

fly ash – Flugasche

flying embers – Flugfeuer

flying sparks – Funkenflug

flywheel – Exzenter, Schwungrad

foam – Schaum, Löschschaum

foam blanket – Schaumdecke, Schaumteppich

foam branch pipe (FBP) – Schaumstrahlrohr, Schaumrohr

foam compatibility – Schaumverträglichkeit

foam concentrate – Schaummittel, Schaumbildner

foam concentrate container – Schaummittelbehälter

foam constancy – Schaumstabilität, Schaumbeständigkeit

foam decomposition – Schaumzerfall

foam distribution unit – Abrollbehälter Wasser/Schaum (AB-Schaum)

foam equipment – Schaumausrüstung, Schaumlöschgerät

foam extinction – Schaumzerfall

foam extinguisher – Schaumlöscher

foam extinguishing method – Schaumlöschverfahren

foam fire-fighting vehicle – Schaumlöschfahrzeug (SLF)

foam generator (FG) – Leichtschaumgenerator (LG), Schaumgenerator

foam gun – Schaumkanone

foam monitor – Schaumwerfer

foam pod – Abrollbehälter Wasser/Schaum (AB-Schaum)

foam solution – Schaummittel-Wasser-Gemisch

foam stability – Schaumstabilität, Schaumbeständigkeit

foam stabilizer – Schaumstabilisator

foam steadiness – Schaumstabilität, Schaumbeständigkeit

foam tank tender – Schaumlöschtankfahrzeug

foam tender (FoT) – Schaumlöschfahrzeug (SLF)

foam trailer (FTr) – Schaumanhänger

foam-compatible – schaumverträglich

foamed plastic – Schaumstoff

foaming – Verschäumung, Schaumerzeugung

foaming agent – Schaumbildner, Schaummittel

foaming device – Schaumerzeuger

foaming index – Verschäumungszahl (VZ)

foaming range – Verschäumungsbereich

foaming ratio – Verschäumungszahl (VZ)

foaming the landing runway/strip – Landebahnbeschäumung

foam-inhibiting – schaumhemmend

focal point of the mission – Einsatzschwerpunkt

focus of mission – Einsatzschwerpunkt

fog – Nebel, Schleier, Dunst

fog extinction – Nebellöschverfahren

fog nail – Löschnagel

fog nozzle, fog nozzle branch pipe – Hohlstrahlrohr (HSR)

foil – Folie; auch: Florett

folding ladder – Klappleiter

folding ruler – Gliedermaßstab, „Zollstock“

folding scoop stretcher – Schaufeltrage

folding step – Klapptritt

folding stretcher – Klapptrage

folding yardstick – Gliedermaßstab, „Zollstock“

follow-up check – Nachkontrolle

follow-up examination – Nachuntersuchung

follow-up extinguishing – Nachlöschen

follow-up extinguishing work – Nachlöscharbeiten

foot brake – Fußbremse

footing – Fundament, Sockel, Basis

footpath – Fußweg

footstep – Schritt, Fußtritt; auch: Trittstufe beim Einsatzfahrzeug

for internal use only (FIUO) – Nur für Dienstgebrauch (NfD)

for official use only (FOUO) – Nur für Amtsgebrauch

forbidden – verboten, untersagt, nicht erlaubt

force effect – Krafteinwirkung

force of gravity – Schwerkraft, Erdanziehungskraft

force of law – Rechtskraft, Gesetzeskraft

forced ventilation – Druckbelüftung

forced ventilation system – Druckbelüftungsanlage

forced-air system – Zwangslüftung

forecast – Prognose, Vorhersage

foreign body – Fremdkörper

forest fire – Waldbrand

forest fire cause – Waldbrandursache

forest fire defence – Waldbrandabwehr

forest fire forecast – Waldbrandvorhersage, Waldbrandprognose

forest fire risk level – Waldbrandgefahrenstufe

forest fire warning – Waldbrandwarnung

forest fire warning level – Waldbrandwarnstufe, Waldbrandgefahrenstufe

forest fire-fighting – Waldbrandbekämpfung

forestry office – Forstamt

foreword – Vorwort

fork spanner – Gabelschlüssel

forklift – Gabelstapler

form of employment – Einsatzform („Beschäftigungsform")

form of operations – Einsatzform („Betriebsform")

formation – Formation, Bildung von ...; auch: Bildung, Ausbildung

formation of a charcoal layer – Bildung einer Holzkohleschicht

formation of fire compartments – Brandabschnittsbildung

forms – Formulare, Vordrucke

formula – Formel

formulary – Formelsammlung

foundation – Fundament, Sockel, Grundlage, Basis; auch: Stiftung

four-eyes principle – Vier-Augen-Prinzip

four-stretcher ambulance – 4-Tragen-Krankentransportwagen (4KTW)

four-wheel drive – Allradantrieb

fracture – Bruch (Material, Knochen)

fragile – zerbrechlich, brüchig, fragil

fragrance – Duft, Geruch, Wohlgeruch

frame – Rahmen, Gestell; auch: Zarge (Tür, Fenster)

frame construction – Rahmenkonstruktion, Skelettbauweise

framework – Rahmen, Gerippe, Fachwerk, Gerüst

framework recommendation – Rahmenempfehlung

Frankfurt shovel – Frankfurter Schaufel

freeway – Autobahn

freezing point – Gefrierpunkt

freight – Fracht, Frachtgut, Ladung

frequency – Frequenz

frequency adjustment – Frequenzeinstellung

frequency change – Frequenzänderung

frequency modulation (FM) – Frequenzmodulation (FM)

fresh – frisch, neu, ungebraucht

fresh air – Frischluft

fresh air supply – Frischluftzufuhr

fresh air vent – Zuluftöffnung, Frischluftöffnung

friction – Reibung (mechanische)

friction brake – Reibungsbremse

friction energy – Reibungsenergie

friction loss – Reibungsverlust

frictional heat – Reibungswärme

frictional resistance – Reibungswiderstand

fright reaction – Schreckreaktion, Angstreaktion

front monitor – Frontmonitor

front mounted pump – Vorbaupumpe

front passenger airbag – Beifahrerairbag

front side – Vorderseite

frost protection agent – Frostschutzmittel

frostbite – Erfrierung

fuel – Brennstoff, Brennmaterial, Kraftstoff, Treibstoff, Benzin

fuel assembly – Brennelement

fuel canister – Kraftstoffkanister

fuel cell – Brennstoffzelle

fuel cooling station – Abklingbecken (Kerntechnik)

fuel finding paste – Öl- und Kraftstoff-Nachweispaste

fuel rods – Brennstäbe (Kerntechnik)

fuel-lean combustion – brennstoffarme Verbrennung

fuel-rich combustion – brennstoffreiche Verbrennung

fugacity – Flüchtigkeit

full access – Vollzugriff

full alarm – Vollalarm

full arrest – Herz-Kreislauf-Stillstand

full face mask – Vollmaske (Atemschutz)

full facepiece respirator – Atemschutzvollmaske

full jet – Vollstrahl

full jet pipe – Vollstrahlrohr

full load – Volllast

full monitoring – Vollüberwachung

full penetration – vollständiger Durchbruch

full protective clothing – Vollschutz

full storey – Vollgeschoss

full-body fall arrest harness – Absturzsicherungsgeschirr

full-body harness – Komplettgurt, Kombigurt

full-duplex – Vollduplex

full-time firefighter – Berufsfeuerwehrmann/-frau

fully developed fire – Vollbrand

fume – Rauch, Dampf, Ausdünstung, Dunst

fume extraction – Rauchabsaugung

fume gas layer – Rauchgasschicht

fume hood – Abzugshaube, Abzug

fume poisoning – Rauchvergiftung, Rauchgasintoxikation

functional control – Funktionskontrolle

functional endurance – Funktionserhalt

functional principle – Funktionsprinzip

functional supervision – Fachaufsicht

functional test – Funktionsprüfung

fundamental particle – Elementarteilchen

fungi – Pilze

funnel – Trichter

furan compounds – Furane, Furan-Verbindungen

furanics – Furane, Furan-Verbindungen

furnace – Feuerungsanlage

fused salt electrolysis – Schmelzflusselektrolyse

fusible link – Schmelzverbindung, Schmelzlot-Verbindung

fusible solder – Schmelzlot

fusible-link sprinkler – Schmelzlotsprinkler

fusion bomb – Wasserstoffbombe

fusion centre of the German BOS-Organisations – Gemeinsames Melde- und Lagezentrum (GMLZ)

G

gable – Giebel

gable wall – Giebelwand

gaiter – Gamasche

gale – Starkwind, Sturm

gamma decay – Gamma-Zerfall

gamma disintegration – Gamma-Zerfall

gamma dose-rate constant – Gamma-Dosisleistungskonstante

gamma emission – Aussendung von Gamma-Strahlung

gamma emitter – Gamma-Strahler

gamma photon – Gamma-Photon

gamma quant – Gamma-Quant

gamma radiation – Gamma-Strahlung

gamma ray source – Gamma-Strahlenquelle

gamma transition – Gamma-Übergang

GAMS-rule (approx..: DBRA-rule: recognize danger, barricade, rescue people, alert special forces) – GAMS-Regel (Gefahr erkennen, Absperren, Menschenrettung durchführen, Spezialkräfte alarmieren)

garage – Garage

garbage dump – Müllkippe, Müllabladeplatz

garden hose – Gartenschlauch

garden pump sprayer – Gartenspritze

gas – Gas

gas alert – Gasalarm

gas burner – Gasbrenner

gas cartridge – Druckgaspatrone, Gaskartusche

gas chain saw – Motorsäge, benzinbetrieben

gas chromatography (GC) – Gaschromatographie (GC)

gas cloud – Gaswolke

gas concentration – Gaskonzentration

gas cylinder – Gasflasche

gas cylinder cart – Gasflaschenwagen

gas cylinder dolly – Gasflaschenwagen

gas cylinder valve – Gasflaschenventil

gas density – Gasdichte

gas detection system (GDS) – Gaswarnlage (GWA)

gas detector – Gasmelder, Gasspürgerät

gas detector bellows – handbetriebene Gasspürpumpe (Faltenbalgpumpe)

gas detector device – Gasspürgerät

gas diffusion – Gasdiffusion

gas diffusion layer (GDL) – Gasdiffusionsschicht (GDS)

gas dispersion – Gasausbreitung, Gasdispersion

gas explosion – Gasexplosion

gas extinguisher – Gaslöscher

gas filter – Gasfilter

gas firing – Gasfeuerung

gas impermeability – Gasundurchlässigkeit

gas jet – Gasstrahl

gas law's – Gasgesetze

gas leak detector – Gasspürgerät (Lecks)

T. Schmiermund, *Fachwörterbuch Feuerwehr und Brandschutz*, https://doi.org/10.1007/978-3-662-64120-0_29

gas lighter – Gasanzünder

gas liquefaction – Gasverflüssigung

gas main – Gasleitung

gas mask – Gasmaske

gas mixture – Gasgemisch

gas permeability – Gasdurchlässigkeit

gas pipe – Gasleitung

gas poisoning – Gasvergiftung

gas propagation – Gasausbreitung

gas sampling tube – Gassammelröhrchen

gas sensor – Gassensor

gas supply – Gasversorgung

gas tank – Gasbehälter

gas testing tube – Gasprüfröhrchen

gas-air-mixture – Gas-Luft-Gemisch

gaseous extinguishing agent – Löschgas

gaseous extinguishing system – Gaslöschanlage

gaseous fire-extinguishing agents – gasförmige Löschmittel

gaseous phase – Gasphase, gasförmige Phase

gaseous state – gasförmiger Zustand, Gaszustand

gaseous substances – gasförmige Stoffe

gases – Gase

gasification – Vergasung

gasify – vergasen

gasket – Dichtmanschette

gasoline – Benzin

gasoline canister – Benzinkanister, Kraftstoffkanister

gasoline finding paste – Öl- und Kraftstoff-Nachweispaste

gas-proof – gasdicht

gas-tight – gasdicht

gastrointestinal anthrax – Darmmilzbrand

gatekeeper – Pförtner

gathering information – Informationsgewinnung

gauging – Messen; auch: Eichung

gear oil – Getriebeöl

gearbox oil – Getriebeöl

Geiger-Muller counter – Geigerzähler, Geiger-Müller-Zähler/Zählrohr

gel battery – Gelakkumulator

general anaesthesia – Vollnarkose

general conditions – allgemeine Bedingungen, Rahmenbedingungen

general duty roster – allgemeiner Dienstplan, Rahmendienstplan

general gas constant – allgemeine Gaskonstante

general gas equation – allgemeine Gasgleichung

general order – Gesamtbefehl

general purpose adhesive – Alleskleber

general terms – allgemeine Begriffe

generally – grundsätzlich

generalship – Strategie

generation – Erzeugung, Entwicklung; auch: Generation

generator – Stromerzeuger

genetic damage – genetische Schäden

genetic engineering (GE) – Gentechnik

Genetic Engineering Safety Ordinance – Gentechniksicherheitsverordnung (GenTSV)

genetically modified organism (GMO) – gentechnisch veränderte Organismen (GVO)

geodetic discharge height – geodätische Förderhöhe

geodetic suction height – geodätische Saughöhe

germ – Keim

German Civil Code – Bürgerliches Gesetzbuch (BGB)

German food coupling – Milchrohrverschraubung

German Institute for Standardisation – DIN (Deutsches Institut für Normung e. V.)

German Radiation Protection Ordinance – Strahlenschutzverordnung (StrlSchV)

German Red Cross – Deutsches Rotes Kreuz (DRK)

German Standard – DIN-Norm

ghost flames – Geisterflammen

girder – Balken, Träger, Tragbalken

girth – Umfang; auch: Gurt, Sattelgurt

girth hitch – doppelter Ankerstich

give a shot – Spritze verabreichen

give an injection – Spritze verabreichen

give first aid – Erste Hilfe leisten

give out – aufgeben, Versagen (Bauteile)

give out rope – Seil ausgeben

glacial acetic acid – Eisessig, reine (100 %ige) Essigsäure

gland seal – Stopfbuchse, Stopfbuchs(ab)dichtung

glanders – Rotz

glass block – Glasbaustein

glass breaker – Scheibenzertrümmerer

glass fibre reinforced plastic (GRP) – glasfaserverstärkter Kunststoff (GFK)

glass-bulb sprinkler – Glasfasssprinkler

glasses – Brille

glasspaper – Sandpapier, Schmirgelpapier

glass-wool – Glaswolle

glazing – Verglasung (Baukunde)

gleam – glänzen, schimmern, funkeln, glimmen

global positioning system (GPS) – globales Positionssystem

gloves – Handschuhe

glow – Glut, Glühen, Glimmen, glühen, glimmen, leuchten

glow temperature – Glimmtemperatur

glowing – Glühen

glowing fire – Glutbrand

glue – Klebstoff, Leim

goal – Tor, Ziel, Treffer; auch: Zweck

goal-orientated – zielorientiert

goggles – Schutzbrille

Gone On Arrival (GOA) – Anrufer nicht auffindbar

goniometer – Winkelmesser

governor – Landrat

grade – Klasse, Note, Sorte, Stufe; auch: Grad, Dienstgrad

grade crossing – Bahnübergang

grading system – Bewertungssystem (z. B. bei Prüfungen)

granules – Granulat

graphite – Graphit

grass fire – Grasbrand

grassland – Wiese

gravel – Kies

gravitational force – Erdanziehungskraft, Schwerkraft

grease – Schmierfett, Fett; einfetten, einschmieren

grease fire – Fettbrand

grease gun – Fettpresse

grease nipple – Schmiernippel

great tattoo – großer Zapfenstreich

grid – Raster

grinder – Trennschleifer

grinding sparks – Schleiffunken

groove – Nut (Nut-Feder-Verbindung)

gross calorific value – Brennwert, Bruttoheizwert

gross heat of combustion – Brennwert, Bruttoverbrennungswärme

gross weight – Gesamtgewicht, Bruttogewicht

grossly negligent – grob fahrlässig

ground – Grund, Boden, Erdboden, Gelände, Gebiet; auch: Grundlage

ground anchor – Erdanker

ground fault interrupter (GFI) – Fehlerstromschutzschalter (FI), FI-Schutzschalter

ground floor – Erdgeschoss

ground hydrant – Unterflurhydrant

ground hydrant key – Unterflurhydrantenschlüssel

ground plan – Lageplan, Grundriss

ground rod – Erdungsstange

ground wire – Erdungskabel

grounding – Erdung (Elektrizität)

grounding accessories – Erdungszubehör

ground-state atom – Atom im Grundzustand

group (team-of-nine) – Gruppe (Gr)

group alarm – Gruppenalarm

group alerting – Gruppenalarmierung

group leader – Gruppenführer (GrFü)

group of detectors – Meldergruppe

guard of honour – Ehrenwache

guidance – Orientierungshilfe, Beratung, Leitung, Anleitung, Lenkung, Führung, Unterweisung

guide line – Rückweg-Führungsleine (Leinensicherungssystem, Atemschutzeinsatz)

guideline – Richtlinie (RL)

gully – Gully

gully cover – Gullydeckel

gumboots – Gummistiefel

gurney – fahrbares Krankenbett, mobile Krankenliege, Patiententransportwagen

gutter – Rinne, Dachrinne, Regenrinne

H

Haber's lethality product – Habersches Tödlichkeitsprodukt

haematoma – Bluterguss, Hämatom

haemorrhagic fever – hämorrhagisches Fieber

Hague Convention – Haager Konvention

Hague Land Warfare Convention – Haager Landkriegsordnung (HLKO)

half hitch – Halbschlag

half mask – Halbmaske

half-duplex – Halbduplex

half-life period – Halbwertszeit (HWZ)

half-time radioactive period – Halbwertszeit (HWZ)

half-value layer (HVL) – Halbwertschicht (HWS) bzw. Halbwertschichtdicke (HWS)

hallucinogens – Halluzinogene

haloalkanes – Halogenalkane

halogenated hydrocarbons – Halogenkohlenwasserstoffe

halon extinguisher – Halonlöscher

halons – Halone

hammer – Hammer

hammer handle – Hammerstiel

hammer head – Hammerkopf

hand broom – Handfeger

hand brush – Handbürste

hand lamp – Handlampe

hand motion – Handbewegung

hand pallet truck – Gabelhubwagen

hand searchlight – Handscheinwerfer

hand signals – Handzeichen

hand suction pump – Handabsaugpumpe

hand truck – Sackkarre, Handwagen, Handkarren

hand-controlled branch pipe – handgeführtes Strahlrohr/Mehrzweckstrahlrohr

handcuff knot – Fesselknoten

hand-held fire extinguisher – Handfeuerlöscher

handheld transceiver (HT) – Handfunkgerät (HFuG), Handsprechfunkgerät

handle – Griff

handle for valves – Handrad für ein Ventil, „Ventilgriff"

handling – Handhabung, Umgang

hand-operated siren – Handsirene

hand-operated winch – handbetriebene Winde

handouts – Handzettel, Merkblätter, Flugblätter

handrail – Handlauf

handsaw – Fuchsschwanz

handwheel – Handrad

hangman's knot – Henkersknoten

harbour fire brigade – Hafenfeuerwehr

hard soap – Kernseife

hardly combustible – schwer brennbar

hardly flammable – schwer entzündlich

hardness of water – Wasserhärte

hardwood handle – Hartholzgriff

T. Schmiermund, *Fachwörterbuch Feuerwehr und Brandschutz*, https://doi.org/10.1007/978-3-662-64120-0_30

harmful – gesundheitsschädlich, schädlich, ungesund

harmful substance – Schadstoff

harmless – ungefährlich, unschädlich, harmlos

harvest – Ernte

hatchet – Beil

hatchet bag – Beiltasche

hatchet of honour – Ehrenbeil

HAUS-rule (approx..: ODBS-rule: obstacles – distances – base – safety) – HAUS-Regel (Hindernisse – Abstände – Untergrund – Sicherheit)

have an injection – Spritze erhalten

hay – Heu

hay fork – Heugabel

hay self-ignition – Heuselbstentzündung

hazard – Gefahr; auch: Gefährdung

hazard analysis (HA) – Gefährdungsanalyse (GA)

hazard area – Gefahrenbereich

hazard bonus – Gefahrenzulage

hazard characteristics – Gefahreneigenschaften

hazard classes – Gefahrklassen, Gefahrgutklassen, Klassen gefährlicher Güter

hazard control – Gefahrenabwehr

hazard control centre – Gefahrenmeldezentrale (GMZ)

hazard diamond – Gefahrendiamant

hazard divisions – Untergruppen der Gefahrklassen

hazard group – Gefahrengruppe

hazard identification number – Nummer zur Kennzeichnung der Gefahr, Kemler-Zahl

hazard information (HI) – Gefahreninformation, Gefahrenhinweise

hazard labelling – Gefahrstoffkennzeichnung

hazard labels – Gefahrzettel, Gefahrenschilder

hazard level – Gefahrenstufe, Gefährdungsgrad

hazard number – Gefahrnummer

hazard potential – Gefährdungspotential, Gefahrenpotential

hazard source – Gefahrenquelle

hazard symbol – Gefahrensymbol

hazard triangle – Warndreieck

hazard warning panel (on trucks) – Warntafel (Lkw-Kennzeichnung)

hazard warning triangle – Warndreieck

hazard zone – Gefahrenbereich

hazardous – gefährlich, riskant, unsicher

hazardous characteristics – gefährliche Eigenschaften, Gefährlichkeitsmerkmale

hazardous event – gefährliches Ereignis

hazardous goods – gefährliche Güter

Hazardous Goods Ordinance – Road – Gefahrgutverordnung Straße (GGVS)

Hazardous Goods Ordinance – Road, Rail – Gefahrgutverordnung Straße, Eisenbahn (GGVSE)

hazardous materials (hazmat) – gefährliche Stoffe und Güter (GSG)

hazardous materials type „NBC" (nuclear, biological, chemical) – ABC-Gefahrstoffe

hazardous substances – Gefahrstoffe

hazardous substances combat vehicle – Gerätewagen Gefahrgut (GW-G)

Hazardous Substances Ordinance – Gefahrstoffverordnung (GefStoffV)

hazardous waste – Sondermüll, gefährlicher Abfall

hazards – Gefahren

hazmat (= hazardous materials) – gefährliche Stoffe und Güter (GSG)

hazmat accident – Gefahrgutunfall

hazmat class – Gefahrengruppe

hazmat operation – Gefahrguteinsatz

hazmat quantity – Menge an Gefahrstoff, Gefahrstoffmenge

hazmat suit – Chemikalienschutzanzug (CSA)

hazmat transfer pump – Gefahrgutumfüllpumpe (GUP)

hazmat transportation – Gefahrguttransport

hazmat truck – Gefahrguttransporter, Gefahrgut-Lkw

H-bomb – Wasserstoffbombe

head harness – Kopfbebänderung

head lamp – Stirnlampe

head of operations – Einsatzleiter (ELtr); auch: Betriebsleiter

head-tilt and chin-lift manoeuvre (HTCL-manoeuvre) – lebensrettender Handgriff = Überstrecken des Kopfes (Beatmung)

headtorch – Stirnlampe

health – Gesundheit

health care – Gesundheitsvorsorge

health examination – Gesundheitsuntersuchung

health hazard – Gesundheitsgefährdung

health protection – Gesundheitsschutz

health risk – Gesundheitsgefahr

hearing aid – Hörgerät

hearing protection – Gehörschutz

hearing protection plugs – Gehörschutzstöpsel

heart – Herz

heart attack – Herzinfarkt

heart massage – Herzmassage

heart rate (HR) – Pulsfrequenz, Herzfrequenz (HF)

heat – Hitze, Wärme

heat absorption – Wärmeabsorption

heat accumulation – Wärmestau

heat balance – Wärmebilanz

heat budget – Wärmebilanz

heat build-up – Wärmestau

heat capacity – Wärmekapazität

heat conduction – Wärmeleitung

heat conductor – Wärmeleiter

heat consumption – Wärmeverbrauch

heat convection – Wärmeströmung, Konvektion, Wärmemitführung

heat detector – Wärmemelder, Thermomelder

heat development – Hitzeentwicklung

heat dissipation – Wärmeableitung

heat exchange – Wärmeaustausch

heat exchanger – Wärmetauscher

heat exhaust venting system – Wärmeabzugsanlage

heat exposure – Hitzeeinwirkung

heat flow – Wärmestrom, Wärmefluss

heat flow density – Wärmestromdichte

heat flow rate – Wärmestrom, Wärmefluss

heat flux – Wärmestrom, Wärmefluss

heat generation – Wärmeentwicklung, Wärmeerzeugung

heat gun – Heißluftpistole

heat ignition – Wärmezündung

heat input – Wärmeeintrag

heat insulating foil – Rettungsdecke (Folie, Gold/Silber)

heat insulation – Wärmedämmung, Wärmeisolierung

heat loss – Wärmeverlust

heat of activation – Aktivierungswärme

heat of combustion – Verbrennungswärme

heat of condensation – Kondensationswärme

heat of fire – Brandwärme

heat of formation – Bildungsenergie, Bildungswärme

heat of gasification – Vergasungswärme, Verdampfungswärme

heat of melting – Schmelzwärme

heat of neutralization – Neutralisationswärme

heat of reaction – Reaktionswärme

heat of sublimation – Sublimationswärme

heat of vaporization – Verdampfungswärme

heat outlet – Wärmeabgabe, Wärmeableitung

heat output – Heizleistung

heat passage – Wärmedurchgang

heat protection – Wärmeschutz, Hitzeschutz

heat protection gloves – Wärmeschutzhandschuhe

heat protective clothing – Wärmeschutzkleidung

heat protective hood – Wärmeschutzhaube

heat protective suit – Wärmeschutzanzug, Hitzeschutzanzug

heat quantity – Wärmemenge

heat radiation – Wärmestrahlung

heat recovery – Wärmerückgewinnung

heat release – Wärmefreisetzung

heat release rate (HRR) – Wärmefreisetzungsrate, Brandleistung

heat removal – Wärmeabfuhr

heat resistance – Wärmebeständigkeit, Hitzeverträglichkeit

heat resistant plastics – wärmebeständige Kunststoffe

heat sealer – Folienschweißgerät

heat sensor – Wärmefühler

heat shock – Hitzeschock

heat spread – Wärmeausbreitung

heat stress – thermische Beanspruchung, „Hitzestress"

heat stroke – Hitzschlag

heat supply – Wärmeversorgung

heat transfer – Wärmetransport, Wärmeübergang, Wärmeübertragung, Wärmedurchgang

heat transfer medium – Wärmeträger, Wärmeüberträger

heat transition – Wärmeübergang, Wärmeübertragung

heat transmission – Wärmeübergang, Wärmeübertragung, Wärmedurchgang

heat transmission coefficient – Wärmedurchgangskoeffizient

heating – Heizung; auch: Erwärmung

heating equipment – Heizgerät

heating installation – Heizungsanlage

heating oil – Heizöl

heating value – Heizwert

heat-insulation foil – Wärmedämmfolie

heat-transmission agent – Wärmeüberträger

heavy current – Starkstrom

heavy gasoline – Schwerbenzin

heavy hydrogen – Deuterium (^{2}H, D), schwerer Wasserstoff

heavy water reactor (HWR) – Schwerwasserreaktor

heavy-gas cloud – Schwergaswolke

heavy-gas effect – Schwergaseffekt

heavy-gas propagation – Schwergasausbreitung

hedge – Hecke

heel – Ferse

heel a ladder – aufrichten einer Leiter

Heidelberger extension – Heidelberger Verlängerung

height difference – Höhenunterschied, Höhendifferenz

height of building – Gebäudehöhe

height of fall – Fallhöhe

height of jet – Wurfhöhe (z. B. Wasserstrahl)

height rescue – Höhenrettung

height rescuer – Höhenretter

helicopter – Hubschrauber

helicopter bucket – Löschwasser-Außenlastbehälter

helium core – Heliumkern

helmet – Helm

helmet lamp – Helmlampe

helmet marking – Helmkennzeichnung

helmet visor – Helmvisier

Helmets off! – Helm ab!

help – Hilfe

help period – Hilfsfrist

helper – Helfer

helve – Griff (eines Werkzeugs)

hepatitis A vaccination (HAV vaccination) – Impfung gegen Hepatitis A (HAV-Impfung)

hepatitis B vaccination (HBV vaccination) – Impfung gegen Hepatitis B (HBV-Impfung)

herbicide – Unkrautvernichtungsmittel, Herbizid

Hess theorem – Hess'scher Satz

heterogeneous catalysis – heterogene Katalyse

heterogeneous inhibition – heterogene Inhibition

heterogenous mixture – Gemenge, heterogenes Gemisch

hex key – Innensechskantschlüssel, Inbusschlüssel

hex wrench – Innensechskantschlüssel, Inbusschlüssel

hierarchy – Hierarchie, Rangordnung, Rangfolge

high active waste (HAW) – hochradioaktiver Abfall

high command – Oberkommando, Führungsstab

high expansion foam (HX) – Leichtschaum

high flammability – Leichtentflammbarkeit

high light torch – Starklichtfackel

high pressure – Hochdruck

high radiation area – strahlungsgefährdeter Bereich

high voltage (HV) – Hochspannung

high voltage cable – Hochspannungsleitung

high voltage distribution – Hochspannungsverteilung

high voltage system – Starkstromanlage

high-altitude rescue – Höhenrettung

high-altitude rescuer – Höhenretter

high-efficiency particulate absorbing (air) filter (HEPA filter) – Schwebstofffilter (SSF)

high-level water tank – Wasserhochbehälter

highly combustible – leichtbrennbar

highly concentrated – hochkonzentriert

highly enriched uranium – hochangereichertes Uran

highly flammable – leichtentflammbar, leichtentzündbar

highly reactive – reaktionsfreudig, hochreaktiv

highly volatile – leicht flüchtig

high-pressure cleaner – Hochdruckreiniger

high-pressure hose – Hochdruckschlauch

high-pressure line – Hochdruckleitung

high-pressure pump – Hochdruckpumpe

high-rise building fire – Hochhausbrand

high-visibility clothing – Warnkleidung

high-visibility vest (HV-vest) – Warnweste

hip roof – Walmdach

historical building – Baudenkmal

historical monument – Baudenkmal

hitches – Stiche (Knoten)

hoisting gears – Hebezeuge

holding basin – Abklingbecken (Kerntechnik)

holding fixture – Halterung, Haltevorrichtung

hole punch – Locheisen

hollow – hohl, Hohlraum, Mulde

hollow space – Hohlraum

home fire – Wohnungsbrand, Wohnhausbrand, Hausbrand

homeless shelter – Obdachlosenheim

homogeneous catalysis – homogene Katalyse

homogenous inhibition – homogene Inhibition

homogenous mixture – homogenes Gemisch

homopolar bond – homöopolare Bindung, Atombindung

honorary – ehrenamtlich

honorary division – Ehrenabteilung

honorary member – Ehrenmitglied

honorary post – Ehrenamt

honourable discharge – ehrenhafter Abschied

hood – Abzugshaube, Abzug; auch: Kapuze, Haube

hook – Haken

hook belt – Hakengurt

hook ladder – Hakenleiter

hook wrench – Hakenschlüssel

hooter – Hupe

hopeless situation – ausweglose Situation

horn – Hupe

horsepower (HP) – Pferdestärke (PS)

hose – Schlauch

hose basket – Schlauchtragekorb, Schlauchkorb

hose bib – Wandwasseranschluss

hose burst – Schlauchplatzer

hose carrier basket – Schlauchtragekorb, Schlauchkorb

hose clamp – Schlauchschelle

hose connection – Schlauchverbindung

hose couplings – Schlauchkupplungen

hose crack – Schlauchriss, Schlauchplatzer

hose crew – Schlauchtrupp (S-Tr)

hose crew leader – Schlauchtruppführer, -in (S-TrFü, S-TrFü'in)

hose crew member – Schlauchtruppmann, -frau (S-TrM, S-TrFr)

hose diameter – Schlauchdurchmesser

hose drying tower – Schlauchturm

hose fitting – Feuerlöscharmatur

hose gaiter – Schlauchbinde

hose laying unit – Abrollbehälter Schlauch (AB-Schlauch)

hose line – Schlauchleitung

hose operator – Strahlrohrführer

hose package – Schlauchpaket

hose pod – Abrollbehälter Schlauch (AB-Schlauch)

hose ramp – Schlauchbrücke

hose reel – Schlauchhaspel

hose reel cabinet – Wandhydrant

hose replacement – Schlauchwechsel

hose squadron – Schlauchtrupp (S-Tr)

hose squadron leader – Schlauchtruppführer, -in (S-TrFü, S-TrFü'in)

hose squadron member – Schlauchtruppmann, -frau (S-TrM, S-TrFr)

hose strap – Schlauchhalter

hose support – Stützkrümmer

hose team – Schlauchtrupp (S-Tr)

hose team leader – Schlauchtruppführer, -in (S-TrFü, S-TrFü'in)

hose team member – Schlauchtruppmann, -frau (S-TrM, S-TrFr)

hose tower – Schlauchturm

hose trailer – Schlauchanhänger

hose-holder – Schlauchhalter

hose-layer (HL) – Schlauchwagen (SW)

hospital – Krankenhaus

hot fire training system – Brandsimulationsanlage

hot spots – Glutnester

hot surface – heiße Oberfläche

house connection – Hausanschluss

house connection box – Hausanschlusskasten

house fire – Wohnhausbrand, Hausbrand

house service connection – Hausanschlusskasten

house-building permission – Baugenehmigung

household – Haushalt

household waste – Hausmüll

housing – Gehäuse; auch: Unterkunft, Behausung

human dignity – Menschenwürde

human error – menschliches Fehlverhalten

human rescue – Menschenrettung

human right's – Menschenrechte

humidify – anfeuchten, befeuchten

hurricane lamp – Sturmlaterne

hurt – Schmerz, Verletzung

hut – Hütte, Baracke, Bude

hydrant – Hydrant (H)

hydrant cover – Hydrantendeckel

hydrant sign – Hydrantenschild

hydrant wrench – Hydrantenschlüssel

hydraulic cutter – hydraulische Rettungsschere

hydraulic jack – hydraulischer Wagenheber

hydraulic line – Hydraulikleitung

hydraulic liquid – Hydraulikflüssigkeit

hydraulic oil – Hydrauliköl

hydraulic platform – Hubsteiger, Gelenkmast (GM)

hydraulic power unit – Hydraulikaggregat

hydraulic rescue apparatus – hydraulisches Rettungsgerät

hydraulic rescue cutter – hydraulische Rettungsschere

hydraulic rescue cylinder – hydraulischer Rettungszylinder

hydraulic rescue device – hydraulisches Rettungsgerät, Rettungssatz

hydraulic rescue spreader – hydraulischer Spreizer/Rettungsspreizer

hydraulics pump – Hydraulikpumpe

hydrocarbons – Kohlenwasserstoffe

hydrochloric acid – Salzsäure

hydrochlorofluorocarbons (HCFC) – Fluorchlorkohlenwasserstoffe, teilhalogeniert (H-FCKW)

hydrofluorobromocarbond (HFBC) – Fluor-Brom-Kohlenwasserstoffe, teilhalogeniert (HFBKW)

hydrofluorocarbons (HFC) – Fluorkohlenwasserstoffe, teilhalogeniert (H-FKW)

hydrogen – Wasserstoff

hydrogen bridge linkage – Wasserstoffbrückenbindung

hydrogen chloride – Chlorwasserstoff

hydrogen cyanide – Blausäure

hydrogen sulphide – Schwefelwasserstoff

hydrogenation – Hydrierung

hydrophilic – hydrophil

hydrophobically – hydrophob

hydrophobicity – Hydrophobierung (Ergebnis)

hydrophobisation – Hydrophobierung (Vorgang)

hydro-shield – Hydroschild

hydrostatic pressure – hydrostatischer Druck

hygiene – Hygiene

hypertension – Bluthochdruck

hypothermia – Unterkühlung

hypovolemic shock – Volumenmangelschock

I

ice accident – Eisunfall

ice load – Eislast

ice rescue sled – Eisrettungsschlitten

ice sledge – Eisschlitten, Eisrettungsschlitten

icon – Bildzeichen

ideal gas – ideales Gas

ideal gas law – allgemeine Gasgleichung

ideal gases – ideale Gase

identical – identisch

identification limit – Nachweisgrenze

ignitability – Zündfähigkeit, Entzündlichkeit

ignitable – entzündlich, zündfähig

ignite – entzünden, zünden, anzünden

ignition – Zündung, Entzündung, Anzünden, Entzünden

ignition characteristics – Zündeigenschaften, Zündverhalten, Zündcharakteristik

ignition danger – Zündgefahr

ignition delay time – Zündverzugszeit

ignition energy – Zündenergie

ignition flame – Zündflamme

ignition lag – Zündverzug

ignition limit pressure – Zündgrenzdruck

ignition phase – Zündphase

ignition point – Zündpunkt

ignition probability – Zündwahrscheinlichkeit, Entzündungswahrscheinlichkeit

ignition protection type – Zündschutzart

ignition range – Zündbereich

ignition source – Zündquelle

ignition spark – Zündfunke

ignition temperature – Zündtemperatur, Entzündungstemperatur

ignition time – Zündzeit

ill person – erkrankte Person, kranke Person

illumination – Beleuchtung

immediate alarm – sofortiger Alarm

immediate information – Sofortinformation, sofortige Information

immediate report – Sofortmeldung, sofortiger Bericht

immediate rescue vehicle – Vorausrüstwagen (VRW)

immediate treatment (IT) – Sofortbehandlung

immediately – unverzüglich, sofort, direkt, gleich

imminent danger to life and health – unmittelbare Gefahr für Leben und Gesundheit

imminent danger to life and health (IDLH) – unmittelbare Gefahr für Leben und Gesundheit

immiscible – unvermischbar

immission – Immission

immission control – Immissionsschutz

immobilization – Immobilisation, Ruhigstellung

immobilize – immobilisieren

T. Schmiermund, *Fachwörterbuch Feuerwehr und Brandschutz*, https://doi.org/10.1007/978-3-662-64120-0_31

immovable – unbeweglich, fest, ortsfest

immune – immun

immunity – Immunität

immunization – Immunisierung, Impfung

impact force – Fangstoß (Höhenrettung)

impact sparks – Schlagfunken

impact wave – Stoßwelle

impermeable – undurchlässig

implosion – Implosion

impregnating – Imprägnieren

impregnation – Imprägnierung

improvement – Verbesserung

impure – unrein, verunreinigt

in case of doubt – im Zweifelsfall

in groups – gruppenweise

inactive – inaktiv

incandescent – weißglühend

incapacitating concentration-time product – kampfunfähig machendes Konzentrations-Zeit-Produkt

incapacitation – Handlungsunfähigkeit

incendiary – Brandstifter; auch: Brand

incendiary agent – Brandmittel, Brandsatz

incendiary composition – Brandsatz

incendiary material – Brandmaterial, Brandsatz

incidence – Inzidenz

incident – Störfall, Ereignis, Zwischenfall, Ereignisfall; auch: Schadenslage

incident command – Einsatzleitung (EL) (Funktion)

incident command organisation – Einsatzleitung (Begriff)

incident commander – Einsatzleiter (ELtr)

incident commander fire service – Einsatzleiter Feuerwehr

incident planning – Einsatzplan

incident room – Einsatzzentrale

incipient fire – Entstehungsbrand, beginnendes Feuer

incision – Schnittwunde

inclinometer – Neigungsmesser

incorporation – Inkorporation, Aufnahme in den Körper

increased radioactivity – erhöhte Radioaktivität

incubation – Inkubation

incubation period (ICP) – Inkubationszeit (IKZ)

indemnity – Schadensersatz, Entschädigung

independent water supply – unabhängige Löschwasserversorgung

indicating device – Anzeigeeinrichtung

indication sign – Hinweisschild, Hinweiszeichen

indicator lamp – Kontrollleuchte, Anzeigelampe

indicator tube – Prüfröhrchen

indirect attack – indirekter (Lösch)Angriff

indirect cooling – indirekte Kühlung

indirect fire attack – indirekter Löschangriff

individual dosimeter – Personendosimeter

individual order – Einzelbefehl

indoor air quality (IAQ) – Raumluftqualität

induction – Einberufung

induction paper – Einberufungsbescheid

induction time – Induktionszeit

industrial buildings – Industriebauten

industrial chemicals – Industriechemikalien

industrial water – Brauchwasser

industry – Gewerbe, Industrie

inert gas – Schutzgas, Inertgas

inert gases – inerte Gase

inert liquids – inerte Flüssigkeiten

inerting – Inertisierung

inexhaustible water supply for fire-fighting – unerschöpfliche Löschwasserversorgung

infection control suit – Infektionsschutzanzug

infection protective suit – Infektionsschutzanzug

infectiosity – Infektiosität

infectious – ansteckend, infektiös

infectious disease – Infektionskrankheit

infectious dose (ID) – Infektionsdosis (ID)

infectious substance – ansteckungsgefährlicher Stoff

infectious waste – infektiöser Abfall

infectivity – Infektiosität

infestation – Verseuchung, Befall (durch Mikroorganismen)

infested – verseucht, befallen (durch Mikroorganismen)

inflammability – Entflammbarkeit, Entzündbarkeit

inflammability limits – Entflammbarkeitsgrenzen

inflammable – entflammbar, brennbar, entzündbar

inflammation – Entzündung, Entflammung; auch: Entzündung (medizinisch)

inflammation point – Brennpunkt

inflammation temperature – Entflammungstemperatur

inflatable life raft – Rettungsinsel

influence – Einfluss

influence of temperature – Temperatureinfluss

influenza vaccination – Impfung gegen Grippe

information and communication – Information und Kommunikation (IuK)

information and communication technology (ICT) – Informations- und Kommunikationstechnik

information and control centre (ICC) – Leitstelle

information assessment – Informationsverarbeitung

information sign – Hinweisschild

informative message – Rückmeldung

infrared detector – Infrarotmelder

infrared radiation (IR-radiation) – Infrarotstrahlung (IR-Strahlung)

infusion – Infusion

infusion bag – Infusionsbeutel

infusion bottle – Infusionsflasche

infusion catheter – Infusionskatheter

infusion device – Infusionsvorrichtung

infusion period – Infusionsdauer

infusion pump – Infusionspumpe

infusion rate – Infusionsgeschwindigkeit

infusion set – Infusionsbesteck

infusion stand – Infusionsständer

ingest – einnehmen, aufnehmen

ingestion – Einnahme, Aufnahme, Nahrungsaufnahme

inhalable – atembar

inhalation anthrax – Lungenmilzbrand

inhalation of radon – Radoninhalation

inhalation resistance – Einatemwiderstand

inhalation valve – Einatemventil

inhale – einatmen

inhaled air – Einatemluft

inhibiting agent – Hemmstoff, Inhibitor

inhibiting effect – antikatalytischer (Lösch) Effekt, inhibitorischer (Lösch)Effekt

inhibition – Inhibition

inhibition effect – Inhibitionseffekt (Löscheffekt)

inhibitor – Inhibitor, Antikatalysator

initial actions – Erstmaßnahmen

initial attack – Erstangriff

initial fire – Entstehungsbrand, anfängliches Feuer

initial phase – Entstehungsphase

initial phase of fire – Anfangsphase eines Brandes

initiator – Initialzünder

injection – Injektion; auch: Impfung

injectional anthrax – Injektionsmilzbrand

injector pump – Injektorpumpe

injury – Verletzung, Schaden

inlet – Eingang

inlet hose – Eingangsschlauch

inlet pressure – Eingangsdruck (Pumpe)

inline inductor – Zumischer

inner mask – Innenmaske

inner room – gefangener Raum

inoperable – nicht betriebsbereit

inoperative – außer Betrieb, funktionsunfähig, wirkungslos

inorganic – anorganisch

inorganic chemistry – anorganische Chemie

inpatient – stationär, stationärer Patient

input pressure – Eingangsdruck (Pumpe)

inquiry of fire cause – Brandursachenermittlung

in-rack sprinkler – Regalsprinkler

insecticide – Insektenvernichtungsmittel

insensitive to moisture – feuchtigkeitsunempfindlich

inside fire – Innenbrand

insignia – Abzeichen, Rangabzeichen, Dienstgradabzeichen

insolating breathing apparatus (BA) – Isoliergerät

insolation – Sonneneinstrahlung

insolation encephalopathy – Sonnenstich

insoluble in water – wasserunlöslich

inspect – prüfen, kontrollieren, überprüfen, inspizieren

inspection – Prüfung, Kontrolle, Überprüfung, Inspektion

inspection duty – Inspektionspflicht, Kontrollpflicht

inspection plate – Prüfplakette

installation – Einbau, Montage, Installation, Anschluss

installation for delivery of extinguishant – Anlage zur Löschmittelförderung

installation for water removal – Entwässerungsvorrichtung

instant information – Sofortinformation, sofortige Information

instant report – Sofortmeldung, sofortiger Bericht

instant-picture-camera – Sofortbildkamera

institutions and organisations in the field of safety and security – Behörden und Organisationen mit Sicherheitsaufgaben (BOS)

instruction – Anweisung, Befehl, Instruktion, Ausbildung, Unterweisung, Belehrung, Unterricht

instruction for use – Gebrauchsanweisung, Gebrauchsanleitung

instruction leaflet – Gebrauchsanweisung

instruction manual – Bedienungsanleitung

instructions – Anleitung, Anweisung, Unterweisung

instructor – Ausbilder

insulated – isoliert

insulating layer – Dämmschicht, Isolierschicht

insulating materials – Dämmstoffe

insulating tape – Isolierband (elektr.)

insulation – Isolierung, Dämmung

insulation layer – Dämmschicht, Isolierschicht

insurable value – versicherbarer Wert, Versicherungswert

insurance – Versicherung

insurance value – Versicherungswert

insured person – Versicherter

insurer – Versicherer

intake – Aufnahme, Einlass

intake socket – Einlassstutzen, Ansaugstutzen

intensity – Intensität

intensity of extinguishing agent supply – Intensität der Löschmittelzufuhr

intensive care helicopter (ICH) – Intensivtransporthubschrauber (ITH)

intent – Absicht, Intention, Vorsatz

intention – Absicht, Intention, Vorsatz

interaction – Wechselwirkung

intercom system – Sprechanlage

interface – Schnittstelle

interim security – Zwischensicherung

interior attack – Innenangriff

intermediate tank – Zwischenbehälter

intermixture – Vermischung, Gemisch

intermolecular attraction – zwischenmolekulare Anziehung

internal checking – Eigenkontrolle, Selbstüberwachung

internal irradiation – innere Bestrahlung

internal thread – Innengewinde; auch: interner Beitrag

internal wall – Innenwand

International Certificate of Vaccination or Prophylaxis (ICVP) – Impfpass, internationaler

International Red Cross (IRC) – Internationales Rotes Kreuz (IRK)

international unit system (SI) – internationales Einheitensystem (SI)

interpersonal area – zwischenmenschlicher Bereich

interrupt – unterbrechen

intertisation – Inertisierung

intervention – Eingriff, Eingreifen, Intervention, Einsatz

intervention planning – Einsatzplan, Alarmplan

intervention preparedness of forces – Einsatzfähigkeit

intervention time – Eingreifzeit

intravenous (i.v.) – intravenös

intravenous bag (i.v. bag) – Infusionsbeutel

intravenous bottle (i.v. bottle) – Infusionsflasche

intravenous bottle holder (i.v. bottle holder) – Infusionsständer

intravenous drip bottle – Infusionsflasche, Infusionstropfflasche

intravenous line (i.v. line) – Infusionsschlauch

intravenous pole (i.v. pole) – Infusionsständer

intrinsic radiation – Eigenstrahlung

intrinsic safety (IS) – Eigensicherheit

intubation – Intubation

intumescent – intumeszierend

intumescent coating – Anstrich, der eine Dämmschicht bildet

inventory – Inventar, Vorräte

inverse-square law – Abstandsgesetz, umgekehrtes (inverses) Gesetz der quadrierten Werte

inversion temperature – Inversionstemperatur

investigation time – Erkundungszeit, Untersuchungszeit

involved forces – eingesetzte Kräfte/Einsatzkräfte

inward opening door – nach innen öffnende Tür

ion – Ion

ion dose – Ionendosis

ion exposure dose – Ionendosis

ionic – ionisch

ionic attraction – Anziehung von Ionen

ionic bond – Ionenbindung

ionisation – Ionisation

ionisation chamber (IC) – Ionisationskammer

ionisation counter – Ionisationszähler

ionisation density – Ionisationsdichte

ionisation detector – Ionisationsmelder

ionisation smoke detector – Ionisationsrauchmelder

ionised atom – ionisiertes Atom

ionising radiation – ionisierende Strahlung („radioaktive“ Strahlung)

ion-mobility spectrometer (IMS) – Ionenmobilitätsspektrometer (IMS)

irradiance – Bestrahlungsintensität

irradiation – Bestrahlung

irradiation dose – Bestrahlungsdosis

irradiation intensity – Bestrahlungsintensität

irregular – unregelmäßig, ungleichmäßig; auch: ungesetzlich, regelwidrig

irritant – Reizstoff, reizend

irritant gases – reizende Gase, Reizgase

irritants – Reizstoffe

irritated skin – gereizte Haut

irritating the eye – augenreizend

irritation – Reiz, Reizung, Irritation

isolated line – Stichleitung

isolation (from a mixture) – Isolation, Abtrennung

isotonic – isotonisch

isotope – Isotop

isotope rule – Isotopenregel

issue of orders – Auftragserteilung, Befehlsgebung

issue orders – Anweisungen erteilen, Aufträge erteilen

Italian hitch – Halbmastwurfsicherung (HMS)

J, K

jack – Hebevorrichtung, Heber, Wagenheber, Klinke

jack knife – Kappmesser

jacking pads, jacks – hydraulische Stützen (Drehleiter, Kran)

jack-lift – Hubwagen

jammed person – eingeklemmte Person

janitor – Pförtner

jaw thrust manoeuvre – Esmarch-Handgriff, Kopf-Kiefer-Handgriff

jaw vice – Schraubstock

Jaws of Life® – hydraulisches Rettungsgerät, Rettungssatz

jet of flame – Stichflamme („Flammenstrahl")

jet pipe mouth piece – Strahlrohrmundstück

jib – Kranausleger

jigsaw – Stichsäge

jimmy – Brechstange

job – Job, Arbeit, Auftrag, Stelle, Tätigkeit

joist hanger – Balkenschuh

Joule-Thomson effect – Joule-Thomson-Effekt

jump start – Starthilfe

jumping blanket – Sprungtuch

jumping cushion – Sprungpolster

jumping sheet – Sprungtuch

junction – Kreuzung, Knotenpunkt, Verbindung, Verbindungsstelle

junior firefighters – Jugendfeuerwehrangehörige

jurisdiction – Zuständigkeit

justify – einstellen, justieren; auch: rechtfertigen

Keep clear! – Abstand halten!; Frei halten!

keep free of smoke – Rauchfreihaltung

Kemler number – Nummer zur Kennzeichnung der Gefahr, Kemler-Zahl

kerma (kinetic energy released per unit mass) – Kerma (kinetische Energie, die je Masseneinheit freigesetzt wird)

kernmantle rope – Kernmantelseil

key – Schlüssel (für Türen)

key cylinder – Schließzylinder, Türzylinder

key-operated switch – Schlüsselschalter

killick hitch – Zimmermannsstich mit Halbschlag

kindling – Anzündholz

kindling temperature – Zündtemperatur

kinetic energy – Bewegungsenergie, kinetische Energie

kink protection – Knickschutz (z. B. Kabel)

kitchen paper towel – Küchenrolle

kitchen towel – Küchenhandtuch

knot of a tie – Krawattenknoten

knots – Knoten

T. Schmiermund, *Fachwörterbuch Feuerwehr und Brandschutz*, https://doi.org/10.1007/978-3-662-64120-0_32

L

labels – Etiketten; auch: Gefahrzettel

laboured breathing – Atemnot

lachrymator – Augenreizstoff, Tränengas

lachrymatory – augenreizend, zu Tränen reizend

lack of air – Luftmangel

lack of oxygen – Sauerstoffmangel

ladder – Leiter (Steiggerät)

ladder belt – Hakengurt

ladder head – Leiterkopf

ladder heel – Leiterfuß

ladder mechanism – Leitergetriebe (Drehleiter)

ladder paws – Fallhaken

ladder shoe – Leiterfuß

ladder stringer – Leiterholm

ladder truck – Leiterfahrzeug

laminar – laminar

laminar flame – laminare Flamme

laminar flow – laminare Strömung

landfill – Mülldeponie, Deponie, Halde

landfill fire – Müllhaldenbrand, Deponiebrand

landfill gas (LFG) – Deponiegas

landing valve – Niederschraubventil

lanyard – Tragegurt; auch: Verbindungsmittel (Absturzsicherung)

large axe – Holzaxt, Fällaxt

large fire – großes Feuer, Großbrand

large scale disaster – Großschadensereignis

large spring hook – Rohrkarabiner

large tank fire tender – Großtanklöschfahrzeug (GTLF)

large tank fire-fighting vehicle – Großtanklöschfahrzeug (GTLF)

lark's head/foot – doppelter Ankerstich

Lassa fever – Lassa-Fieber

latched – eingerastet, verriegeln, einklinken

late sequelae – Spätfolgen

late shift – Spätschicht

latency time – Latenzzeit

latent heat – latente Wärme

latent period – Latenzzeit

lateral recumbent position – stabile Seitenlage (Erste Hilfe)

lath – Latte

lathwork – Lattung

latitude – Ermessensspielraum

laundry – Wäscherei

law – Recht, Gesetz

Law for the Protection against Contagious Disease – Infektionsschutzgesetz (IfSG)

law giver – Gesetzgeber

law maker – Gesetzgeber

law of conservation of energy – Energieerhaltungssatz, Satz von der Erhaltung der Energie

law of partial pressures – Gesetz der Partialdrücke

lawful – rechtmäßig, legal, gesetzlich

T. Schmiermund, *Fachwörterbuch Feuerwehr und Brandschutz*, https://doi.org/10.1007/978-3-662-64120-0_33

lay a hose line – eine Schlauchleitung verlegen

lay helper – Laienhelfer

lay person – Laie

lay rescuer – Laienhelfer, Laienretter

layer – Schicht (Oberfläche)

lead – Blei

lead accumulator – Bleiakkumulator („Autobatterie")

lead apron – Bleischürze

lead climber – Vorsteiger (Höhenrettung)

lead climbing – Vorstieg (Höhenrettung)

lead shielding – Bleiabschirmung

lead storage battery – Bleiakkumulator („Autobatterie")

lead structure – Leitstruktur, Leitsubstanz

lead substance – Leitsubstanz

lead wool – Bleiwolle

lead-acid battery – Bleiakkumulator („Autobatterie")

leader – Führer, Leiter, Anführer

leadership – Führung

leadership style – Führungsstil

leading by orders – Führen durch Befehl

leading fire-fighter (LFf) – Truppführer (TrFü)

leaflets – Merkblätter, Prospekte, Broschüren

leak – Leck

leakage – Leckage

leakage current – Ableitung (von elektr. Strom), Stromverlust, Spannungsverlust

leakage loss – Leckverlust, Verlust durch ein Leck

leakage test – Dichtheitsprüfung

leaky – undicht

leather boots – Lederstiefel

ledge – Leiste, Sims, Fenstersims

legal – gesetzlich, legal, gesetzmäßig, rechtmäßig

legal basis – Rechtsgrundlage, rechtliche/gesetzliche Grundlage

legal entity – juristische Person

legal force – Rechtskraft, rechtliche Handhabe

legal regulations – gesetzliche Regelungen

legal requirement – gesetzliche Bestimmung/Vorschrift

legal supervision – Rechtsaufsicht, rechtliche Aufsicht

legally valid – rechtskräftig, rechtsgültig

legislation – Gesetzgebung

legislator – Gesetzgeber

length of escape route – Fluchtweglänge

length of the ladder – Leiterlänge

less volatile – schwer flüchtig

lesson plan – Lehrplan

lethal concentration – letale Konzentration (LC)

lethal dose (LD) – letale Dosis (LD), tödliche Dosis

lethal exposure dose – letale Expositionsdosis

lethality – Letalität, Tödlichkeit

lethality product – Tödlichkeitsprodukt

level control – Niveauregulierung

level crossing – Bahnübergang

level indicator – Füllstandsanzeige

level of command – Führungsebene

level of intervention – Einsatzstufe

level of knowledge – Kenntnisstand

level of responsibility – Verantwortungsebene

lever – Hebel, Brechstange

lever principle – Hebelgesetz

lever-operated ball valve – Kugelhahn mit Hebelbetätigung

lewisite – Lewisit

liability insurance – Haftpflichtversicherung

liaison officer (LNO) – Verbindungsoffizier (VO)

liaison-personnel – Verbindungspersonal (z. B. zu anderen Dienststellen)

liberating heat – wärmeabgebend, wärmefreisetzend

licence – Erlaubnis, Genehmigung, Zulassung

life belt – Rettungsgürtel; auch: Rettungsring

life hammer – Nothammer, Rettungshammer

life jacket – Rettungsweste, Schwimmweste

life ring – Rettungsring

life saver – Lebensretter, Rettungsschwimmer; auch: Rettungsring

lifeboat – Rettungsboot (RTB)

lifeguard – Rettungsschwimmer

lifeline – Rettungsleine

life-saving measures – Lebensrettende Maßnahmen

life-threatening – lebendbedrohend

lifetime – Lebensdauer

lifetime (cumulative) occupational radiation exposure – Berufslebensdosis (Radioaktivität)

lift – Aufzug, Fahrstuhl; auch: Heben, Auftrieb (in Luft)

lift cage – Aufzugskabine

lift point – Anhebepunkt

lift shaft – Aufzugsschacht

lifting bag (‚air-bag') – Lufthebekissen, Luftheber (LH), Hebekissen

lifting beam – Hebebaum, Hebebalken

lifting device – Hebevorrichtung

lifting height – Hubhöhe

lifting jack – hydraulischer Heber; auch: Wagenheber

lifting means – Anschlagmittel

lifting rescue vehicles – Hubrettungsfahrzeuge

light – Licht, Lampe, leuchten, beleuchten; auch: entzünden, anzünden

light bar – Lichtbalken

light barrier – Lichtschranke

light beam – Lichtstrahl

light bulb – Glühbirne

light corpuscles – Lichtquanten, Lichtteilchen; auch: leichte Teilchen

light dome – Lichtkuppel

light equivalent – Lichtäquivalent

light gasoline – Leichtbenzin

light mast – Lichtmast; oder: leichter Mast

light pole – Lichtmast

light pole vehicle – Lichtmastfahrzeug

light portable pump (LPP) – Lenzpumpe

light quanta – Lichtquanten, Lichtteilchen

light speed – Lichtgeschwindigkeit

light steel support – leichte Stahlstütze

light water reactor (LWR) – Leichtwasserreaktor (LWR)

light well – Lichtschacht

lightning – Blitz

lightning apparatus – Beleuchtungsgerät

lightning arrester (LA) – Blitzableiter

lightning bolt – Blitzschlag

lightning device – Beleuchtungsgerät

lightning discharge – Blitzentladung

lightning installation – Beleuchtungsanlage

lightning protection – Blitzschutz

lightning protection measures – Blitzschutzmaßnahmen

lightning protector – Blitzableiter

lightning strike – Blitzunfall

lightweight concrete – Leichtbeton

lime – Kalk

lime soap – Kalkseife

limestone – Kalkstein

limit – Grenze, Grenzwert; auch: begrenzen, beschränken

limit of detection – Bestimmungsgrenze

limit time of exposure – Expositionszeit begrenzen

limit value monitoring – Grenzwertüberwachung

limited concentration – Grenzkonzentration

limiting factor – begrenzender Faktor

limiting oxygen concentration – Sauerstoffgrenzkonzentration

linchpin – Dreh- und Angelpunkt, Stütze

line – Linie, Zeile, Verbindung, Leitung (für Energien)

line securing system – Leinensicherungssystem

linear burning rate – lineare Abbrandgeschwindigkeit

linear thermal expansion – Längenausdehnung

linear thermal expansion coefficient – Längenausdehnungskoeffizient

line-up order – Antreteordnung

link up – verbinden, ankoppeln, koppeln

linseed oil – Leinöl

lintel – Türsturz, Sturz, Querbalken

liquefied natural gas (LNG) – verflüssigtes Erdgas

liquid fire – Flüssigkeitsbrand

liquid fire-extinguishing agents – flüssige Löschmittel

liquid level – Flüssigkeitsstand, Füllstand

liquid level indication – Füllstandsanzeige

liquid manure – Gülle

liquid nitrogen – flüssiger Stickstoff

liquid oxygen (LOX) – flüssiger Sauerstoff

liquid state – flüssiger Zustand

liquid substances – flüssige Stoffe

liquids – Flüssigkeiten

liquified gases – Flüssiggase

liquified petroleum gas (LPG) – Flüssiggas

litmus – Lackmus

litmus paper – Lackmuspapier

litter – Wurf; auch: Streu, Stroh

live (saving) raft – Rettungsfloß

load – Last, Belastung, Bürde; auch: Last, Ladung, Fracht

load bearing capacity – Tragfähigkeit

load compartments – Laderaum

load handling device – Lastaufnahmemittel

load hook – Kranhaken, Traghaken

load lifting device – Lastaufnahmemittel, Lasthebegerät

load lifting means – Lastaufnahmemittel, Lasthebemittel

load of ruins – Trümmerlast

loadable – belastbar, ladbar

loading – Belastung, Beladung

loading capacity – Belastbarkeit, Ladekapazität

local area network (LAN) – Netzwerk, lokales

local dose – Ortsdosis

local dose rate – Ortsdosisleistung

local fire brigade – Ortsfeuerwehr

local galvanic element – Lokalelement

local health authority – Gesundheitsamt

locality – Ort

lock – Schloss, Verschluss; auch: sperren, verriegeln, arretieren

locked-in – eingerastet, eingeschlossen

locker – Schließfach, Spind, Aufbewahrungskiste; auch: Geräteraum

locking – Verschließen

locking brake – Feststellbremse

locking cylinder – Schließzylinder, Türzylinder

locking mechanism – Verschlussmechanismus, Feststellanlage, Verschluss

loft – Dachboden, Speicher

logistical support – logistische Unterstützung

logistics – Versorgung

long line – Feuerwehrleine

long-distance delivery of water – Löschwasserförderung über lange Wegstrecken

long-living – langlebig (Isotope)

long-nose pliers – Spitzzange

long-term damages – Langzeitschäden, Spätschäden

loop – Schleife, Schlaufe, Öse; auch: Ringleitung

loose – schlaff (z. B. Seil)

loose connection – Wackelkontakt

Loschmidt number – Loschmidt-Konstante

loss factor – Verlustfaktor

loss in activity – Aktivitätsverlust

loss of activity – vollständiger Aktivitätsverlust

low active waste (LAW) – schwach radioaktiver Abfall

low areas – tiefliegende Bereiche

low current – Schwachstrom

low expansion foam (LX) – Schwerschaum

low expansion foam branch (pipe) – Schwerschaumrohr

low in oxygen – sauerstoffarm

low voltage – Niederspannung

low voltage distribution – Niederspannungsverteiler

low voltage installation – Niederspannungsanlage

lower explosion limit (LEL) – untere Explosionsgrenze (UEG)

lower explosion point – unterer Explosionspunkt

lower water authority – untere Wasserbehörde

lowering – ablassen (jemanden herunterlassen)

lower-off point – Abseilpunkt

low-radiation – strahlungsarm

low-smoke layer – raucharme Schicht

low-vibration – schwingungsarm

lubricate – einfetten, einölen, einschmieren

lubricating oil – Schmieröl

lubrication – Schmierung

lumberman's knot – Zimmermannsstich (Knoten)

luminescence – Lumineszenz

lump hammer – Fäustel

lung anthrax – Lungenmilzbrand

lung demand valve – Lungenautomat

lye – Lauge

lye-proof – laugenbeständig, alkalibeständig

M

machinist – Maschinist

magnesium torch – Magnesiumfackel

magnetic field – Magnetfeld

magnetic field strength – magnetische Feldstärke

magnetic valve – Magnetventil

main attack direction – Hauptangriffsrichtung

main attack path/route – Hauptangriffsweg

main extinguishing effect – Hauptlöscheffekt

main objective – Hauptaufgabe

main point fire brigade – Stützpunktfeuerwehr

main shut-off device – Hauptabsperrschieber

main slide – Hauptschieber

main spool – Hauptschieber

main stop tab – Hauptabsperrschieber

main switch – Hauptschalter

main tap – Haupthahn

main valve – Hauptventil, Haupthahn

main water pipe – Hauptwasserleitung

mains tester – Phasenprüfer

maintaining – Wartung, Aufrechterhaltung

maintenance – Wartung, Instandhaltung

maintenance-free – wartungsfrei

major emergency response vehicle (MERV) – Großraumrettungswagen (G-RTW)

major fire – Großbrand

major incident – Großschadensereignis

make an inventory – inventarisieren

make tight – abdichten, dichten

makeshift shelter – Behelfsunterkunft

male thread – Außengewinde

malfunction – Fehlfunktion

malfunctioning – Fehlfunktion, Störung

maliasmus – Rotz

malice – Mutwille, Vorsatz, Bosheit

malicious false alarm – böswilliger Alarm

management – Leitung, Führung, Betriebsleitung, Verwaltung

manager – Leiter, Manager, Geschäftsführer

mandatory signs – Gebotszeichen

manhole cover – Gullydeckel, Schachtdeckel

manhunt – Fahndung

manlift – Arbeitsbühne

man-made disaster – menschliches Versagen

man-made fibres – Kunstfasern

manner – Art und Weise

manometer – Manometer, Druckmesser, Druckmessgerät

manometric discharge height – manometrische Förderhöhe

manometric pressure height – manometrische Druckhöhe

T. Schmiermund, *Fachwörterbuch Feuerwehr und Brandschutz*, https://doi.org/10.1007/978-3-662-64120-0_34

manometric suction height – manometrische Saughöhe

manual – Bedienungsanleitung

manual alarm point – Druckknopfmelder

manual fire siren – Handsirene

manual mode – Handbetrieb

manual release – Handauslösung

manufacturer – Hersteller, Produzent, Erzeuger

manure – Mist, Dung

manure fork – Mistgabel

manure hoe – Dunghaken

map exercise – Planübung, ‚Planspiel'

maps – Karten (Kartenkunde)

maps of hydrants – Hydrantenpläne

Marburg virus – Marburg-Virus

marking – Markierung

marsh fire – Moorbrand

mask – Maske

mask glasses – Maskenbrille

masking tape – Kreppband

masonry – Mauerwerk

mason's hammer – Maurerhammer

mass burning rate – Massenabbrandgeschwindigkeit, massenbezogene Abbrandrate

mass casualty incident (MCI) – Massenanfall von Verletzten (MANV)

mass concentration – Massenkonzentration

mass defect – Massendefekt

mass exchange – Stoffaustausch

mass fire – Massenbrand

mass flow – Massestrom

mass fraction – Massenanteil

mass of atom – absolute Atommasse

massive – massiv, fest

master alarm – Hauptalarm

master chimney sweeper – Schornsteinfegermeister

master key – Hauptschlüssel

mastering of the disaster – Bewältigung der Schadenslage

match – Spiel, Begegnung; auch: Streichholz, Zündholz

material asset – Sachwert, materieller Vermögenswert

material assets insurance – Sachwertversicherung

material difficult to ignite – schwer entzündbarer Stoff

material exchange – Stoffaustausch, Materialaustausch

material flammable with difficulty – schwer entflammbarer Stoff

material safety data sheet (MSDS) – Sicherheitsdatenblatt (SDB)

matter waves – Materiewellen

maximum concentration – Höchstkonzentration

maximum detector – Maximalmelder

maximum experimental safe gap (MESG) – Grenzspaltweite

maximum explosion-pressure – maximaler Explosionsdruck

maximum heart rate (MHR) – maximale Herzfrequenz

maximum permissible workplace concentration – maximale Arbeitsplatzkonzentration (MAK)

maximum permitted concentration – zulässige Höchstkonzentration

maximum pressure-rise – maximaler Druckanstieg

maximum rate of pressure – zeitlicher maximaler Druckanstieg

maximum temperature detector – Wärmemaximalmelder

maximum voluntary ventilation (MVV) – Atemzeitvolumen, Atemgrenzwert

mayor – Bürgermeister

mean – Mittelwert, Durchschnitt

mean burning rate – mittlere Abbrandgeschwindigkeit

mean concentration-time product – mittleres Konzentrations-Zeit-Produkt

mean lethal concentration-time product – mittleres letales Konzentrations-Zeit-Produkt

means – Mittel

means of coercion – Zwangsmittel

means of command and control – Führungsmittel

means of communication – Kommunikationsmittel

means of escape – Fluchtwege

means of production – Arbeitsmittel

meantime between failures (MTBF) – Ausfallsicherheit

measles-mumps-rubella vaccination (MMR vaccination) – Impfung gegen Masern-Mumps-Röteln (MMR-Impfung)

measured value – Messwert, gemessener Wert

measurement equipment – Messgerät, Messwerkzeug

measures to fence off – Absperrmaßnahmen

measuring blood pressure – Blutdruckmessung

measuring chamber – Messkammer

measuring cup – Messbecher

measuring equipment – Messgeräte, Messinstrument

measuring tool – Messgerät, Messwerkzeug

measuring uncertainty – Messunsicherheit

mechanic's level – Wasserwaage

mechanical – mechanisch

mechanical energy – mechanische Energie

mechanical strength – mechanische Festigkeit

medal – Medaille, Orden

medal ribbon – Ordensband

median lethal concentration (LC_{50}) – mittlere letale Konzentration (LC_{50})

median lethal dose (LD_{50}) – mittlere letale Dosis (LD_{50})

medical call – Notfalleinsatz, Notfall (medizinischer)

medical gases – medizinische Gase

medical gloves – medizinische Handschuhe

medical ID alert bracelet – Notfallarmband

medical personnel – Sanitätspersonal

medical service – Sanitätsdienst

medical vehicle – Sanitätskraftwagen (Sanka)

medical ventilator – Beatmungsgerät

medical-grade petroleum spirit – Wundbenzin

medicaments – Medikament, Arzneimittel

medium expansion foam (MX) – Mittelschaum

medium expansion foam branch (pipe) – Mittelschaumrohr

medium fire – Mittelbrand

medium pressure – Mitteldruck

medium pressure line – Mitteldruckleitung

medium-density fibreboard (MDF) – Holzfaserplatte, mitteldichte

medivac chopper – Rettungshubschrauber (RTH)

melioidosis – Melioidose

melting – schmelzen

melting behaviour – Schmelzverhalten

melting point (mp) – Schmelzpunkt (Fp)

melting process – Schmelzvorgang

melting range – Schmelzbereich

membrane – Membran, Folie, Häutchen

meningitis – Meningitis, Hirnhautentzündung

Mercator projection – Mercator-Projektion (Kartenkunde)

mercury – Quecksilber

message – Nachricht, Botschaft, Meldung, Mitteilung

metal fire – Metallbrand

metal fire extinguishing powder – Metallbrandpulver

metal saw – Eisensäge

metallic bond – Metallbindung

metering valve – Dosierventil

methane – Methan

METHANE-mnemonic: mission details – exact location – time and type of incident – hazards in the area – approach routes – number of casualties – expected respons – METHANE-Merkschema: Einsatzdetails – exakte Ortsangabe – Art und Zeit des Vorfalls – Gefahren in der Umgebung – Anfahrtswege – Verletztenanzahl – erwartete Reaktion

method of extinguishing by aerosol – Aerosollöschverfahren

method of extinguishing by water – Wasserlöschverfahren

metric graph paper – Millimeterpapier

metric horsepower (mhp) – metrische Pferdestärke = Pferdestärke (PS)

mezzanine – Zwischengeschoss

mid ship mounted pump – Mitteleinbaupumpe

millimetre paper – Millimeterpapier

million – Million (1 000 000 = 10^6)

mineral fibre – Mineralfaser

mineral oil – Mineralöl

mineral wool – Mineralwolle

mineral-oil resistant – mineralölbeständig

minimal damage – Bagatellschaden

minimal ignition-hazardous oxygen concentration – minimaler zündgefährlicher Sauerstoffgehalt

minimal treatment (MT) – minimale Behandlung

minimization requirement – Minimierungsgebot

minimum air requirement – Mindestluftbedarf

minimum autoignition temperature – Mindestzündtemperatur

minimum burning temperature – Mindestverbrennungstemperatur

minimum detection limit (MDL) – untere Nachweisgrenze

minimum fire resistance – Mindestfeuerwiderstandsdauer

minimum gate-trigger current – Mindestzündstrom

minimum ignition energy – Mindestzündenergie

minimum ignition temperature – Mindestzündtemperatur

minimum ignition time – Mindestzünddauer

minimum operating pressure – Mindestbetriebsdruck

minimum operating time – Mindestbetriebszeit

minimum oxygen requirement – Mindestsauerstoffbedarf

minimum pressure – Mindestdruck

minimum requirement – Mindestanforderung

minor damage – Bagatellschaden

minute respiratory volume (MRV) – Atemminutenvolumen (AMV)

misbehaviour – Fehlverhalten

miscellaneous – verschieden, vermischt

miscibility – Mischbarkeit

miscibility gap – Mischungslücke

miscibility with water – Mischbarkeit mit Wasser

misconduct – Fehlverhalten, Verfehlung im Amt, Amtsvergehen

mission – Auftrag, Einsatz, Aufgabe, Mission

mission control – Einsatzleitung (EL) (Funktion)

mission diary – Einsatztagebuch

mission order – Einsatzauftrag

mission preparation – Einsatzvorbereitung (im Vorfeld)

mission report – Einsatzbericht

mission situations – Einsatzsituationen

mission type A (= atomic) – A-Einsatz, Einsatz mit radioaktiven Stoffen

mission type B (= biological) – B-Einsatz, Einsatz mit Biostoffen

mission type C (= chemical) – C-Einsatz, Einsatz mit Chemikalien

mission-concentrated cooperation – auftragsbezogene Zusammenarbeit

missions – Aufgaben, Aufträge, Missionen

mission-tactics – Auftragstaktik

mist – Nebel, Dunst, Trübung

mist mask – Staubmaske, Feinstaubmaske

mist respirator mask – Feinstaubmaske

mixed phases – Mischphasen

mixed wood – Mischwald

mixing element – Mischelement

mixing temperature – Mischungstemperatur

mixture – Mischung, Gemisch

mixture formation – Gemischbildung

mixture of substances – Stoffgemisch

mobile command post – bewegliche Befehlsstelle

mobile crane – Kranwagen (KW)

mobile hose reel – fahrbare Schlauchhaspel

mobile intensive care unit (MICU) – Intensivtransportwagen (ITW)

mobile kitchen – Feldküche, mobile Küche

mobile radio – Funkgerät (FuG)

mode of operation (MO) – Arbeitsweise, Funktionsweise

model for effects with toxic gases – Modell für Effekte mit toxischen Gasen (MET)

model town exercise – Planübung, ‚Planspiel'

moderating block – Bremsblock (Kerntechnik)

moderator – Moderator, Bremssubstanz (Kerntechnik)

modifications – Änderungen, Veränderungen, Abänderungen

modular construction – Modulbauweise, modularer Aufbau

modular system – Baukastensystem

moisten – benetzen, befeuchten, anfeuchten

moisture-sensitive – feuchtigkeitsempfindlich

molality – Molalität

molar concentration – Stoffmengenkonzentration

molar mass – molare Masse, Molmasse

molar volume – molares Volumen

mole (mol) – Mol (mol); auch: Maulwurf, Mole (z. B. Hafenmole)

mole fraction – Stoffmengenanteil

molecular attraction – zwischenmolekulare Anziehung

molecular mass – Molekülmasse, Molmasse, molare Masse

molecular motion – Molekularbewegung

molecule – Molekül

Molotov-cocktail (an incendiary composition) – Molotow-Cocktail

moment of momentum – Drehimpuls

momentum – Impuls, Schwung

momentum conservation – Impulserhaltung

monitor – Monitor, Wasserwerfer, Schaum-Wasser-Werfer

monitoring – Überwachung (von Geräten)

monkey grip – Affengriff

mono-constituent substance – Reinstoff, nur aus einer Komponente bestehender Stoff

mononuclidic element – Reinelement

moor fire – Heidebrand

mop up – aufräumen, aufwischen, säubern

morbidity – Morbidität

morning shift – Frühschicht

mortal danger – Lebensgefahr

mortality – Mortalität, Sterberate

mortar – Mörtel; auch: Granatwerfer

motor circuit breaker/switch – Motorschutzschalter, Zündunterbrecher

motor power – Motorleistung

motor vehicle – Kraftfahrzeug

motor vehicle accident (MVA) – Verkehrsunfall (VU)

motorway – Autobahn

mountain rescue – Bergrettung

mounted pump – Einbaupumpe

mounting – Montage, Einbau ; auch: Halterung

mounting device – Haltevorrichtung

mounting point – Befestigungspunkt

mouth piece – Strahlrohrmundstück

mouth-to-mouth resuscitation – Mund-zu-Mund-Beatmung

mouth-to-nose resuscitation – Mund-zu-Nase-Beatmung

movability – Beweglichkeit, Transportfähigkeit

movable – beweglich, mobil, ortsbeweglich, verschiebbar

movement area – Bewegungsfläche

movement order – Verlegungsbefehl

movement time – Bewegungszeit; auch: Fluchtzeit

moving stairway/staircase – Rolltreppe

mud-flap – Schmutzfänger (Kfz)

mullion – Pfosten

multi stage pump – mehrstufige Pumpe

multi-chamber container – Mehrkammerbehälter

multi-function ladder – Multifunktionsleiter

multi-grade oil – Mehrbereichsöl

multimeter – Vielfachmessgerät

multiple trauma – Polytrauma

multiple-casualty incident (MCI) – Massenanfall von Verletzten (MANV)

multi-purpose branch pipe – Mehrzweckstrahlrohr

multi-purpose foam (compound) – Mehrbereichsschaummittel (MBS)

multi-purpose jet pipe type BM/CM – BM-/CM-Strahlrohr

multi-resistant staphylococci – Staphylokokken, multiresistente

multi-stored – mehrgeschossig

multitool – Multifunktionswerkzeug

multi-use branch (pipe) – Mehrzweckstrahlrohr

multi-user band radio – Bündelfunk

Munter hitch – Halbmastwurfsicherung (HMS)

mushrooms – Pilze

mustard gas – Lost, Senfgas, S-Lost, Schwefel-Lost, Yperit

mutagen – erbgutverändernd

mutual aid – nachbarliche Hilfe

mutual solubility curve – Löslichkeitskurve, gegenseitige

mycotoxin – Pilzgift, Pilzgift

N

name badge – Namensschildchen (z. B. an der Kleidung)

name plate – Namensschild, Türschild

name tag – Namensschild

nape – Nacken, Genick

narcotic – Narkotikum, betäubend wirkende Substanz

nasopharyngeal agents – Nasen-Rachen-Reizstoffe

nasopharyngeal tube – Wendl-Tubus, Nasopharyngealtubus

national contact point – Alarmspitze

natural background radiation – natürliche Hintergrundstrahlung

natural gas – Erdgas

natural hazard – Elementarschaden (Versicherungen); Naturgefahren

natural safe – ausfallsicher, eigensicher

natural sciences – Naturwissenschaften

NBC (nuclear, biological, chemical) – ABC (atomar, biologisch, chemisch)

NBC emergency forces – ABC-Einsatzkräfte

NBC equipment – ABC-Ausrüstung

NBC forces – ABC-Einsatzkräfte

NBC hazard situation – ABC-Gefahrenlage

NBC hazards – ABC-Gefahren

NBC mission – ABC-Einsatz

NBC operation – ABC-Einsatz

NBC protection suit – ABC-Schutzanzug

NBC task forces – ABC-Einsatzkräfte

NBC training – ABC-Ausbildung

necessary stair – notwendige Treppe

necessity – Notwendigkeit

neck – Hals, Nacken, Genick

neck brace – Halskrause

neck cover – Nackenschutz, Nackenabdeckung

neck guard – Nackenschutz

neck protection – Nackenschutz

need of extinguishing agent – Löschmittelbedarf

need of extinguishing water – Löschwasserbedarf

needed information – benötigte Information

needle – Nadel; auch: Kanüle

negative pressure – Unterdruck

negligence – Fahrlässigkeit

negligent – fahrlässig

negligent arson – fahrlässige Brandstiftung

neighbourhood – Nachbarschaft

neighbourhood protection – Nachbarschaftsschutz

neighbourly help – Nachbarschaftshilfe

Nernst equation – Nernstsche Gleichung

nerve agents – Nervenkampfstoffe

nerve gas – Nervengas

net heat of combustion – Heizwert, Nettoverbrennungswärme

nettle agent – Nesselstoff (Kampfstoff, Rotkreuz)

neutral point – Neutralpunkt

neutral wire – Nullleiter

T. Schmiermund, *Fachwörterbuch Feuerwehr und Brandschutz*, https://doi.org/10.1007/978-3-662-64120-0_35

neutralization – Neutralisation

neutralization agent – Neutralisationsmittel

neutralizer – Neutralisationsmittel

neutralizing heat – Neutralisationswärme

neutron – Neutron

neutron activation analysis – Neutronenaktivierungsanalyse

neutron counter – Neutronenzähler

neutron flux – Neutronenfluss

neutron radiation – Neutronenstrahlung

night drill – Nachtübung

night lighting – Nachtbeleuchtung

night observation device (NOD) – Nachtsichtgerät

nightshift – Nachtschicht

night-vision device – Nachtsichtgerät

nine-eleven call center, 911 call center {US} – Rettungsleitstelle

nine-eleven call, 911 call {US} – Notruf

nine-eleven-operator, 911-operator {US} – Leitstellendisponent (Rettungsdienst)

nine-nine-nine call, 999 call {GB} – Notruf

nitric acid – Salpetersäure

nitric acid ester – Salpetersäureester

nitric acid ester poisoning – Salpetersäureestervergiftung

nitrogen – Stickstoff

nitrogen mustard gas – Stickstoff-Lost (N-Lost)

nitrogen oxides – Stickoxide (NO_x)

nitrous gases – nitrose Gase

no decompression limit (NDL) – Nullzeit (Tauchen)

no smoking – Rauchverbot

noble gas – Edelgas

noble metal – Edelmetall

no-go area – Sperrgebiet

no-go zone – Sperrgebiet, Sperrzone

noise prevention – Schallschutz, Lärmschutz

noise protection – Lärmschutz

noise suppression – Rauschunterdrückung

nominal capacity – Nennleistung, Nennkapazität

nominal condition – Sollzustand, Nennzustand

nominal delivery rate – Nennfördermenge

nominal diameter – Nenndurchmesser, Nennweite (NW)

nominal discharge – Nennförderleistung

nominal discharge height – Nennförderhöhe

nominal displacement – Nennförderleistung

nominal opening temperature – Nennöffnungstemperatur (Sprinkler)

nominal power – Nennleistung

nominal rescue height – Nennrettungshöhe

nominal speed – Nenndrehzahl

nominal value – Nennwert, Nominalwert, Sollwert

nominal width – Nennweite (NW)

non-aging – alterungsbeständig, nicht alternd

non-aqueous – nicht wässrig

non-combustible – nicht brennbar

non-dimensional – dimensionslos

non-emergency patient – Nichtnotfallpatient

non-ferrous metal – Buntmetall, Nichteisen-Metall (NE-Metall)

non-governmental sanitary organizations – Hilfsorganisationen (HiOrg)

non-inflammable – nicht entflammbar

non-inflammable roof-covering – harte Bedachung

non-ionic – nichtionisch

non-ionizing radiation – nichtionisierende Strahlung

nonjudicial punishment – Disziplinarstrafe, nichtgerichtliche Bestrafung

non-negligent – nicht nachlässig, vorsätzlich, nicht fahrlässig

nonnegligent arson – vorsätzliche Brandstiftung

non-public fire brigade – nichtöffentliche Feuerwehr

non-self-ignitable – nicht selbstentzündlich

non-stationary – nicht ortsfest

non-usable – nicht verwendbar

non-volatile – nicht flüchtig

norm – Regel

normal atom – normales Atom = Atom im Grundzustand

normal conditions – Normalbedingungen

normal hydrogen electrode (NHE) – Normalwasserstoffelektrode

normal pressure pump – Normaldruckpumpe

normal state – Normalzustand

normally combustible – normal brennbar

normally flammable – normal entzündlich

normally inflammable – normal entflammbar, normal entzündbar

normally inflammable – normal entflammbar

no-smoking area – Rauchverbotsbereich

not dangerous – nicht gefährlich, ungefährlich

not harmful – ungefährlich, unschädlich, harmlos

not safe – nicht sicher

notice of departure – Abmeldung

notional yield – Nennausbeute

Novichok – Nowitschok

nozzle – Düse

nozzle diameter – Düsenweite

nozzle discharge – Ausfluss (aus einem Strahlrohr)

nuclear (power) phase-out – Atomausstieg

nuclear accident – Atomunfall, nuklearer Unfall

nuclear binding energy – Kernbindungsenergie

nuclear bomb – Atombombe

nuclear chain reaction – nukleare Kettenreaktion, Kernkettenreaktion

nuclear charge – Kernladung

nuclear chemistry – Kernchemie

nuclear disaster – Atomkatastrophe

nuclear disintegration – Kernzerfall, Atomzerfall

nuclear energy – Kernenergie, Atomenergie; auch: Kernkraft

nuclear fission – Atomspaltung, Kernspaltung

nuclear fusion – Kernfusion

nuclear medicine – Nuklearmedizin

nuclear meltdown – Kernschmelze

nuclear plant – Atomkraftwerk (AKW), Kernkraftwerk (KKW)

nuclear power – Kernenergie, Atomenergie

nuclear power engineering – Kerntechnik, Kernkrafttechnik

nuclear power plant (NPP) – Atomkraftwerk (AKW), Kernkraftwerk (KKW)

nuclear reaction – Kernreaktion

nuclear reactor – Atomreaktor, Kernreaktor

nuclear reactor accident – Reaktorunfall

nuclear research – Nuklearforschung

nuclear technology – Kerntechnik

nuclear waste – radioaktiver Abfall, Atommüll

nuclear-powered – atomgetrieben

nucleus – Kern, Atomkern

nuclides – Nuklide

number of revolutions (rpm – revolutions per minute) – Drehzahl (Upm – Umdrehungen pro Minute)

number of storeys – Geschossanzahl

nurse – Krankenschwester

nursing home – Altersheim

O

obey an order – einen Befehl befolgen, einem Befehl gehorchen

objective of the mission – Auftragsinhalte, Einsatzziel, Ziel des (Einsatz-)Auftrags

obligation – Verpflichtung, Pflicht

obscuration (by smoke) – Sichtminderung (durch Rauch)

observations – Beobachtungen

obstacles – Hindernisse

obstructions – Behinderungen (Hindernisse)

obstructive disease – obstruktive Erkrankung

obstructive disorder – obstruktive Erkrankung

occupational accident – Arbeitsunfall

occupational disease – Berufskrankheit

occupational health check – arbeitsmedizinische Untersuchung

occupational physician – Arbeitsmediziner

occupational safety and health – Arbeitsschutz

octan number – Oktanzahl

octane rate – Oktanzahl

octet rule – Oktettregel

odorant – Odoriermittel

odorization – Odorierung

odour – Geruch, Duft, Gestank

odour of fire – Brandgeruch

off duty – außer Dienst

offence against regulations of the fire service – Dienstvergehen (Feuerwehr)

off-gas temperature – Abgastemperatur

office puncher – Bürolocher

officer in charge (OiC) – Einsatzleiter (ELtr), verantwortliche Führungskraft

officer in charge (OiC) of the volunteer fire brigade – Wehrführer der Freiwilligen Feuerwehr (WeFü), Wehrleiter

official calibration – Eichung

official expert – Sachverständiger

official order – dienstlicher Befehl

offset overhand bend (OOB) – Sackstich, Schlaufenknoten, Bandschlingenknoten

Ohm's law – Ohm'sches Gesetz

oil – Öl, einölen, schmieren

oil accident – Ölunfall

oil adsorbent – Ölbinder, Ölbindemittel

oil barrier – Ölsperre

oil binder – Ölbinder, Ölbindemittel

oil binding agent – Ölbinder, Ölbindemittel

oil boom – Ölsperre

oil burner – Ölbrenner

oil can – Ölkanne

oil central heating – Ölzentralheizung

oil change – Ölwechsel

oil consumption – Ölverbrauch

oil drain plug – Ölablassschraube

oil finding paste – Ölnachweispaste

oil heating – Ölheizung

oil level check – Ölstandkontrolle

oil level warning light – Ölstandkontrollleuchte

oil on road – Ölspur

T. Schmiermund, *Fachwörterbuch Feuerwehr und Brandschutz*, https://doi.org/10.1007/978-3-662-64120-0_36

oil separator – Ölabscheider

oil skimmer – Ölskimmer

oil spill – Ölverschmutzung

oil spill barrier – Ölsperre

oil trap – Ölabscheider („Ölfalle")

oil-fired heating system – Ölheizungsanlage

oil-spill tender – Rüstwagen Öl (RW-Öl)

ointment – Salbe

old building – Altbau

on duty – im Dienst

on orders from above – auf höheren/übergeordneten Befehl

on-board equipment – Bordausrüstung

on-call room – Bereitschaftsraum (Zimmer)

oncogenic – krebserzeugend

one-helper method – Ein-Helfer-Methode (Erste Hilfe, HLW)

one-one-two call, 112 call {EU} – Notruf

one-shift-personnel (of a fire station) – Wachabteilung

one-time use filtering escape device – Einweg-Fluchtfiltergerät

one-way gloves – Einmalhandschuhe, Einweghandschuhe

ongoing fire – andauernder Brand

on-ramp – Autobahnauffahrt, Auffahrt auf die Schnellstraße

on-scene care – Erstversorgung

on-site fire brigade – Betriebsfeuerwehr (BtF)

opacity (of smoke) – Lichtschwächung (durch Rauch)

open fire – offener Brand

open flames – offene Flammen

open water – offenes Gewässer

open wounds – offene Wunden

openable – aufklappbar

open-end wrench – Gabelschlüssel

open-hand knot – Sackstich, Schlaufenknoten, Bandschlingenknoten

operable – durchführbar, praktikabel, betriebsbereit

operating instructions – Gebrauchsanweisung

operating manual – Bedienungsanleitung

operating method – Arbeitsmethode

operating plan – Ablaufplan

operating pressure – Arbeitsdruck

operating principles – Funktionsprinzipien; auch: Einsatzgrundsätze

operating time – Betriebszeit

operating voltage – Betriebsspannung

operation – Betrieb, Operation, Einsatz, Arbeitsgang

operation abroad – Auslandseinsatz

operation instructions – Bedienungsanleitung

operation mode – Betriebsart

operation sector – Einsatzabschnitt (EA)

operation staff – Einsatzleitung (Team)

operation with dangers due to biological substances – B-Einsatz, Einsatz mit Biostoffen

operation with dangers due to chemical substances – C-Einsatz, Einsatz mit Chemikalien

operation with dangers due to radioactive substances – A-Einsatz, Einsatz mit radioaktiven Stoffen

operational ability – Einsatzfähigkeit, Betriebsfähigkeit

operational command – Einsatzleitung (EL) (Funktion)

operational conditions – Einsatzbedingungen

operational crew – Einsatztrupp

operational duty – operativer Dienst

operational equipment – Einsatzmittel

operational fire protection – betrieblicher Brandschutz

operational forces – Einsatzkräfte

operational means – Einsatzmittel

operational measures – Einsatzmaßnahmen

operational order – Einsatzbefehl

operational personnel – Einsatzpersonal

operational procedure – Einsatzablauf

operational readiness – Funktionsbereitschaft

operational safety – Arbeitssicherheit, Betriebssicherheit

operational scenarios – Einsatzszenarien

operational situations – Einsatzsituationen

operational success – Einsatzerfolg

operational tactics – Einsatztaktik

operational tasks – Einsatzaufgaben

operational threshold – Einsatzschwelle

operational threshold values – Einsatzschwellenwerte

operational tolerable value – Einsatztoleranzwert (ETW)

operational units – Einsatzkräfte

operational value – Einsatzwert

operative-tactical address (OPTA) – operativ-taktische Adresse (OPTA)

operative-tactical-measures – operativtaktische Maßnahmen

operator – Betreiber, Bediener, Veranstalter

optical density of smoke – optische Rauchdichte

optical signals – optische Signale

optional equipment – Sonderausrüstung

oral order – mündlicher Befehl

order – Anordnung, Befehl

order and report system – Befehls- und Meldesystem

order of events – Ablauf, zeitliche Abfolge von Geschehnissen

order system – Befehlssystem

ordinance – Verordnung, Anordnung

ordnance – Nachschub

organ dose – Organdosis

organic chemistry – organische Chemie

organic peroxides – organische Peroxide

organic solvents – organische Lösemittel

organisation – Gliederung, Organisation

Organisation for the Prohibition of Chemical Weapons (OPCW) – Organisation für das Verbot chemischer Waffen

organophosphates – Phosphorsäureester

oriel – Erker

origin – Ursprung, Herkunft, Abstammung

oropharyngeal airway (OPA) – Guedeltubus

oropharyngeal tube – Oropharyngealtubus

outbreak of fire – Brandausbruch

outcome of measures – Ergebnis der Maßnahmen

outer wall – Außenwand

outlet pressure – Ausgangsdruck

outline – Umriss

outpatient – ambulanter Patient

output – Ausgabe, Leistungsabgabe, Ausstoß, Produktionsleistung

output pressure – Ausgangsdruck

outside (fire) attack – Außenangriff

outside fire – offener Brand, Brand im Freien

outside stair – Außentreppe

outside temperature – Außentemperatur

outward and return (route) – Hin- und Rückweg

outward opening door – nach außen öffnende Tür

outward route – Hinweg

overall command – Gesamteinsatzleitung

overall intention – Gesamtabsicht

overall political responsibility – politisch Gesamtverantwortlicher

overall situation – Gesamtlage

overflow – Überlauf

overhand knot – Überhandknoten

overhead cable – Freileitungskabel

overhead transmission line – Freileitung

overhead-projector, OH-projector – Tageslichtprojektor

overheat – überhitzen

overheating – Überhitzung

overload – Überlastung

overpotential – Überspannung

overrun brake – Auflaufbremse

overshoot – übersteuern

overstated – überbewertet

overvoltage – Überspannung

ownership – Eigentum, Besitz

oxidable – oxidierbar

oxidant – Oxidationsmittel, Sauerstoffträger (chem.)

oxidation – Oxidation

oxidation degree – Oxidationsgrad

oxidation inhibitor – Oxidationsverhinderer

oxidation number – Oxidationszahl

oxidation process – Oxidationsvorgang

oxidation rate – Oxidationsgeschwindigkeit

oxidation state – Oxidationsstufe

oxidation-sensitive – oxidationsempfindlich

oxidizing (substance) – brandfördernd (Substanz)

oxidizing agent – Oxidationsmittel, Sauerstoffträger (chem.)

oxidizing catalyst – Oxidationskatalysator

oxidizing substances – oxidierend wirkende Stoffe

oxygen – Sauerstoff

oxygen carrier – Sauerstoffträger (med.)

oxygen cylinder – Sauerstoffflasche, Sauerstoffdruckgasflasche

oxygen deficiency – Sauerstoffmangel, Sauerstoffentzug

oxygen demand – Sauerstoffbedarf

oxygen equivalent – Sauerstoffäquivalent

oxygen index (OI) – Sauerstoff-Index (OI)

oxygen intoxication – Sauerstoffvergiftung, Sauerstoffintoxikation

oxygen limit – Sauerstoffgrenze

oxygen mask – Sauerstoffmaske

oxygen meter, ox-meter – Sauerstoffmessgerät

oxygen requirement – Sauerstoffbedarf

oxygen resuscitation – Sauerstoffbeatmung

oxygen starvation – Sauerstoffmangel

oxygen starved – sauerstoffarm

oxygen supply – Sauerstoffzufuhr

oxygen tank – Sauerstoffbehälter

oxygen toxicity – Sauerstoffvergiftung, Sauerstoffintoxikation

oxygen toxicity syndrome – Sauerstofftoxikose

oxygen-acetylene cutting – Autogenschneiden, Brennschneiden (Acetylen-Sauerstoff)

oxygen-acetylene welding – Autogenschweißen, Schweißen (Acetylen-Sauerstoff)

oxyhydrogen – Knallgas

P, Q

pacemaker (PM) – Schrittmacher, Herzschrittmacher (HSM)

packaging – Verpackung

packing materials – Verpackungsmaterialien

paddles – Paddel; auch für: EKG-Elektroden

padlock – Vorhängeschloss

pager – Funkmeldeempfänger, Meldeempfänger, „Piepser"

paging device – Funkmeldeempfänger (FME), „Piepser"

paid fire brigade – Berufsfeuerwehr (BF), „bezahlte Feuerwehr"

pairing – Paarbildung

pairing energy – Paarbildungsenergie

pallet store – Regalstapellager

palpate – tasten, abtasten, betasten

pan – Pfanne, Trog, Schale, Wanne, Becken, Mulde

panelling – Wandverkleidung, Täfelung

panic bar – Panikstange

parallel connection – Parallelschaltung

paralysis of the sense of smell – Lähmung des Geruchssinns

paralytic shellfish poisoning (PSP) – Muschelvergiftung (durch STX verursacht)

paramedic – Rettungsassistent (RettAss, RA), Notfallsanitäter (NotSan)

paramedic instructor – Lehrrettungsassistent

parameter – Kenngröße, Parameter

parameters of state – Zustandsgrößen

parking position – Parkposition

partial discharge – Teilflutung; auch: Teilentladung

partial failure – teilweiser Ausfall, Teilausfall

partial malfunction – teilweise Fehlfunktion, Teilausfall

partial monitoring – Teilüberwachung

partial pressure – Partialdruck, Teildruck

partial protection – Teilschutz

partial vapour pressure – Partialdampfdruck, Teildampfdruck

partial-body dose – Teilkörperdosis

partial-body exposure – Teilkörperbestrahlung

participants – Teilnehmer, Beteiligte

particleboard – Spanplatte

particles radiation – Teilchenstrahlung

particularities – Besonderheiten; auch: Festlegungen

particulate respirator – Partikelfilter (Atemschutzmaske)

partition – Abtrennung (räumlich)

partition wall – Trennwand

passive leg-raising (PLR) – Schocklagerung

passivity – Passivität

pasty – pastös

pathogen – Krankheitserreger, Erreger

pathogenicity – Pathogenität

T. Schmiermund, *Fachwörterbuch Feuerwehr und Brandschutz*, https://doi.org/10.1007/978-3-662-64120-0_37

patient – Patient

patient transfer – Patiententransport

patient transport ambulance (PTA) – Krankentransportwagen (KTW)

patient transport unit (PTU) – Patienten-Transport-Einheit (PTE)

peavy – Wendehaken, Fällheber

pebbles – Kieselsteine

pedal – Pedal

pedal brake – Fußbremse

pedal cutter – Pedalschneider

pedal cutting device – Pedalschneidgerät

pedestal – Podest

pencil – Bleistift

pencil sharpener – Bleistiftspitzer

pendulum breathing – Pendelatmung

penetration – Durchbruch, Durchdringung

penetration time – Durchbruchszeit, Durchdringungszeit

pent roof – Pultdach

per-/poly-fluorinated carbons (PFC) – per-/polyfluorierte Chemikalien (PFC)

perceptions – Wahrnehmungen

percussion drilling machine – Schlagbohrmaschine

perfect gas equation – allgemeine Gasgleichung

per-fluorinated tensids (PFT) – Perfluortenside

perfluorooctanesulfonic acid (PFOS) – Perfluoroctansulfonsäure (PFOS)

perfluorooctanoic acid (PFOA) – Perfluoroctansäure

Periodic Table of the Elements (PTE) – Periodensystem der Elemente (PSE)

peristaltic pump – Schlauchpumpe

permanent – ständig, permanent, dauerhaft, unbefristet

permanent disposal site – Endlager, dauerhafte Entsorgungsstelle

permanent gas – permanentes Gas

permanent hardness of water – permanente Härte des Wassers

permanently available to intervence – ständig alarmierbar

permanently connected – fest angeschlossen

permanently installed – fest eingebaut

permanently manned – ständig besetzt

permanent-pressure fire extinguisher – Dauerdruck-Feuerlöscher

permeability – Durchlässigkeit

permeable – durchlässig

permissible axle load – zulässige Achslast

permissible pressure hose length – zulässige Druckschlauchlänge

permissible total weight – zulässige Gesamtmasse

permission – Erlaubnis, Genehmigung, Zulassung

permission to smoke – Raucherlaubnis

peroxides – Peroxide

peroxyacetic acid (PAA) – Peroxyessigsäure (PES)

persistent – beständig, schwer abbaubar, persistent; bei Kampfstoffen: sesshaft

person in command – Einsatzleiter (ELtr), Befehlshaber

person who caused a fire – Brandverursacher

personal alert safety system (PASS) – Notsignalgeber, Totmanneinrichtung (Atemschutzeinsatz)

personal distress alarm (PDA) – Notsignalgeber, Totmanneinrichtung (Atemschutzeinsatz)

personal dose – Personendosis

personal dose equivalent – Personenäquivalentdosis

personal equipment – persönliche Ausrüstung

personal injury – Personenschaden

personal line – persönliche Sicherheitsleine (Atemschutzeinsatz)

personal observations – eigene Beobachtungen

personal perceptions – eigene Wahrnehmungen

personal protection – persönlicher Schutz, Körperschutz

personal protective equipment (PPE) – Persönliche Schutzausrüstung (PSA)

personal special equipment – Persönliche Sonderausrüstung

personnel – Personal

personnel car – Mannschaftstransportfahrzeug (MTF)

perspiration – Schweiß, Schwitzen

perspire – schwitzen

pesticide – Pflanzenschutzmittel, Pestizid

petrol station – Tankstelle

petrol-engined saw – Motorsäge, benzinbetrieben

petroleum jelly – Vaseline

pH scale – pH-Skala, pH-Wert-Skala

pharyngeal tube – Pharyngealtubus

phase tester – Phasenprüfer

phased evacuation – kontrollierte/schrittweise Evakuierung

pH-meter – pH-Meter, pH-Wert-Messgerät

phonetic alphabet – Buchstabieralphabet, Buchstabiertafel

phosgene oxim (CX) – Phosgenoxim

phosgene – Phosgen

phosphorescence – Phosphoreszenz

phosphorescent – nachleuchtend

phosphoric acid – Phosphorsäure (H_3PO_4)

phosphoric acid ester poisoning – Phosphorsäureestervergiftung

phosphoric acid esters – Phosphorsäureester

photoelectric effect (PE) – Photoeffekt (PE)

photoionization detector (PID) – Photoionisationsdetektor (PID)

photon energy – Photonenenergie

photon equivalent – Photonenäquivalent

photonuclear reaction – Kernphotoeffekt

photovoltaic system – Photovoltaik-Anlage

photovoltaics (PV) – Photovoltaik

photovoltaics module – Photovoltaikelement

pH-value – pH-Wert

physical – physisch, physikalisch, körperlich, mechanisch

physical condition – Gesundheitszustand, physische (körperliche) Verfassung

physical containment level (L1-L4) – Sicherheitsstufe (Labor, L1-L4)

physical explosion – physikalische Explosion

physical fire model – physikalisches Brandmodell

physical law – physikalisches Gesetz

physical safeguarding measure – bauliche Sicherungsmaßnahme

physical state – Aggregatzustand

physical work – körperliche Arbeit

physiological saline solution – physiologische Kochsalz-Lösung, Ringer-Lösung

phytotoxic – pflanzenschädlich

pick axe – Pickel, Spitzhacke

pick-up hose for extinguishing agent – Zumischerschlauch

pick-up point – Abholpunkt

pictorial symbol – Bildzeichen, Bildsymbol

pier – Pfeiler

pig (a nickname) – Bleiblock (Radioaktivität)

pillar – Säule, Stütze

pillar hydrant – Überflurhydrant

pillar hydrant key – Überflurhydrantenschlüssel

pillar of smoke – Rauchsäule

pilot flame – Zündflamme, Pilotflamme

pilot line – Steuerleitung

pilot valve – Steuerventil

piloted ignition – Fremdzündung

pinchers – Beißzange

pipe – Rohr, Leitung; auch: Pfeife, Flöte

pipe cross-section – Rohrquerschnitt

pipe cutter – Rohrschneider

pipe diameter – Rohrdurchmesser

pipe elbow – Krümmer, Knie, Rohrbogen

pipe leader – Strahlrohrführer

pipe suspension – Rohraufhängung

pipe wrench – Rohrzange

pipeline – Rohrleitung, Pipeline

piping network – Rohrnetz

pistol type branch – Pistolenstrahlrohr

pit saw – Zugsäge

pitchfork – Heugabel

place of fire origin – Brandausbruchsstelle (BA-Stelle)

place of incident – Unfallstelle, Ereignisort

place of operations – Einsatzstelle (ESt)

place of work – Arbeitsplatz

plague – Pest

plan – Plan, Vorhaben, Konzept, Programm

plan of action – Einsatzplan, Aktionsplan

Planck's quantum – Planck'sches Wirkungsquantum

plank – Bohle

planning – Planung

planning application – Bauantrag

planning documents – Planungsunterlagen

plant – Anlage, Werk, Produktionsanlage; auch: Pflanze, Gewächs

plant fire brigade – Werkfeuerwehr (WF)

plant pest – Pflanzenschädling

plant protection – Pflanzenschutz

plant safety – Anlagensicherheit

plant-protective agent – Pflanzenschutzmittel, Pestizid

plasma cutting device – Plasmaschneidgerät

plaster – Innenputz, Verputz

plastic container – Kunststoffbehälter

plastic foam – Schaumstoff

plastic waste – Kunststoffabfall

plastics – Kunststoffe

plastics fire – Kunststoffbrand

pleasant smell – angenehmer Geruch

pleasure craft – Sportboot

pliers – Zange

plumber – Klempner

plumber's helper – Pümpel (Gummi-Saugglocke)

plunger – Pümpel (Gummi-Saugglocke)

plywood – Sperrholz

pneumatic brake – Druckluftbremse

pneumatic chisel – Druckluftmeißel

pneumatic hammer – Drucklufthammer

pneumonic plague – Lungenpest

pocket dosimeter – Stabdosimeter

pocket-calculator – Taschenrechner

pockets of embers – Glutnester

point of origin – Ursprungspunkt, Ausgangspunkt

point of view – Gesichtspunkte

poison control centre – Vergiftungszentrale

poison gas – Giftgas

poisoning – Vergiftung

poisonous – giftig

poisonous if inhaled – giftig beim Einatmen

poisonousness – Giftigkeit

pole climbers – Steigeisen

police – Polizei (POL)

police radio – Polizeifunk

polish – Politur

polishing – polieren

polishing agent – Poliermittel

pollutant – Schadstoff

pollute – verschmutzen, verunreinigen, verpesten, vergiften

pollution – Verschmutzung, Umweltverschmutzung, Verunreinigung

pollution control – Umweltschutz

pollution level – Schadstoffbelastung

polymerizing substances – polymerisierbare Stoffe

pompier ladder – Hakenleiter

pool – Lache, Pfütze; auch: Becken, Schwimmbad

pool fire – Lachenbrand

population (pop.) – Bevölkerung

porch – Veranda, Vordach

porosity – Porosität

portable – tragbar, ortsbeweglich

portable fire extinguishers – tragbare Feuerlöscher

portable fire ladders – tragbare Feuerwehrleitern

portable fire pump normal pressure (PFPN) – tragbare Feuerlöschkreiselpumpe, Tragkraftspritze (TS)

portable fire pump vehicle – Tragkraftspritzenfahrzeug (TSF)

portable fire-extinguishers – tragbare Feuerlöschgeräte

portable flashing light – tragbare Warnblinkleuchte

portable generator – tragbarer Stromerzeuger

portable hose reel – tragbare Schlauchhaspel

portable ladders – tragbare Leitern

portable lightning apparatus – tragbares Beleuchtungsgerät

portable pump trailer – Tragkraftspritzenanhänger (TSA)

portable radio device – Handfunkgerät (HFuG), Handsprechfunkgerät

portable smoke blocker – mobiler Rauchverschluss

portable smoke curtain – mobiler Rauchverschluss

porter – Pförtner

porter's lodge – Pförtnerloge

porter's office – Pförtnerloge

position – Position, Dienststellung

positional change – Stellungswechsel (räumlich)

positive pressure – Überdruck

positive pressure breathing apparatus – Überdruck-Pressluftatmer, Atemschutzgerät mit Überdruck

possibility – Möglichkeit

possibility for escape – Fluchtmöglichkeit

post – Pfosten, Ständer; auch: Post, Posten, Amt, Stellung

post-alerting – Nachalarmierung

post-extinguishing work – Nachlöscharbeiten

potassium – Kalium

potassium lye – Kalilauge

potential energy – potentielle Energie, Lageenergie

potential equalization – Potentialausgleich

pour point – Stockpunkt

pourability – Rieselfähigkeit

powder – Pulver

powder extinguisher – Pulverlöscher

powder fire-fighting vehicle – Pulverlöschfahrzeug (PLF)

powder monitor – Pulverwerfer

powder nozzle – Pulverstrahlrohr

power – Leistung, Kraft, Strom, Energie; auch: Macht, Befugnis, Stärke

power consumption – Leistungsaufnahme

power current – Starkstrom

power current tool box – Starkstrom-Werkzeugkasten

power drain – Stromverbrauch, Leistungsabgabe

power failure – Stromausfall

power installation – Strominstallation, meist Starkstrom

power source – Energiequelle

power supply – Energieversorgung

power supply company (PSC) – Energieversorgungsunternehmen (EVU)

power take-off (PTO) – Nebenantrieb

practicable – möglich, ausführbar, durchführbar

practice alarm – Probealarm

practice tower – Übungsturm

practitioner – Fachmann, Praktiker, Anwender

precaution – Vorsicht, Vorkehrung, Vorsichtsmaßnahme, Vorsorge

precautional measure – Vorsichtsmaßnahme

precautionary measure – Vorsichtsmaßnahme

precleaning – Vorreinigung

pre-determined attendance – Ausrückeordnung, Abmarschfolge

predetermined breaking point – Sollbruchstelle

prediction – Voraussage

preface – Vorwort, Einleitung; auch: Vorderseite

pre-fire planes – Feuerwehrpläne

preliminary investigation (PI) – Ermittlungsverfahren (EV)

premixed flame – vorgemischte Flamme

preparation – Bereitstellung

preparation of action – Einsatzvorbereitungen (aktuelle)

prepared – vorbereitet

preselect command – Vorbefehl

presenting a fire hazard – brandgefährlich, Brandgefahr

preservation – Konservierung

preservative – Konservierungsmittel

press button fire alarm box – Druckknopfmelder

pressure – Druck

pressure at the input-side – Eingangsdruck (Pumpe)

pressure balance – Druckbilanz

pressure control – Druckregelung

pressure control valve – Druckregelventil

pressure coupling – Druckkupplung

pressure drop – Druckabfall

pressure fluctuation – Druckschwankung

pressure gauge – Manometer, Druckmessgerät, Druckanzeige

pressure gauge device – Druckmesseinrichtung

pressure gradient – Druckgefälle

pressure head – Druckhöhe

pressure hose – Druckschlauch

pressure hose length – Druckschlauchlänge

pressure increase – Druckerhöhung

pressure limiting valve – Druckbegrenzungsventil

pressure loss – Druckverlust

pressure measurement – Druckmessung

pressure measurement unit – Druckmessgerät, Druckmesseinheit

pressure meter – Druckmessgerät

pressure monitoring – Drucküberwachung

pressure reducer – Druckminderer

pressure reducing valve – Druckreduzierventil, Reduzierventil

pressure regulator – Druckregler

pressure release – Druckentspannung (Druck ablassen)

pressure relief – Druckentlastung

pressure relief flap – Druckentlastungsklappe

pressure relief valve – Überdruckventil, Druckentlastungsventil

pressure relief vent – Druckentlastungsöffnung

pressure resistance – Druckfestigkeit

pressure resistant – druckfest

pressure rise – Druckanstieg

pressure rise at to time – zeitlicher Druckanstieg

pressure shock – Druckstoß

pressure socket piece – Druckstutzen (Pumpe)

pressure surge – Druckstoß

pressure tank – Druckbehälter, Drucktank

pressure test – Druckprüfung

pressure valve – Druckventil

pressure variation – Druckschwankung

pressure vessel – Druckbehälter, Druckkessel

pressure vessel burst – Druckbehälterzerknall

pressure vessel rupture – Druckbehälterzerknall, Druckbehälterbruch

pressure wave – Druckwelle

pressure-dependent – druckabhängig

pressure-less – drucklos

pressure-resistant – druckbeständig

pressure-tight – druckdicht

pressure-vacuum gauge – Druck-/Unterdruck-Messgerät, Manometer für Über- und Unterdruck

pressurized water fire-extinguisher – Dauerdruck-Wasserlöscher

pressurized water reactor (PWR) – Druckwasserreaktor (DWR)

prestressed concrete – Spannbeton

pre-treatment – Vorbehandlung

prevailing – vorherrschend

preventer – Einreißhaken

prevention – Verhütung

prevention of accidents – Unfallverhütung

preventive – vorbeugend

preventive action – vorbeugende Maßnahme

preventive explosion protection – vorbeugender Explosionsschutz

preventive fire protection – Brandverhütung

pre-warning – Vorwarnung

pre-warning duration – Vorwarndauer

pre-warning time – Vorwarnzeit

prewet – vorbenetzen, anfeuchten

primal pump – Entlüftungspumpe

primary coolant – Primärkühlmittel

primary coolant circuit – Primärkühlmittelkreislauf

primary explosion – Primärexplosion

primary key – Hauptschlüssel

primary survey – Erstversorgung

principle of proportionality – Grundsatz der Verhältnismäßigkeit

principle of reasonableness – Grundsatz der Verhältnismäßigkeit

prion disease (PD) – Prionenerkrankung

prions – Prionen

probability – Wahrscheinlichkeit

probability of occurrence – Eintrittswahrscheinlichkeit

procedure – Prozedur, Verfahren, Arbeitsmethode

process – Prozess, Verfahren, Ablauf

process of action – Handlungsablauf

process of ignition – Zündvorgang

process of thinking and acting – Denk- und Handlungsablauf

process water – Brauchwasser

producer – Produzent, Erzeuger, Hersteller

production requirements – betriebliche Voraussetzungen

profession – Beruf

professional – Fachmann, Profi

professional fire brigade – Berufsfeuerwehr (BF)

professional fire service – Berufsfeuerwehr (BF)

profiled steel – Profilstahl

progressive smouldering – fortschreitendes Schwelen

prohibited – verboten

prohibited area – Sperrbereich

prohibition – Verbot

prohibition sign – Verbotszeichen

prohibition to return to a particular place – Platzverweis

prop – Abstützung

propagation – Vermehrung, Ausbreitung

propagation limit – Ausbreitungsgrenze

propagation of flames – Flammenausbreitung

propagation velocity – Ausbreitungsgeschwindigkeit

propagation velocity of electromagnetic waves – Ausbreitungsgeschwindigkeit elektromagnetischer Wellen

propellant gas – Treibgas

proper angle of a ladder – richtiger Leiteranstellwinkel, Anleiterwinkel

proper mass – Ruhemasse

property – Eigentum, Besitz; auch: Eigenschaft

property protection – Sachwertschutz, Eigentumsschutz

property value – Sachwert, Eigentumswert

property value protection – Sachwertschutz, Schutz des Eigentumswerts

proportion – Proportion, Anteil, Verhältnis, Mengenverhältnis

proportional counter – Proportionalzähler

proportional counter tube – Proportionalzählrohr

proportionality – Verhältnismäßigkeit

proprietary – Eigentum

protected area – geschützter Bereich, Schutzbereich

protection – Schutz, Sicherung

protection against contact – Berührungsschutz

protection against toxic gases – Gasschutz, Schutz vor giftigen Gasen

protection area – Schutzbereich

protection category – Schutzklasse

protection cover – Schutzabdeckung, Schutzhülle, Abdeckplane

protection goals – Schutzziele

protection of animals – Schutz von Tieren

protection of material assets – Schutz von Sachwerten

protection of persons – Schutz von Personen

protection of persons and material assets – Schutz von Personen und Sachwerten

protection of persons, animals and material assets – Schutz von Personen, Tieren und Sachwerten

protection of place of operation – Sicherung der Einsatzstelle

protection of property – Eigentumssicherung

protection of the flank – Flankenschutz

protection of the operation – Einsatzsicherung

protection team – Sicherheitstrupp (SiTr)

protection zone – Schutzzone

protective arrangement – Schutzvorkehrungen

protective class – Schutzklasse

protective clothing – Schutzkleidung

protective curtain – Schutzvorhang, „Eiserner Vorhang" (Bühne, Theater)

protective effect – Schutzwirkung

protective equipment – Schutzausrüstung

protective gear – Schutzausrüstung

protective gloves – Schutzhandschuhe

protective grounding – Schutz-Erdung

protective hood – Schutzabdeckung, Schutzhaube

protective immunization – Schutzimpfung

protective insulating gloves – Elektrohandschuhe, Elektroschutzhandschuhe

protective measure – Schutzmaßnahme, Sicherungsmaßnahme

protective shield – Schutzschirm

protective spectacles – Schutzbrille

protective suit – Schutzanzug

protective system – Schutzsystem

protective toecap – Zehenschutzkappe („Stahlkappe" im Sicherheitsschuh)

protective value – Schutzwert

protein foam compound – Proteinschaummittel (PS)

protolysis – Protolyse

proton – Proton

proton number – Ordnungszahl, Kernladungszahl

prove – nachweisen, beweisen, prüfen

provision – Bestimmung, Bereitstellung

prusik knot – Prusikknoten

prussic acid – Blausäure

pry bar – Brechstange, Stemmeisen, Nagelzieher

psychosocial emergency care – psychosoziale Notfallversorgung (PSNV)

psycho-toxic agents – psychotoxische Kampfstoffe

public assembly room – Versammlungsstätte, öffentlicher Versammlungsraum

public fire brigade – öffentliche Feuerwehr

public health authority – Gesundheitsbehörde

public health officer – Amtsarzt

public medical officer – Amtsarzt

public water main – öffentliche Wasserleitung

public water supply – öffentliche Wasserversorgung

publication – Veröffentlichung

public-safety answering point (PSAP) – Notrufabfragestelle

puddle – Pfütze, Lache

Pulaski – Pulaski-Axt

pull down hook – Einreißhaken

pulley – Umlenkrolle

pulley block – Flaschenzug

pulling cable – Zugseil

pull-out from the nuclear energy – Kernenergie-Ausstieg

pull-rope – Zugseil

pulmonary agents – Lungenkampfstoffe

pulse rate – Zählrate, Impulsrate; auch: Pulsfrequenz

pump – Pumpe

pump capacity – Pumpenleistung

pump casing – Pumpengehäuse

pump characteristic (line) – Pumpenkennlinie

pump inlet – Saugeingang

pump operator – Maschinist, Pumpenbediener

pump outlet pressure – Pumpendruck

pump truck – Gabelhubwagen

pump water tender – Tanklöschfahrzeug (TLF)

pump-outlet pressure control – Pumpendruckregelung

puncher – Locher

pupils – Pupillen

pure – pur, rein, unverfälscht

pure substance – Reinstoff

purity – Reinheit

purlin – Pfette

purpose – Zweck, Ziel, Bestimmung

push button alarm – Druckknopfmelder

put at risk – einem Risiko aussetzen, gefährden

putty – Kitt

pyrexia – Fieber

pyrolysis – Pyrolyse

pyrolysis gases – Pyrolysegase

pyrolysis products – Pyrolyseprodukte

pyrophoric – pyrophor

pyrotechnic articles – Feuerwerkskörper, pyrotechnische Erzeugnisse/Artikel

pyrotechnics – Feuerwerkskörper, pyrotechnische Erzeugnisse; auch: Feuerwerkerei

Q-fever – Q-Fieber

quadrillion – Billiarde (1 000 000 000 000 000 = 10^{15})

qualified person – fachkundige Person, Experte

quality factor (QF) – Qualitätsfaktor (QF), Bewertungsfaktor

quantity of dangerous material – Menge an Gefahrstoff, Gefahrstoffmenge

quantity of extinguishing agent – Löschmittelmenge

quantity of extinguishing water – Löschwassermenge

quantity of fires and fire looses – Brandgeschehen

quantity of water – Wassermenge, Löschwassermenge

quantity per pack – Füllmenge, Packungsinhalt

quantity ratio – Mengenverhältnis

quick attack – Schnellangriff, Schnellangriffseinrichtung

quick attack hose – Schnellangriffsschlauch

quick check – Einsatzkurzprüfung (Atemschutz)

quick coupling – Schnellkupplung

quick coupling handle – Schnellkupplungsgriff

quick reference – Kurzinformation

quick-acting valve – Schnellschlussventil

quilt – Decke (zum Wärmen)

R

rabbit fever – Hasenpest, Tularämie

rabies – Tollwut

rack storage – Regallager

radar beams – Radarstrahlen

radiancy – Strahlen

radiant flux – Strahlungsleistung

radiant heat flux – Strahlungswärmestromdichte

radiant-flux density – Bestrahlungsdichte

radiate heat – Wärme abstrahlen

radiation – Strahlung

radiation accident – Strahlenunfall

radiation biology – Strahlenbiologie

radiation damages – Strahlenschäden

radiation detector – Strahlenspürgerät

radiation dose/dosage – Strahlendosis

radiation dosimeter – Strahlungsdosimeter

radiation effect – Strahlenwirkung

radiation energy – Strahlungsenergie

radiation exposure – Strahleneinwirkung, Strahlenbelastung

radiation hazard – Strahlengefährdung

radiation heat – Strahlungswärme

radiation injuries – Strahlenschäden

radiation limits – Strahlengrenzwerte, Grenzwerte für ionisierende Strahlung

radiation load – Strahlenbelastung

radiation measuring devices – Strahlungsmessgeräte

radiation plume – Strahlungswolke, radioaktive Wolke

radiation poisoning – Strahlenvergiftung

radiation power – Strahlungsleistung

radiation protection – Strahlenschutz

radiation protection area – Strahlenschutzbereich

radiation protection measures – Strahlenschutzmaßnahmen

radiation protection officer (RPO) – Strahlenschutzbeauftragter (SSB)

radiation protection suit – Strahlenschutzanzug

radiation protection supervisor – Strahlenschutzverantwortlicher

radiation safety officer (RSO) – Strahlenschutzbeauftragter (SSB)

radiation sensitivity – Strahlenempfindlichkeit

radiation shielding – Abschirmung gg. Strahlung, Strahlenabschirmung

radiation sickness – Strahlenkrankheit (leichter Verlauf), „Strahlenkater"

radiation source – Strahlenquelle

radiation suit – Strahlenschutzanzug

radiation toxicity – Strahlenvergiftung

radiation weighting factor – Strahlenwichtungsfaktor

radiation-level check – Messung der Strahlenbelastung

radiation-resistant – strahlungsbeständig

radiative heat transfer – Wärmestrahlungsübergang

radical chain reaction – Radikalkettenreaktion

T. Schmiermund, *Fachwörterbuch Feuerwehr und Brandschutz*, https://doi.org/10.1007/978-3-662-64120-0_38

radicals – Radikale

radio alarm system – Funkmeldesystem (FMS)

radio call – Funkruf

radio call-sign – Funkrufname

radio channel – Funkkanal

radio communication – Sprechfunkverkehr, Funkverbindung (grundsätzliche)

radio communication (of scene of operation) – Einsatzstellenfunk

radio contact – Funkverbindung

radio device – Funkgerät (FuG)

radio failure – Ausfall der Funkverbindung

radio link – Funkverbindung

radio message – Funkspruch

radio operator – Funker

radio path – Funkstrecke

radio relay station – Relaisfunkstelle

radio shadow – Funkschatten

radio silence – Funkstille

radio station – Funkstelle (FuSt)

radio unit – Funkanlage

radio waves – Radiowellen

radioactive decay – radioaktiver Zerfall

radioactive disintegration – radioaktiver Zerfall

radioactive material – radioaktive Stoffe, radioaktives Material

radioactive plume – Strahlungswolke, radioaktive Wolke

radioactive waste – radioaktiver Abfall, Atommüll

radioactivity – Radioaktivität

radioautographic – Autoradiographie

radiobiological – strahlenbiologisch

radiocarbon method – Radiocarbon-Methode, C-14-Methode (Altersbestimmung)

radiodense – strahlenundurchlässig

radiographic – Radiographie

radiological agent – radiologischer Kampfstoff

radiological weapon – radiologische Waffe

radiolucent – strahlendurchlässig

radiolysis – Radiolyse

radiolysis products – Radiolyseprodukte

radioman (RM) – Funker

radiometer – Radiometer

radionuclide – Radionuklid

radionuclide generator – Radionuklidgenerator

radiopaque – strahlenundurchlässig

radios – Rundfunkgeräte, Radios

radiosensitivity – Strahlenempfindlichkeit

radius – Ausladung (z. B. beim Kran)

radon – Radon

radon concentration – Radonkonzentration

radon exposure – Radonbelastung

radon gas – Radongas

rafters – Sparren

rail tank – Eisenbahn-Kesselwagen

railway overbridge – Bahnüberführung

railway rescue vehicle – Rüstwagen Schiene (RW Schiene, RW-S)

rain clothes – Regenbekleidung

rain jacket – Regenjacke

rain trousers – Regenhose

rake – Rechen

ram pressure – Staudruck

ramified water main – Verästelungsnetz

ramp planks – Auffahrbohlen

random sample – Stichprobe

range – Reichweite; auch: Entfernung, Angebot, Umfang

range of jet – Wurfweite (z. B. Wasserstrahl)

range of temperature – Temperaturbereich

range of use – Einsatzbreite

rank – Rang, Dienstgrad, Rangstufe

rank badges – Dienstgradabzeichen, Rangabzeichen

rank insignias – Dienstgradabzeichen, Rangabzeichen

rank marking – Dienstgradabzeichen, Rangabzeichen

rank order – Rangordnung

ranking – Rangfolge, Einstufung

rapid intervention vehicle (RIV) – Schnellangriffsfahrzeug

rapid sequence intubation (RSI) – Schnell-Intubation

rappelling – Abseilen

rare earths (elements) – Seltene Erden

rasp – Raspel (Werkzeug)

rate of delivery – Förderstrom

rate of explosion – Explosionsgeschwindigkeit

rate of fire spread – Brandausbreitungsgeschwindigkeit

rate of flame spread – Flammenausbreitungsgeschwindigkeit

rate of heat release – Wärmefreisetzungsrate

rate of pressure – zeitlicher Druckanstieg

rate of rotation – Umdrehungsgeschwindigkeit, Rotationsgeschwindigkeit

rate-of-rise – Anstiegsgeschwindigkeit

rate-of-rise detector – Differentialmelder, Thermodifferentialmelder

rate-of-rise temperature detector – Wärmedifferentialmelder, Thermodifferentialmelder

rating basis – Bewertungsgrundlage, Bemessungsgrundlage

rating values – Beurteilungswerte

ratio of air – Luftverhältnis

ratio of densities – Dichteverhältnis

ratio of vapour densities – Dampfdichteverhältnis

Rautek grip – Rautek-Griff

ray – Strahl

reach of ruins – Trümmerschatten, Trümmerbereich

reaction – Reaktion

reaction equation – Reaktionsgleichung

reaction mechanism – Reaktionsmechanismus

reaction rate – Reaktionsgeschwindigkeit

reaction time – Reaktionszeit, Ausrückezeit

reaction velocity – Reaktionsgeschwindigkeit

reaction zone – Reaktionszone

reactor life – Reaktorlaufzeit

reactor pressure vessel (RPV) – Reaktordruckbehälter (RDB)

read out – ablesen, Ablesung (Gerät, Messwert); auch: vorlesen

readily flammable – leichtentflammbar, leichtentzündbar

readiness for action – Einsatzbereitschaft

reading – lesen, Lesung; auch: Ablesung (Gerät, Messwert)

ready for use – betriebsbereit, einsatzbereit, gebrauchsfertig

ready-message – Bereitmeldung

real burning rate – reale Abbrandgeschwindigkeit/Abbrandrate

real gas – reales Gas

realization – Durchführung, Verwirklichung

real-time mode – Echtzeitbetrieb

reanimation – Reanimation, Wiederbelebung

rear-mounted pump – Heckpumpe

reasonableness – Zumutbarkeit

rebreather – Kreislaufgerät (Atemschutz)

receiver – Empfänger (Gerät)

recipient – Empfänger (Person)

reciprocating primer – Kolbenpumpe

reciprocating pump – Kolbenpumpe

reckless – grob fahrlässig, rücksichtslos, leichtsinnig

recognized contractor – anerkannte Fachfirma, anerkannter Auftragnehmer

reconnaissance – Erkundung, Aufklärung

reconnaissance flight – Erkundungsflug, Aufklärungsflug

reconnaissance helicopter – Erkundungshubschrauber

reconnaissance plane – Erkundungsflugzeug

reconnaissance troop – Erkundungstrupp

reconstruction – Wiederaufbau, Rekonstruktion, Sanierung

recover – retten, bergen, Rettung, Bergung

recovery – Erholung, Wiederherstellung

recovery crane – Bergungskran

recovery time – Erholungszeit

recreation room – Aufenthaltsraum

recycling – Wiederverwertung

red crescent – Roter Halbmond

red-fuming nitric acid (RFNA) – Salpetersäure, rotrauchende

red-glowing – rotglühend

redox equation – Redox-Gleichung

redox pair – Redox-Paar

redox reaction – Redox-Reaktion

redox system – Redox-System

reducer – Reduktionsmittel, Reduzierer

reducing – Abmagern (Löscheffekt), Reduzieren

reducing adapter – Reduzierstück

reducing agent – Reduktionsmittel

reductant – Reduktionsmittel

reduction – Reduktion, Verringerung, Ermäßigung, Rückgang, Minderung

reduction valve – Reduzierventil, Druckreduzierventil

redundance – Redundanz

redundancy – Redundanz

reef bend – Kreuzknoten

reef knot – Weberknoten, Kreuzknoten

reel – Haspel, Rolle, Spule

re-examination – Nachuntersuchung

reference books – Nachschlagewerke

refill – Nachfüllung

refuelled – betankt

refurbishment – Instandsetzung

regenerative breathing apparatus – Regenerationsgerät (RG) (Atemschutz)

regenerator – Regenerationsgerät (RG) (Atemschutz)

registers – Verzeichnisse

regular – regelmäßig, regulär

regulation – Vorschrift, Verordnung, Regelung; auch: Regulierung, Steuerung

regulatory offence – Ordnungswidrigkeit

reignition – Rückzündung, Wiederentzündung

reinforcement – Verstärkung

relative atomic mass (RAM) – relative Atommasse

relative burning rate – relative Abbrandgeschwindigkeit

relative molecular mass (RMM) – relative Molekülmasse

relaxation – Erschlaffung

relay – Relais; auch: Ablösung

relay operation – Relaisbetrieb

relay station – Relaisstelle

release – Veröffentlichung, Freigabe; auch: Freisetzung

release element – Freischaltelement (FSE)

relief – Ablösung

relief of forces – Ablösung von Einsatzkräften

relief-forces – Reservekräfte, ablösende Kräfte

relieve – entlasten, ablassen (Druck reduzieren)

remedial action – Schadensbehebung

remote control – Fernbedienung (Gerät)

remote operation – Fernbedienung (Tätigkeit)

remote transmission – Fernübertragung

removable – abnehmbar, herausnehmbar, demontierbar

render inert – inertisieren

renovation – Renovierung, Sanierung

repair – Reparatur

repeater – Verstärker (Funktechnik)

repellent – abweisend, abstoßend

replacement – Ersatz, Austausch, Ablösung, Nachfolger

replacement power supply – Ersatzstromversorgung

replacement power supply system – Ersatzstromversorgungsanlage

replenishing – Nachfüllung

report – Bericht, Meldung, Mitteilung

report system – Meldesystem

representation – Darstellung, Wiedergabe; auch: Lagedarstellung

representation of the situation – Lagedarstellung

reprocessing plant – Wiederaufbereitungsanlage

request for proposal (RFP) – Angebotsanfrage, Ausschreibung

requirement – Anforderung, Voraussetzung, Bedarf, Erfordernis

requirement of extinguishing agent – Löschmittelbedarf

rescue – retten, bergen, Rettung, Bergung

rescue aircraft – Rettungsflugzeug

rescue airplane – Rettungsflugzeug

rescue and salvage operations – Rettungs- und Bergungsarbeiten

rescue bid – Rettungsversuch

rescue blanket – Rettungsdecke (Stoff-/Wolldecke)

rescue cage – Rettungskorb (Drehleiter)

rescue chain – Rettungskette

rescue chut – Rettungsrutsche

rescue costs – Rettungskosten

rescue crew – Rettungsmannschaft

rescue cruiser – Seenotrettungskreuzer

rescue cushion – Sprungkissen

rescue device – Rettungseinrichtung

rescue diver – Rettungstaucher

rescue dog – Rettungshund

rescue dog handler – Rettungshundeführer

rescue dog squadron – Rettungshundestaffel

rescue dog team – Rettungshundestaffel

rescue dummy – Rettungspuppe

rescue effort – Rettungsmaßnahmen, Rettungsbemühungen

rescue equipment – Rettungsausrüstung, Rettungsgeräte, Rettungsmittel

rescue flight – Rettungsflug

rescue forces – Rettungskräfte

rescue height – Rettungshöhe

rescue helicopter – Rettungshubschrauber (RTH)

rescue knife – Rettungsmesser

rescue ladder – Rettungsleiter

rescue line – Rettungsleine

rescue measure – Rettungsmaßnahme

rescue mission – Rettungseinsatz, Rettungsarbeiten, Rettungsaktion

rescue of animals – Tierrettung (Tätigkeit)

rescue of people – Menschenrettung

rescue operation – Notoperation; auch: Rettungseinsatz, Rettungsaktion

rescue organization – Rettungsorganisation

rescue panel – Rettungstafel

rescue party – Rettungsmannschaft

rescue plane – Rettungsflugzeug

rescue platform – Rettungsplattform

rescue pumper – Hilfeleistungslöschfahrzeug (HLF)

rescue resources – Rettungsmittel

rescue rope – Feuerwehrleine, Rettungsseil, Fangleine

rescue scissors – Rettungsschere (med.)

rescue service – technische Hilfeleistung (TH)

rescue services – Rettungswesen, Rettungsdienst (RD)

rescue sledge – Rettungsschlitten

rescue spreader – Rettungsspreizer, Spreizer (SP)

rescue squad – Rüstzug

rescue stretcher – Rettungstrage

rescue team – Rettungstrupp, Rettungsmannschaft

rescue technology – Rettungstechnik

rescue tool – Rettungswerkzeug

rescue unit – Rettungseinheit

rescue winch – Rettungsseilwinde, Rettungswinde

rescue work – Rettungsarbeiten, Rettungseinsatz

rescue-rope bag – Fangleinenbeutel

reserve – Reserve

reserve material – Reserve, Reservematerial

reserve of extinguishing water – Löschwasserreserve

reservoir dam – Staudamm

residential area – Wohngebiet

residual current operated circuit breaker (RCCB) – Fehlerstromschutzschalter (FI), FI-Schutzschalter

residual current protective device (RCD) – Fehlerstromschutzschalter (FI), FI-Schutzschalter

residual heat – Restwärme

residual radiation – Reststrahlung

residual risk (RR) – Restrisiko (RR)

resistance – Resistenz, Beständigkeit

resistance list – Beständigkeitsliste

resistant to ageing – alterungsbeständig

resolve – Entschluss, Entschlossenheit

respiration – Atmung, Beatmung

respiration apparatus – Beatmungsgerät

respiration mask – Beatmungsmaske, Atemmaske

respirator – Beatmungsgerät; auch: Atemschutzmaske

respirator training facility – Atemschutzübungsanlage

respiratory aid – Beatmungshilfe, Atemhilfe

respiratory arrest – Atemstillstand

respiratory centre – Atemzentrum

respiratory poison – Atemgift

respiratory protection – Atemschutz

respiratory protection monitoring – Atemschutzüberwachung (ASÜ)

respiratory protection monitoring system – Atemschutzüberwachungssystem

respiratory protective device – Atemschutzgerät (ASG)

respiratory rate – Atemfrequenz (AF)

respiratory resistance – Atemwiderstand

respiratory system – Atmungsorgane

respiratory tract burn – Atemwegsverätzung

respire – atmen

response behaviour – Ansprechverhalten

response group – Eingreiftruppe

response pressure – Ansprechdruck

response temperature – Ansprechtemperatur

response time – Ausrückezeit, Reaktionszeit, Eingreifzeit, Ansprechzeit

responsibility – Verantwortung

rest energy – Ruheenergie

rest mass – Ruhemasse

resting time – Ruhezeit

restoration costs – Wiederherstellungskosten

restoration of operational readiness – Wiederherstellung der Einsatzbereitschaft

restricted access – Zutrittsbeschränkung

restricted area – Sperrgebiet

restricted zone – Sperrgebiet, Sperrzone

resuscitation – Reanimation, Wiederbelebung

resuscitation room – Reanimationsraum (im Krankenhaus)

retainability – Rückhaltevermögen

retainer – Halter, Haltevorrichtung, Fangvorrichtung

retaining wall – Stützmauer

retainment – Rückhaltung

retainment capacity – Rückhaltevermögen

retarder – Verzögerer

retention – Rückhaltung

retention pond – Rückhaltebecken, Auffangwanne

retention sump – Auffangwanne

retirement – Pensionierung, Verabschiedung aus dem Dienst

retirement division – Altersabteilung

retractor – „Rückholer" (z. B. bei Sicherheitsgurten); auch: Wundhaken

retreat – Rückzug

retreat path – Rückzugsweg

retreat way – Rückzugsweg

return route – Rückweg

reusable – wiederverwendbar

reuse – Wiederverwendung

revitalisation – Wiederbelebung

revolutions – Umdrehungen; auch: Revolutionen

revolving light – Rundumkennleuchte (RKL)

RICE-principle (rest – ice – compression – elevation) – PECH-Regel (Pause – Eis – Compression – Hochlagern)

ricin – Rizin

ricin poisoning – Rizin-Vergiftung

rickettsia – Rickettsien

ridge – First

right means – richtige Mittel (Einsatzmittel)

rights – Rechte

rights and obligations – Rechte und Pflichten

rigid – starr, steif, unbeweglich, strikt, zäh

rigidity – Starrheit, Steifheit, Stabilität, Zähigkeit

ring wrench – Ringschlüssel

Ringer solution – Ringer-Lösung, physiologische Kochsalz-Lösung

ring-pipe system – Ringleitungssystem

riot control agents – Reizstoffe, Mittel um Unruhen/Aufstände unter Kontrolle zu bringen

rising main – Steigleitung

risk – Risiko, Gefahr

risk analysis – Risikoanalyse

risk assessments – Risikobewertung

risk calculation – Risikoberechnung

risk estimation (RE) – Risikoabschätzung (RA)

risk evaluation – Risikobewertung

risk factor – Risikofaktor

risk groups – Risikogruppen

risk management – Risikomanagement

risk of asphyxiation – Erstickungsgefahr

risk of burns by corrosion – Verätzungsgefahr

risk of contamination – Kontaminationsgefahr

risk of corrosion – Korrosionsgefahr

risk of deformation – Verformungsgefahr

risk of explosion – Explosionsgefahr

risk of fire – Brandgefahr

risk of fire spread – Gefahr der Brandausbreitung

risk of implosion – Implosionsgefahr

risk of infection – Infektionsgefahr, Ansteckungsgefahr, Infektionsrisiko

risk of intoxication – Vergiftungsgefahr

risk of life – Lebensgefahr

risk of panic – Panikgefahr

risk of reaction – Reaktionsgefahr

risk of unlawful interception – Abhörgefahr

risk potential – Gefährdungspotential, Risikopotential
risk provisions – Risikovorsorge
river – Fluss
road map – Straßenkarte
road sweeper – Kehrmaschine
road traffic accident (RTA) – Verkehrsunfall (VU)
road traffic regulations – Straßenverkehrsordnung (StVO)
roadside emergency telephone – Notrufsäule
rock wool – Steinwolle
rodent plague – Hasenpest, Tularämie
role play actor – Verletztendarsteller
roll cage – Rollcontainer
roll of bandage – Verbandpäckchen
rolled fire hose – Rollschlauch
roller container transport system (ACTS – abroll container transport system) – Abrollbehälter-Transportsystem
roller container vehicle – Wechselladerfahrzeug (WLF)
roller door – Rolltor
roll-over – Rauchgasdurchzündung
roof – Dach
roof batten – Dachlatte
roof construction – Dachkonstruktion
roof covering – Bedachung, Dach(be)deckung
roof fire – Dachstuhlbrand
roof identification – Dachkennzeichnung (Fahrzeuge)
roof ladder – Dachleiter
roof surface – Dachfläche
roof tile – Dachziegel
roof truss – Dachstuhl
roof vent – Dachentlüftung
roofing – Überdachung, Bedachung
roofing felt – Dachpappe
roofing hammer – Latthammer
room – Zimmer, Raum
room air – Raumluft
room enclosure – Raumabschluss
room of fire origin – Brandentstehungsraum
room temperature – Raumtemperatur
rope – Seil
rope access – seilunterstützte Zugangstechniken (SZT)
rope control device – Seilstoppgerät
rope descend – Seilabstieg, Abseilen
rope down – Abseilen
rope hose holder – Seilschlauchhalter
rope ladder – Strickleiter
rope tension – Seilspannung
rope-down equipment – Abseilgeräte
rotating beacon – Rundumkennleuchte (RKL)
rotational force – Drehmoment
rotodynamic pump – Kreiselpumpe
rotor – Laufrad (Pumpe)
rough guide – Faustformel
roundabout – Kreisverkehr
rubber boots – Gummistiefel
rubber gasket – Gummidichtung
rubber lining – Gummierung
rubber mallet – Gummihammer
rubber seal – Gummidichtung
rubberized apron – Gummischürze
rubber-lined deliver hose – gummierter Druckschlauch
rubbing alcohol – Alkohol für Desinfektionszwecke
rubbish – Abfall
rubbish bag – Abfallsack
rubble – Trümmer, Schutt, Geröll
rule – Regel, Vorschrift, Gebot
rule of thumb – Faustregel
ruler – Lineal
rules of conduct – Verhaltensregeln

rules of engagement (RoE) – Einsatzregeln

ruling – Entscheidung, Regelung

run command – Befehl ausführen!; auch: Fahrbefehl

rung – Sprosse

rung spacing – Sprossenabstand

runner – Läufer; auch: Bandschlinge

running gear – Fahrgestell (allgemein)

runway foam laying device – Landebahnbeschäumungsfahrzeug

runway foaming tender – Landebahnbeschäumungsanhänger

rust – Rost

rust dissolver – Rostlöser

rust-free – rostfrei

rustling flame – rauschende Flamme

S

sabotage – Sabotage

sabotage toxin – Sabotagegift

sabre saw – Säbelsäge

saddle roof – Satteldach

Safar airway – Safar-Tubus

safe load indicator – Belastungsanzeige (Drehleiter)

safeguarding – Absicherung

safeguarding of scene of operation – Absicherung der Einsatzstelle

safeguards – Schutzmaßnahmen, Schutzvorkehrungen

safety – Sicherheit

safety belt – Sicherheitsgurt

safety cable – Fangseil

safety can – Sicherheitskanne

safety catcher – Absturzsicherung

safety clearance – Sicherheitsabstand

safety criteria – Sicherheitskenndaten

safety data sheet (SDS) {GB} – Sicherheitsdatenblatt (SDB)

safety distance – Sicherheitsabstand

safety equipment – Schutzausrüstung, Sicherheitsausrüstung

safety fuse – Sicherung, Schmelzsicherung (elektr.)

safety glass – Sicherheitsglas

safety glasses – Schutzbrille

safety hazard – Sicherheitsgefährdung, Sicherheitsrisiko

safety helmet – Schutzhelm, Sicherheitshelm

safety indexes – sicherheitstechnische Kennzahlen

safety installation – Sicherheitseinrichtung

safety lamp – Sicherheitsleuchte

safety level (L1-L4) – Sicherheitsstufe (Labor)

safety lighting – Sicherheitsbeleuchtung

safety marking – Sicherheitskennzeichnung

safety officer – Sicherheitsbeauftragter

safety precautions – Sicherheitsmaßnahmen, Sicherheitsvorkehrungen; auch: Sicherheitshinweise

safety regulations – Sicherheitsbestimmungen, Sicherheitsvorschriften

safety risk – Sicherheitsrisiko

safety rope – Sicherheitsseil

safety shoes – Sicherheitsschuhe

safety shower – Notdusche

safety signs – Sicherheitszeichen

safety stair – Sicherheitstreppe

safety staircase – Sicherheitstreppenraum

safety trough for tank – Tanktasse

safety valve – Sicherheitsventil

safety-related values – sicherheitstechnische Kennzahlen

salary – Gehalt, Lohn

salts – Salze

salvage – bergen, Bergung

salvage cask – Bergefass

salvage charges – Bergungskosten, Bergegebühren

T. Schmiermund, *Fachwörterbuch Feuerwehr und Brandschutz*, https://doi.org/10.1007/978-3-662-64120-0_39

salvage costs – Bergungskosten

salvage crew – Bergungsmannschaft

salvage devices – Bergungsgeräte

salvage drum – Bergefass

salvage effort – Bergungseinsatz, Bergungsaktion

salvage experts – Bergungsexperten, Bergungsspezialisten

salvage mission – Bergungseinsatz, Bergungsaktion

salvage operations – Bergungsarbeiten

salvage receptacle – Bergebehälter

salvage ship – Bergungsschiff

salvage team – Bergungstrupp

salvage unit – Bergungseinheit

salvage winch – Bergungswinde, Bergungsseilwinde

salvage work – Bergungsarbeiten

sample taking – Probenahme

sampling – Probenahme

sampling date – Probenahmedatum

sampling jar – Probengefäß, Probenflasche

sampling vial – Probenahmefläschchen, Probenflasche

sand – Sand

sand bucket – Sandeimer

sandbag – Sandsack

sandbag barricade – Sandsackbarrikade

sandpaper – Sandpapier, Schmirgelpapier

sanitary facilities – sanitäre Einrichtungen

sanitary installations – sanitäre Einrichtungen

sanitary room – Sanitärraum

sanitation care – Hygienevorsorge

sarin (GB) – Sarin

saturation concentration – Sättigungskonzentration

save – speichern, sparen; auch: retten, sichern, bergen

saving human life – Menschenrettung

saving lives – Menschenrettung

saw – Säge

saw blade – Sägeblatt

saw chain – Sägekette

saw guide – Sägeschiene

sawtooth roof – Sheddach

saxitoxin (STX) – Saxitoxin

scaffold – Gerüst

scald – verbrühen, Verbrühung

scale – Skala, Maßstab (Kartenkunde)

scale of charges – Gebührenordnung

scaling – Skalierung

scaling ladder – Steckleiter

scattered-light detector – Streulichtmelder

scattered-light smoke detector – optischer Rauchmelder, Streulichtrauchmelder

scattering – Streuung

scattering experiment – Streuversuch

scene of fire – Brandstelle (BSt)

scene of operations – Einsatzstelle (ESt)

scent – Duft, Geruch, Aroma, Parfüm

scent of smoke – Rauchgeruch

schedule of fees – Gebührenordnung

sciences – Wissenschaften, Naturwissenschaften

scientific paper – Fachaufsatz

scintigraphy – Szintigraphie

scintillation counter – Szintillationszähler

scoop stretcher – Schaufeltrage

scope – Anwendungsbereich, Umfang, Kompetenzbereich

scope of protection – Schutzumfang

scope of work – Arbeitsbereich

scorch – verschmoren

scouring agent – Scheuermittel

SCRAM (safety cut rope axe man) – Reaktorschnellabschaltung (RESA)

scrap (metal) – Schrott

scratchproof – kratzfest

screening – Abschirmung

screw carabiner – Schraubkarabiner

screw clamp – Schraubzwinge

screw driver – Schraubendreher

screw karabiner – Schraubkarabiner

screw-cap vial – Probenflasche mit Schraubverschluss

screw-on coupling – Schraubkupplung

screw-on valve – Niederschraubventil

scrubbing brush – Scheuerbürste

SCUBA set – Drucklufttauchgerät (DTG)

scuncheon – Laibung

sea damage (SD) – Havarie

seal – abdichten, Dichtung, Abdichtung, Siegel

seal bag – Kanaldichtkissen

sealant – Abdichtmasse, Dichtmasse

sealed tight – dicht verschlossen

sealing – Dichtung, Abdichtung, Versiegelung

sealing compound – Abdichtmasse, Dichtmasse

sealing cushion – Dichtkissen

sealing pad – Dichtkissen

sealing plug – Dichtstopfen

sealing wedge – Dichtkeil

search and rescue helicopter (SAR helicopter) – Such- und Rettungshubschrauber

search and rescue team (SAR-team) – Such- und Rettungsteam

search crew – Suchtrupp

search party – Suchtrupp

seat belt – Sicherheitsgurt (Kfz)

seat belt cutter – Gurtschneider

seat of fire – Brandherd

seating order – Sitzordnung

second alarm – zweiter Alarm; auch: zweiter Abmarsch

second climber – Nachsteiger (Höhenrettung)

second climbing – Nachstieg (Höhenrettung)

second escape route – zweiter Rettungsweg

second fire escape – zweiter Rettungsweg

secondary damage – Folgeschaden

secondary drive – Nebenantrieb

secondary effect of extinguishing – Nebenlöscheffekt

secondary fire – Folgebrand, Sekundärbrand

secondary infection – Sekundärinfektion

second-degree burn – Verbrennung zweiten Grades

second-degree frostbite – Erfrierung zweiten Grades

second-order conductor – Leiter zweiter Ordnung

section – Bereich, Sektion, Abschnitt, Abteilung

section chief – Abteilungsleiter

section of ladder – Leiterteil

sectional model – Schnittmodell

sector – Bereich, Sektor, Abschnitt

sector-commander – Einsatzabschnittsleiter (EALtr)

secure the retreat way – Rückzugsweg sichern

security – Sicherheit

security against interception – Abhörsicherheit

security door system – Sicherheitsschleuse

security regulations – Sicherheitsbestimmungen

security seal – Sicherheitssiegel, Sicherheitsprüfsiegel

sedentariness – Bewegungsmangel; auch: Sesshaftigkeit von Kampstoffen

self-acceleration – selbstbeschleunigt

self-acceleration decomposition – selbstbeschleunigte Zersetzung

self-acceleration decomposition temperature (SADT) – Temperatur der selbstbeschleunigte Zersetzung (z. B. bei Peroxiden)

self-catalysis – Autokatalyse, Selbstkatalyse

self-contained breathing apparatus (SCBA) – umluftunabhängiges Atemschutzgerät, Pressluftatmer (PA)

self-contained underwater breathing apparatus (SCUBA) – Drucklufttauchgerät (DTG)

self-containment suit – Chemikalienschutzanzug (CSA)

self-endangerment – Eigengefährdung, Selbstgefährdung

self-extinguishing – selbstverlöschend, selbstlöschend

self-extinguishing material – schwer brennbarer Stoff

self-heating – Selbsterwärmung

self-heating temperature – Selbsterwärmungstemperatur

self-ignitability – Selbstentzündlichkeit

self-ignitable – selbstentzündlich

self-ignition – Selbstentzündung

self-ignition point – Selbstentzündungstemperatur

self-ignition temperature – Selbstentzündungstemperatur

self-inflaming – selbstentzündlich, selbstentflammend

self-priming – selbstansaugend

self-propagation of flame – selbstständiges Weiterbrennen

self-protection – Selbstschutz

self-reactive substances – selbstzersetzliche Stoffe

self-rescue – Selbstrettung

self-sustaining – selbsterhaltend

semipermeable – halbdurchlässig

senior officer – Wachabteilungsleiter, („Rangältester")

sense of smell – Geruchssinn

sensing – Wahrnehmung; auch: Spüren (i. S. v. Nachweisen)

sensitive to heat – wärmeempfindlich

sensitive to impact – stoßempfindlich

sensitive to moisture – feuchtigkeitsempfindlich

sensitive to shocks – schlagempfindlich

sensitivity – Empfindlichkeit

sensivity of response – Ansprechempfindlichkeit

sensor – Fühler, Messsonde, Messwertaufnehmer

sensory perception – Sinneswahrnehmungen

separated protected premises – abgesetzter Sicherungsbereich

separating effect – Trenneffekt (Löscheffekt)

separation – Trennung, Abtrennung

separator – Abscheider, Separator

sepsis – Blutvergiftung, Sepsis

septic shock – septischer Schock

septicaemia – Blutvergiftung, Sepsis

septicaemic plague – Pestsepsis

sequence – Reihenfolge

sequence of measures – Reihenfolge der Maßnahmen

series circuit – Reihenschaltung (elektr.)

series connection of fire pumps – Reihenschaltung (von Feuerlöschpumpen)

series of fires – Brandserie

serious illness – ernste Erkrankung

service entrance room – Hausanschlussraum (HAR)

service grade – Dienstgrad; auch: Service-Qualität

service pressure – Betriebsdruck

service ribbon – Ordensspange, Bandschnalle

service station – Tankstelle

serving fire-fighter – Feuerwehrmann im aktiven Dienst

set point – Sollwert

setting-up of the vehicles – Aufstellung der Fahrzeuge, Fahrzeugaufstellung

setting-up time – Rüstzeit

set-up area – Aufstellfläche

severe acute respiratory syndrome coronavirus type 2 (SARS-CoV-2) – schweres-akutes-Atemwegssyndrom-Coronavirus Typ 2

severe thunderstorm – schweres Gewitter, starkes Unwetter

severe weather – Unwetter, raues Wetter

sewage – Abwasser, Schmutzwasser

sewage system – Kanalisation

sewage treatment plant – Abwasserreinigungsanlage (ARA)

shackle – Schäkel

shaft – Welle, Schaft, Siel; auch: Schacht

shank – Schaft, Schenkel; auch: Seilverkürzung

sharps – scharfkantige Gegenstände

shattering capability – Brisanz

shearing action – Scherwirkung

sheepshank – Trompetenknoten

sheet bend – Schotenstich, Schotstek

sheet pile wall – Spundwand

shelf – Regal, Fach, Einschub

shell – Schale

shell model – Schalenmodell

shellfish poison – Saxitoxin (ein durch Muscheln übertragenes Gift)

shelter – Schutz; auch: Unterkunft, Obdach, Unterstand, Unterschlupf

shelter against wind and weather – Wind- und Wetterschutz

shielded radiation source – abgeschirmte Strahlenquelle

shielding – Abschirmung

shift – Verschiebung, Umschalttaste; auch: Schicht, Arbeitsschicht

shock – Schock

shock absorber – Stoßdämpfer; auch: Falldämpfer

shock wave – Schockwelle, Druckwelle

shoe – Schuh, auch: Leiterfuß

shoe covers – Überschuhe

shoe protectors – Überschuhe

shopping centre – Einkaufszentrum

shore – Ufer, Stand; auch: Strebe, Stützbalken

short circuit – Kurzschluss (elektr.)

short data service (SDS) – Kurznachricht (Digitalfunk)

short operational check – Einsatzkurzprüfung (Atemschutz)

short-lived – kurzlebig (z. B. Isotope)

short-living – kurzlebig (z. B. Isotope)

short-term stay – Kurzzeitaufenthalt

short-time dose – Kurzzeitdosis

short-time exposure – Kurzzeitaufnahme

shoulder board – Schulterklappe

shoulder strap – Schulterklappe

shovel – Schaufel

shower – Dusche

shunt – beiseitelegen, verschieben, Stoß, „Bums"; auch: Auffahrunfall

shut-off device – Absperrvorrichtung, Absperrarmatur

shut-off flap – Absperrklappe

shut-off gate – Absperrschieber

shut-off valve – Absperrventil

shuttle water relay – Pendelverkehr (Löschwasser)

SI unit – SI-Einheit

Siamese connection – Doppelschlauchanschluss

sick call – Krankmeldung

sick note – Krankmeldung

sick person – erkrankte Person, kranke Person

sicken – erkranken, krank werden

sickness – Erkrankung, Krankheit

side airbag – Seitenairbag

side group – Nebengruppe

side-group element – Nebengruppenelement

sieve analysis – Siebanalyse

signal communication – Fernmeldeverbindung

signal converter – Signalwandler

signal flag – Warnflagge, Signalflagge

signal generator – Signalgeber

signal lantern – Signallaterne

signal of distress – Notsignal, Notzeichen

signal pipe – Signalpfeife

signal service – Fernmeldedienst

signal transductor – Signalwandler

signal whistle – Signalpfeife

signaller – Fernmelder, Funker; auch: Signalgeber, Einweiser

signalling installation – Signalanlage

sign-out – abmelden

silicon – Silicium (ein Element)

silicone – Silikon (ein Kunststoff)

sill – Schwelle, Fenstersims, Fensterbrett, Türleiste, Sims

simmering – köcheln

single phase system – Einphasensystem

single pole ladder – Anstellleiter

single rope technique (SRT) – Ein-Seil-Technik

single stage pump – einstufige Pumpe

single-point smoke detector – einzelner Rauchmelder

single-use – Einmalgebrauch

single-use gloves – Einmalhandschuhe

single-use protective suit – Einmalschutzanzug

siren signals – Sirenensignale

sit harness – Sitzgurt, Hüftgurt

site fire brigade – Werkfeuerwehr (WF)

site occupancy index – Grundflächenzahl

site of fire – Brandstelle (BSt)

situation – Lage, Situation

situation conference – Lagevortrag

situation map – Lagekarte

situation of impact and damage – Schadenslage (insbes. bei VU)

situation reconnaissance – Lageerkundung

situation report (SITREP) – Lagebericht, Lagemeldung

size of damage – Schadensumfang

skeleton – Skelett

skeleton duty roster – Rahmendienstplan

skeleton formula – Skelettformel

sketch – Skizze, Entwurf

sketch of the situation – Lageskizze

skids – Kufen (z. B. Hubschrauber)

skim off – abschöpfen

skin – Haut

skin anthrax – Hautmilzbrand

skin blisters – Blasen auf der Haut

skin care – Hautpflege

skin contact – Hautkontakt

skin irritation – Hautreizung

skin poison – Hautgift

skin resorption – Hautresorption

skull and crossbones (symbol) – Totenkopf (Symbol)

skylight – Dachfenster

skyscraper – Hochhaus

skyscraper fire – Hochhausbrand

slat – Lamelle, Latte

slate – Schiefer

slated roof – Schieferdach

sledgehammer – Vorschlaghammer

sleeve badge – Ärmelabzeichen

slide-in module – Einschub

sliding door – Schiebetür

sliding pole – Rutschstange

slight risk – kleine Gefahr, geringes Risiko

sling – Schlinge, Riemen; auch: Bandschlinge

slip differential – Differentialsperre (Kfz)

slip road – Zufahrtsstraße, Autobahnauffahrt

slope – Steigung, Neigung, Gefälle

slot screwdriver – Schlitzschraubendreher

sluice valve – Schleusenventil

slurry – Aufschlämmung

small fire – Kleinbrand

small fire engine with portable pump – Tragkraftspritzenfahrzeug (TSF)

small fire-fighting vehicle – Kleinlöschfahrzeug (KLF)

small quantity exception – Kleinmengenregelung

smell – Geruch, Duft, Gestank

smell of fire – Brandgeruch

smell of smoke – Rauchgeruch

smoke – Rauch

smoke alarm – Rauchwarnmelder

smoke and heat exhaust system (SHE) – Rauch- und Wärmeabzugsanlage (RWA)

smoke and heat outlet – Rauch- und Wärmeableitung

smoke barrier – Rauchschürze

smoke clearance – Entrauchung

smoke cloud – Rauchwolke, Rauchschwaden

smoke concentration – Rauchkonzentration

smoke control damper – Entrauchungsklappe (ERK)

smoke control door – Rauchschutztür

smoke damage – Rauchschaden

smoke density – Rauchdichte

smoke detector – Rauchmelder

smoke distribution – Rauchausbreitung, Rauchverteilung

smoke diver – Atemschutzgeräteträger (AGT)

smoke exhaust damper – Entrauchungsklappe (ERK)

smoke exhaust ventilation system – Entrauchungsanlage

smoke exhaust venting equipment – Rauchabzugseinrichtung

smoke flap – Rauchklappe

smoke formation – Rauchbildung

smoke gas layer – Rauchgasschicht

smoke gases – Rauchgase

smoke influence effect – Raucheinwirkung

smoke inhalation injury – Rauchvergiftung, Rauchgasintoxikation

smoke intoxication – Rauchvergiftung, Rauchgasintoxikation

smoke particles – Rauchpartikel

smoke point – Rauchpunkt (Öle)

smoke poisoning – Rauchvergiftung, Rauchgasintoxikation

smoke production – Rauchentwicklung

smoke production rate – Rauchentwicklungsrate

smoke propagation – Rauchausbreitung

smoke protection – Rauchschutz

smoke reduction – Rauchverringerung

smoke section – Rauchabschnitt

smoke sensitivity – Rauchempfindlichkeit

smoke spreading – Rauchausbreitung

smoke vent – Rauchklappe

smoke vent opening – Rauchabzugsöffnung

smoke venting – Abführen von Rauch und Wärme, Entrauchung

smoke-blackened – rauchgeschwärzt

smoke-filled – verqualmt, verraucht

smoke-generating device – Nebelgerät

smokeless – rauchschwach

smokeless layer – rauchfreie Schicht

smoke-proof – rauchdicht

smoke-proof closure – rauchdichter Abschluss

smokestack – Schornstein, Rauchfang

smoking corner – Raucherinsel, Raucherecke

smoky – verqualmt, verraucht

smoulder – schwelen, glimmen

smouldering combustion – Schwelen

soldering fire – Schwelbrand, Glimmbrand

smouldering point – Schwelbrand

smoothbore nozzle pipe – Vollstrahlrohr

smother the flames – Flammen ersticken

smothering effect – Stickeffekt, Löscheffekt Ersticken

smouldering combustion – Schwelen

smouldering fire – Schwelbrand, Glimmbrand

smouldering point – Schwelpunkt

snap hook – Karabinerhaken, Schnapp-Haken

snap in, to snap in – einrasten

snapping – Einschnappen

snow load – Schneelast

snow pipe – Schneerohr

snow ploughing service – Räumdienst (im Winter)

snow shovel – Schneeschaufel, Schneeschieber

soap – Seife

socket wrench – Steckschlüssel

soddenness – Durchnässung

sodium – Natrium

sodium lye – Natronlauge

soft soap – Schmierseife

softening point – Erweichungspunkt

solar cell – Solarzelle

solar collector – Solaranlage

solar electricity – Solarstrom

solar energy – Solarenergie

solar panel – Solarpaneel

solar plant – Solaranlage

solar radiation – Sonnenstrahlung

solder – löten

soldering lamp – Lötlampe

solid – fest, massiv

solid carbon dioxide – festes Kohlenstoffdioxid, Kohlendioxidschnee

solid construction – Massivbauweise

solid explosives – feste explosive Stoffe

solid fire-extinguishing agents – feste Löschmittel

solid jet – Vollstrahl

solid rubber – Vollgummi

solid state – fester Zustand

solid stream – Vollstrahl

solid substances – Feststoffe

solidity – Festigkeit, Stabilität

solids – Feststoffe

solubility – Löslichkeit

solubility curve – Löslichkeitskurve

soluble in water – wasserlöslich

solution – Lösung

solvent – Lösemittel

solvent vapour – Lösemitteldampf

soman (GD) – Soman

somatic damages – somatische Schäden

sonic speed – Schallgeschwindigkeit

soot – Ruß

soot fire – Rußbrand

soot formation – Rußbildung

sound detector – Verschütteten-Suchgerät

sound level – Schallpegel

sound pressure – Schalldruck

sound pressure level (SPL) – Schalldruckpegel

sound speed – Schallgeschwindigkeit

sound velocity – Schallgeschwindigkeit

source of damage – Schadensursache

source of heat – Wärmequelle

source strength – Quellenaktivität (radioaktive Quelle)

space – Platz, Raum, Abstand, Zwischenraum, Lücke; auch: Weltraum

space blanket – Rettungsdecke (Stoff-/Wolldecke)

space explosion – Raumexplosion

spacer – Abstandshalter

spacing law – Abstandsgesetz

spacing material – Füllmaterial

spade – Spaten

span width – Spannweite

spanner – Schraubenschlüssel, auch kurz für: Kupplungsschlüssel

spark – Funken

spark discharge – Funkenentladung

spark energy – Funkenenergie

spark plugs – Zündkerzen

sparklers – Wunderkerzen

special arrangement – Sonderregelung

special case – Sonderfall

special constructions – Sonderbauten

special equipment – Sonderausrüstung, Spezialausrüstung

special extinguishing media – Sonderlöschmittel

special fire-fighting vehicle – Sonderlöschfahrzeug

special infection ambulance – Infektionsrettungswagen (I-RTW)

special license – Sondergenehmigung, Sonderlizenz

special operational situations – besondere Einsatzsituationen

special order – besonderer Befehl

special permission – Sondererlaubnis, Sondergenehmigung

special permit – Sondergenehmigung, Sondererlaubnis

special regulation – Sonderregelung

special rights – Sonderrechte

special service – Fachdienst

special surveillance – Überwachung, besondere

special unit – Sondereinheit

special vehicles – Sonderfahrzeuge

specialized authority – Fachbehörde

specialized literature – Fachliteratur

specific activity – spezifische Aktivität

specific caloric value – spezifischer Brennwert

specific heat capacity – spezifische Wärmekapazität

specific heat of combustion – spezifische Verbrennungswärme

specific heat of evaporation – spezifische Verdampfungswärme

specific heat of melting – spezifische Schmelzwärme

specific heat of sublimation – spezifische Sublimationswärme

specific heat of vaporization – spezifische Verdampfungswärme

specific heating value – spezifischer Heizwert

specification – Spezifikation, Beschreibung, (genaue) Angabe

specified condition – Sollzustand

specimen jar – Probenglas, Probenflasche

spectacle wearer – Brillenträger

spectacles – Brille

speculation – Vermutung

speech diaphragm – Sprechmembran

speed limit – Geschwindigkeitsbegrenzung

speed of light – Lichtgeschwindigkeit

speed restriction – Geschwindigkeitsbegrenzung

speedometer – Tacho, Tachometer, Fahrgeschwindigkeitsanzeiger

speedway – Schnellstraße

spherical valve – Kugelhahn

sphygmomanometer – Blutdruckmessgerät

spillage – Austritt, Verschüttetes (Flüssigkeiten)

spilled person – verschüttete Person

spilled substances – ausgelaufene Stoffe

spine board (SB) – Rettungsbrett, Wirbelsäulenbrett, Spineboard

spiral staircase – Wendeltreppe

splint – Schiene (med.)

splinter – Splitter

splitting – Spaltung (mechanisch)

splitting axe – Spaltaxt

splitting maul – Spalthammer

splitting wedge – Spaltkeil

sponge – Schwamm

spontaneous heating – Selbsterwärmung

spontaneous heating temperature – Selbsterwärmungstemperatur

spontaneous ignition – Spontanentzündung, Selbstentzündung

spontaneous oxidation – Autooxidation, Selbstoxidation

spontaneous-ignition temperature (SIT) – Zündtemperatur, Selbstentzündungstemperatur

spontaneously ignitable – selbstentzündlich

spray can – Sprühdose

spray jet – Sprühstrahl

spray nozzle – Sprühdüse, Zerstäuberdüse

spray pattern – Sprühbild

spread of gas clouds – Ausbreitung von Gaswolken

spread test – Ausbreitungsversuch, Streuversuch

spreading – Verbreitung, Ausbreitung

spring – Feder (Drahtspirale)

spring centre punch – Federkörner

springiness – Elastizität, Federung, Sprungkraft

sprinkler – Sprinkler

sprinkler activation area – Wirkfläche (von Sprinklern)

sprinkler control centre – Sprinklerzentrale (SPZ)

sprinkler head – Sprinklerkopf

sprinkler nozzle – Sprinklerdüse

sprinkler pump – Sprinklerpumpe

sprinkler system – Berieselungsanlage

squad – Staffel (St)

squad leader – Staffelführer (StFü)

squadron – Trupp (Tr)

squadron leader – Truppführer (TrFü)

square knot – Kreuzknoten

square spanner – Vierkantschlüssel

square wrench – Vierkantschlüssel

square-box wrench – Vierkantsteckschlüssel

squeegee – Abzieher (Bodenwischer)

St. Johns ambulance (brigade) – Johanniter-Unfall-Hilfe (JUH)

stability – Stabilität, Standsicherheit, Beständigkeit, Haltbarkeit

staff for extraordinary events – Stab für außergewöhnliche Ereignisse

staff room – Aufenthaltsraum; auch: Lehrerzimmer

staff section (S 1 – S 6) – Sachgebiet (S 1 – S 6)

staff section S 1 – personnel/administration – Sachgebiet S 1 – Personal/Innerer Dienst

staff section S 2 – information gathering and assessment – Sachgebiet S 2 – Lage

staff section S 3 – operation – Sachgebiet S 3 – Einsatz

staff section S 4 – logistics – Sachgebiet S 4 – Versorgung

staff section S 5 – media and press – Sachgebiet S 5 – Presse- und Medienarbeit

staff section S 6 – communications and transmission – Sachgebiet S 6 – Information und Kommunikation

stage – Bühne (Theater)

staging area – Bühnenbereich; auch: Bereitstellungsfläche

staircase – Treppenraum

stairway – Treppe

stanchion – Stütze, Strebe, Pfosten; auch: Stiel

standard – Norm

standard combustion enthalpy – Standardverbrennungsenthalpie

standard conditions (0 °C, 1013 mbar) – Normbedingungen, Normzustand (0 °C, 1013 mbar)

standard cubic metre – Normkubikmeter (Gase)

standard decontamination – Standard-Dekontamination

standard intervention process – Standardvorgehen

standard ladder – Anstellleiter

standard of fire resistance – Feuerwiderstandsklasse

standard radioactive source – Prüfstrahler

standard temperature & pressure (STP) – Normzustand (0 °C, 1013 mbar)

standard time-temperature curve – Einheits-Temperatur-Kurve

standardization – Eichung, Normung, Standardisierung

standard-temperature-time-curve – Standard-Temperatur-Zeit-Kurve

standing order – Dienstanweisung; auch: Standard-Einsatz-Regel (SER)

standpipe – Standrohr

Staphylococcal Enterotoxin B (SEB) – Staphylokokken-Enterotoxin B (SEB)

staphylococci – Staphylokokken

staphylococcus infection – Staphylokokken-Infektion

star screwdriver – Torx-Schraubendreher

state (status) – Status

state change – Zustandsänderung

state of aggregation – Aggregatzustand

state of balance – Gleichgewichtszustand

state of calamity – Notstand, Katastrophenfall

state of disaster – Katastrophenfall

state of emergency – Notstand

state of health – Gesundheitszustand

state of matter – Aggregatzustand

static electrification – (elektro)statische Aufladung

static pressure – statischer Druck

statics – Statik

station house – Wache, Feuerwache (FW), Feuerwehrhaus (FwH)

station turn-out area – Wachbereich

stationary – stationär, stehend, ruhend, ortsfest

status message – Statusmeldung

statutory basis – gesetzliche/rechtliche Grundlage

statutory fire brigade – Pflichtfeuerwehr (PF)

steadiness – Beständigkeit, Festigkeit, Stabilität, Gleichmäßigkeit

steady state – stationärer Zustand, gleichbleibender Zustand

steam – Wasserdampf, Dampf

steam explosion – Dampfexplosion

steam extinguishing method – Dampflöschverfahren

steam fire-extinguishing installation – Dampf-Löschanlage

steam pipeline – Dampfleitung

steam-volatile – wasserdampfflüchtig

steam-volatile substances – wasserdampfflüchtige Stoffe

steel – Stahl

steel concrete – Stahlbeton

steel construction – Stahlkonstruktion

steel deck – Stahl-Trapezblech-Dach

steel girder – Stahlträger

steel pipe – Stahlrohr

steeping – Einweichen, Tränkung

stench – Gestank

step ladder – Trittleiter

step voltage – Schrittspannung

sterilization – Entkeimung, Sterilisation, Sterilisierung

stick – Stiel

Stiff Neck™ – HWS-Halskrause

stimulus – Reiz, Anregung, Anreiz

stink – stinken, Gestank

stirrup pump – Kübelspritze

stochastic radiation effect – stochastische Strahlenwirkung

stock – Lager, Vorrat, Bestand

stock of extinguishing agent – Löschmittelvorrat

stoichiometric combustion – stöchiometrische Verbrennung

stoichiometric concentration – stöchiometrische Konzentration

stoichiometric mixture – stöchiometrisches Gemisch

stoichiometric yield – stöchiometrische Ausbeute

stoichiometry – Stöchiometrie

stop button – Stopptaste

stop message: fire under control – Brand unter Kontrolle (BuK)

stop of combustion – Abbruch der Verbrennung

stopgap – Notnagel, Lückenfüller

stopper knot – Stopperknoten

storage – Lager, Lagerung, Vorrat, Speicherung

storage area – Lagerfläche, Lagerbereich

storage battery acid – Akkumulatorsäure, Batteriesäure

storage ring – Speicherring

storage tank – Speichertank, Vorratsbehälter

store section area – Teillagerfläche, Lagerbereich

stored goods – Lagergut

stored quantity – Vorratsmenge, gespeicherte Menge

stored-pressure fire extinguisher – Dauerdruck-Feuerlöscher

storey – Geschoss, Etage

storey ceiling – Geschossdecke

storm – Sturm

storm damages – Sturmschäden

storm force – Windstärke

storm matches – Sturmstreichhölzer, Sturmzündhölzer

Storz couplings – Storz-Kupplungen

stowage place – Stauraum

strain – Stamm, Belastung, Spannung; auch: absieben

strainer – Sieb, Filter; auch: Saugkorb

strainer basket – Saugschutzkorb

strap fall attenuator – Bandfalldämpfer

strategic agents – strategische Kampfstoffe

strategy – Strategie

straw – Stroh

straw roof – Strohdach

stray light – Streulicht

stray light detector – Streulichtmelder

stream – Strom, Zustrom, Strömung; auch: Bach

street map – Straßenkarte

strength – Stärke, Kraft, Festigkeit

stress – Belastungszustand

stretcher – Krankentrage

stretcher chair – Tragestuhl

stretcher-bearer – Krankentragen-Träger

strictly forbidden – streng verboten

strictly prohibited – streng verboten

striker – Stürmer, Streikender; auch: Gasanzünder

string – Schnur, Faden, Saite

strobe light – Blitzlicht

stroke – Schlaganfall

stroke of lightning – Blitzschlag

stroke volume (SV) – Schlagvolumen (SV)

strong gale – starker Sturm (Windstärke 9)

strong interaction – starke Wechselwirkung

structural explosion protection – bautechnischer Explosionsschutz

structural fire protection – baulicher Brandschutz

structural formula – Strukturformel

structural measure – bauliche Maßnahme

structure of an order – Gliederung eines Befehls

strut – Strebe, Spreize

studies of fire causes – Brandursachenforschung

stuffing-box seal – Stopfbuchse, Stopfbuchs(ab)dichtung

style of leadership – Führungsstil

subdivide – unterteilen

sublimation – Sublimation

sublimation point – Sublimationspunkt

sublimation pressure – Sublimationsdruck

sublime – sublimieren

submersible – tauchfähig (Pumpe)

submersible pump – Tauchpumpe (TP)

subordinated forces – unterstellte Einsatzkräfte

subordination – Unterstellungsverhältnis

sub-sector – Unterabschnitt

subsonic noise – Infraschall

substance key – Stoffnummer

substance number – Stoffnummer

substance properties – Stoffeigenschaften

substances liable to spontaneous combustion – selbstentzündliche Stoffe

substances which, in contact with water, emit flammable gases – Stoffe, die in Berührung mit Wasser entzündbare Gase entwickeln

sub-station – Nebenwache

substitute – Ersatz, Ersatzmann, Vertretung

substrate – Untergrund

substructure – Unterkonstruktion, Unterbau

success – Erfolg

suction – Ansaugen, Absaugen

suction capacity – Ansaugleistung

suction check – Saugprüfung

suction collecting head – Sammelstück

suction connection – Saugstutzen

suction coupling – Saugkupplung

suction height – Saughöhe

suction hood – Absaughaube

suction hose – Saugschlauch

suction hose adaptor – Übergangsstück für Feuerwehrsaugschläuche

suction hose connection – Saugstutzen

suction line – Ansaugleitung, Saugleitung

suction operation – Saugbetrieb

suction pipe – Saugleitung, Saugrohr

suction point – Saugstelle

suction support – Saugunterstützung; auch: Ansaugstutzen

suction tube – Ansaugschlauch, Ansaugrohr, Saugrohr

suffocate – ersticken

suffocating effect – Stickeffekt, Löscheffekt Ersticken

suffocation – Erstickung

suitability – Eignung, Tauglichkeit

suitable – geeignet

sulfonamides – Sulfonamide

sulfur dioxide, sulphur dioxide – Schwefeldioxid

sulfur, sulphur – Schwefel

sulfuric acid, sulphuric acid – Schwefelsäure

sulphonic acids – Sulfonsäuren

sulphur mustard gas – Schwefel-Lost (S-Lost)

sunburn – Sonnenbrand

sunstroke – Sonnenstich

super-critically – überkritisch (Gas, Flüssigkeit)

superglue – Sekundenkleber

superheat – überhitzen

superheating – Überhitzung

superior – Vorgesetzter

superiority – Weisungsbefugnis; auch: Überlegenheit

supervised area – Überwachungsbereich

supervisor – Betreuer, Aufseher, Aufpasser

supplemental report – Nachmeldung

supplementary actions – ergänzende Maßnahmen

supplementary education – Weiterbildung, ergänzende Ausbildung

supplementary equipment – zusätzliche Ausrüstung

supply – liefern, versorgen, zuführen, Versorgung, Lieferung, Zuführung, Einspeisung

supply conduit – Versorgungsleitung

supply of extinguishing agent – Löschmittelversorgung, Löschmittelzufuhr

supply of extinguishing water – Löschwasserversorgung

supply pressure – Versorgungsdruck

supply tank – Vorratsbehälter

support – Unterstützung

support fire station – Stützpunktfeuerwehr

supporting combustion – brandfördernd

supporting pillar – Stützpfeiler

surface burn – Brennen der Oberfläche

surface combustion – Abbrennen, Abflammen der Oberfläche

surface fire – Bodenfeuer, Brand der Oberfläche

surface flash – Flammen auf der Oberfläche

surface spread of flame – Ausbreitung von Flammen auf der Oberfläche

surface tension – Oberflächenspannung

surfactants – Tenside

surge suppressor – Überspannungsschutz

surrounding – Umgebung

surrounding fire – Umgebungsbrand

surrounding fire attack – umfassender Löschangriff

surveillance – Überwachung (von Personen)

survey – Erhebung

survival rate – Überlebensrate

susceptible to contamination – kontaminationsgefährdet

suspected toxin – Verdachtsstoff, vermutetes Toxin/Gift

suspension – Suspension

suspension trauma – Hängetrauma

suspicion – Verdacht

suspicion of contamination – Kontaminationsverdacht

sustained combustion – andauerndes Brennen

sustained flame – anhaltende Flammenbildung

sweat – Schweiß, schwitzen

sweating – Schwitzen

swelling – Schwellung, Anschwellung

swift – zügig/rasch

swift water rescue – Rettung aus fließendem Gewässer

switch clock – Zeitschaltuhr, Schaltuhr

switching spark – Schaltfunken

synonym – Synonym

synthesis – Synthese

synthetic resin – Kunstharz

synthetic resin varnish – Kunstharzlack

syringe – Spritze

system component – Anlagenteil

system part – Anlagenteil

system pressure – Netzdruck (Wasserversorgung)

systemic toxic effect – systemische Giftwirkung

T

table of fees – Gebührentabelle

table salt – Kochsalz

tabun (GA) – Tabun

tackle – angehen, angreifen, herangehen, bekämpfen

tacky – zäh (klebrig)

tactical agents – taktische Kampfstoffe

tactical cast – taktischer Auftrag

tactical designation – taktische Bezeichnung

tactical mission cast – taktischer Auftrag

tactical organisation – taktische Gliederung

tactical point of view – taktische Gesichtspunkte

tactical signs – taktische Zeichen

tactical unit – taktische Einheit

tactical withdrawal – taktischer Rückzug

tailor's tape – Bandmaß

take a resolution – Entschlussfassung

take an order – Befehl entgegennehmen

take in rope – Seil einholen

take precautions – Vorkehrungen treffen

tally chart – Strichliste

tamper – manipulieren

tampering – Manipulationen

tamper-proof lock – Sicherheitsschloss

tank car – Tankwagen, Kesselwagen (im Straßentransport)

tank farm – Tanklager

tank fire – Tankbrand

tank fire-fighting vehicle – Tanklöschfahrzeug (TLF)

tank tender – Tanklöschfahrzeug (TLF)

tank truck – Kesselwagen (Straße)

tank vehicle – Kesselwagen (Straße)

tanks – Behälter, Tanks; auch: Panzer

tap water – Leitungswasser

tar – Teer, Bitumen

tare – Tara

tare weight – Tara, Leermasse

target orientated – zielgerichtet

targets – Ziele

tariff schedule – Gebührentabelle

task force – Einsatzgruppe

task fulfilment – Auftragserledigung

tasks – Aufgaben

tattoo – Zapfenstreich; auch: Tätowierung

teaching – Lehren, Unterrichten

teaching aids – Lehrmittel

teaching material – Lehrmittel

team of fire-fighters – Löschmannschaft

team of firemen – Löschmannschaft

team-of-five (squad) – Staffel (St)

team-of-nine (group) – Gruppe (Gr)

team-of-two (squadron) – Trupp (Tr)

tear-gas – Tränengas

technical knowledge – Fachkunde

technical literature – Fachliteratur, technische Literatur

technics of fire protection – Brandschutztechnik

technological disaster – technische Katastrophe

T. Schmiermund, *Fachwörterbuch Feuerwehr und Brandschutz*, https://doi.org/10.1007/978-3-662-64120-0_40

telecommunication – Telekommunikation

telescopic probe – Teleskopsonde

teletype – Fernschreiber

temperature – Temperatur; auch: Fieber

temperature class – Temperaturklasse

temperature coefficient of expansion (TCE) – Wärmeausdehnungskoeffizient (WAK)

temperature conductivity – Temperaturleitfähigkeit

temperature dependence – Temperaturabhängigkeit

temperature of fire – Brandtemperatur

temperature of fire room – Brandraumtemperatur

temperature process by a fire – Temperaturverlauf eines Brandes

temperature range – Temperaturbereich

temperature rise – Temperaturanstieg

temperature scale – Temperaturskala

temperature-dependent – temperaturabhängig

tempests – Stürme, Unwetter

temporal order – zeitliche Reihenfolge

temporally hardness of water – temporäre Härte des Wassers

temporary bridge – Behelfsbrücke, provisorische Brücke

temporary helper – Hilfskraft (Person)

temporary shelter – vorübergehende Unterkunft

tenacity – Tenazität

tender truck – Tanklöschfahrzeug (TLF)

tensids – Tenside

tensile strength – Zugfestigkeit, Nennfestigkeit

tension – Spannung (mechanische)

tent – Zelt

tent hospital – Zeltkrankenhaus

tent village – Zeltdorf

teratogenic – Missbildungen verursachend

term of service – Dienstzeit

terminal atom – Atom, endständiges

terms – Begriffe

terrace – Terrasse

terrace roof – Terrassendach

terrain – Gelände, Terrain, Boden

terrestrial radiation – terrestrische Strahlung

terrestrial trunked radio (TETRA) – Digitalfunk (digitaler Bündelfunk)

terror(ist) attack – Terroranschlag

terroristic background – terroristischer Hintergrund

test centre – Prüfstelle

test certificate – Prüfzeugnis

test intervals – Prüffristen

test method – Prüfmethode, Prüfverfahren

test pressure – Prüfdruck

test report – Prüfprotokoll

test requirements – Prüfbedingungen

test specification – Prüfgrundlage

test specimen – Probekörper

test voltage – Prüfspannung

test, testing – Prüfung (Geräte, Anlagen)

tester – Tester, Testperson; auch: Prüfgerät

testing device – Prüfgerät

testing facility – Prüfeinrichtung

textile hose – Gewebeschlauch

thatched roof – Strohdach

thawing device for hydrants – Hydrantendeckel-Auftaugerät

theoretical air need – theoretischer Luftbedarf

theoretical combustion temperature – theoretische Verbrennungstemperatur

thermal – thermisch

thermal capacity – Wärmekapazität

thermal conduction – Wärmeleitung

thermal conductivity – Wärmeleitfähigkeit

thermal conductivity detector (TCD) – Wärmeleitfähigkeitsdetektor (WLD)

thermal convection – Wärmekonvektion

thermal decomposition – thermische Zersetzung

thermal degradation – Wärmezersetzung, thermische Schädigung

thermal diffusivity – Wärmeleitzahl

thermal dissociation – thermische Dissoziation

thermal energy – Wärmeenergie

thermal equilibrium – Wärmegleichgewicht

thermal expansion – Wärmeausdehnung

thermal explosion – Wärmeexplosion

thermal imager – Wärmebildkamera (WBK)

thermal imaging camera (TIC) – Wärmebildkamera (WBK)

thermal insulating layer – Dämmschicht, Isolierschicht

thermal insulation material – Wärmedämmstoff

thermal lance – Sauerstofflanze

thermal neutrons – thermische Neutronen

thermal properties – thermische Eigenschaften

thermal radiation – Wärmestrahlung

thermal release – Wärmefreisetzung

thermal stability – thermische Stabilität

thermal up current – Wärmeauftrieb, Konvektion

thermal uplift – Wärmeauftrieb, Konvektion

thermal volume expansion – Volumenausdehnung

thermo-conductive – wärmeleitend

thermoconductivity – Wärmeleitfähigkeit

thermocouple – Thermoelement

thermodynamics – Wärmelehre, Thermodynamik

thermoluminescent dosimeter (TLD) – Thermolumineszenzdosimeter (TLD)

thermometer – Thermometer

thermoplastics – Thermoplaste

thermosetting plastics – Duroplaste

thickening – Verdickung

thief knot – Altweiberknoten

third-degree burn – Verbrennung dritten Grades

third-degree frostbite – Erfrierung dritten Grades

This is an exercise. – Das ist eine Übung!

This is *not* a drill! – Das ist *keine* Übung!

This is *not* an exercise! – Das ist *keine* Übung!

thread – Gewinde

three-dimensional – räumlich

three-way distributor – Dreifach-Verteiler

threshold – Schwelle

threshold limit values (TLV) – Grenzwerte

threshold values – Schwellenwerte

throttle valve – Drosselventil

thumb knot – Sackstich, Schlaufenknoten, Bandschlingenknoten

thunderstorm – Gewitter, Unwetter

tidal volume (VT) – Atemzugvolumen (AZV)

tight – fest, dicht, eng

tightening – Festziehen

tilt protection – Kippschutz

timber – Bauholz

timber construction – Holzkonstruktion

timber hitch – Zimmermannsstich (Knoten)

timber hitch with a half stitch – Zimmermannsstich mit Halbschlag

timber preservative – Holzschutzmittel

timbering – Fachwerk, Gebälk

time frame – Zeitrahmen

time monitoring – Zeitüberwachung

time of action – Einsatzzeit; auch: Zeitpunkt des Einsatzes

time of day – Tageszeit

time of departure – Abfahrtszeit, Abmarschzeit

time of evacuation – Evakuierungszeit

time of exposure – Expositionszeit

time of fire growth – Brandentwicklungsdauer

time of readiness – Bereitschaftszeit

time of relief – Ablösezeitpunkt

time of year – Jahreszeit

time slot – Zeitschlitz

time switch – Zeitschaltuhr, Schaltuhr

time to assistance – Hilfsfrist

time-critical – zeitkritisch

timer – Zeitschaltuhr, Schaltuhr

tin opener – Dosenöffner, Büchsenöffner

tin plate – Weißblech

tinplate can – Weißblechdose

tiredness – Müdigkeit, Ermüdung

tissue – Gewebe

tissue retractor – Wundhaken

tissue weighting factor – Gewebewichtungsfaktor

to-and-fro breathing – Pendelatmung

to-and-fro system – Pendelatmungssystem

tolerability limit – Erträglichkeitsgrenze

tolerability threshold – Erträglichkeitsschwelle

tongue – Zunge; auch: Feder (einer Nut-Feder-Verbindung)

tongue and groove joint (TG) – Nut-Feder-Verbindung

tool and gear vehicle – Gerätewagen (GW)

tool case – Werkzeugkoffer

tool-box – Werkzeugkasten

tool-chest – Werkzeugkasten

tools – Werkzeuge

torch – Fackel, Taschenlampe

torque – Drehmoment

torque handle – Drehmomentschlüssel

torque wrench – Drehmomentschlüssel

total breakdown – Gesamtausfall, Totalausfall

total discharge – Gesamtentladung; auch: Vollflutung

total discharge system – Vollflutungsanlage

total hardness – Gesamthärte

total hardness of water – Gesamthärte des Wassers

total heat protection system – Vollwärmeschutzsystem

total loss of pressure – Gesamtdruckverlust

total static head – Gesamtförderhöhe

total weight – Gesamtmasse („Gesamtgewicht")

touch – berühren, anfassen, ertasten, Berührung, Tastgefühl

touch protection – Berührungsschutz

tough – hart, zäh, robust, stark

toughness – Zähigkeit, Härte, Widerstandsfähigkeit

towing rope – Abschleppseil

town gas – Stadtgas

town map – Stadtplan

toxic – giftig

toxic effect – Giftwirkung

toxic gas cloud – Giftgaswolke

toxic gases – giftige Gase

toxic hazard – Gefahr durch giftige Stoffe

toxic shock (TS) – toxischer (giftiger) Schock

toxic substance – Gift, Giftstoff, giftiger Stoff

toxic to reproduction – fortpflanzungsgefährdend

toxic vapours – giftige Dämpfe

toxicant – Gift, Giftstoff, giftig

toxicity – Giftigkeit, Toxizität

toxin – Gift, Giftstoff

tracing – Rückverfolgung, Ermittlung, Fahndung

traction vehicle – Zugfahrzeug

tractive force – Zugkraft

tractive saw – Zugsäge

trade – Gewerbe, Handel, Branche, Markt

trademark – Warenzeichen

traffic accident – Verkehrsunfall (VU)

traffic area – Verkehrsfläche

traffic circle – Kreisverkehr

traffic collision – Fahrzeugkollision

traffic cone – Verkehrsleitkegel

traffic paddle – Winkerkelle

traffic regulations – Straßenverkehrsordnung (StVO)

traffic rotary – Kreisverkehr

traffic route – Verkehrsweg

trailer for fire-fighting equipment – Feuerwehranhänger (FwA)

trailer ladder – Anhängeleiter (AL)

trailer load – Anhängemasse

trailer pump – Anhängerpumpe

train – Zug (Fahrzeug)

train fire – Zugbrand

training – Ausbildung, Schulung, Training

training aids – Lehrmittel

training and supplementary education – Aus- und Weiterbildung

training dummy – Übungspuppe

training for action – Einsatzausbildung

training standards – Ausbildungsnormen

transactional leadership – kooperativer Führungsstil

transfer – Transport

transformation series – Zerfallsreihen, Umwandlungsreihen

transition state – Übergangszustand

transmission device – Übertragungseinrichtung (ÜE)

transmission route – Übertragungsweg

transmission system – Übertragungsanlage

transmit of reports – Meldungen abgeben

transpallet – Gabelhubwagen

transport category – Transportkategorie

transport class – Transportkategorie

transport documents – Beförderungspapiere

transport index (TI) – Transportkennzahl (TKZ)

transport of nuclear waste – Atommülltransport

transport of radioactive waste – Atommülltransport

transport system – Transportanlage

transport unit – Beförderungseinheit

transport vehicle – Transportfahrzeug

transportability – Transportfähigkeit, Transportierbarkeit

Transport-Accident-Information-and-Assistance-System – Transport-Unfall-Informations- und Hilfeleistungssystem (TUIS)

transportation law – Transportrecht

transporter – Transporter, Transportfahrzeug

transporting away – Abtransport

transverse ventilation – Querlüftung

trapezoidal roof – Trapezblechdach

trapezoidal sheet metal – Trapezblech

trapped person – gefangene Person, verschüttete Person

tray – Tablett; auch: Wanne

treatment – Behandlung

treatment process – Behandlungsprozess

treatment station – Behandlungsplatz (BHP)

treetop fire – Wipfelbrand, Wipfelfeuer

trefoil = basic ionizing radiation symbol – Kleeblatt (Trefoil) = Strahlenwarnzeichen

trespass right – Wegerecht

tri jack – Dreibock

triage – Sichtung, Selektierung, sichten, Triagieren (med.)

triage ward – Triageabteilung

triangle wrench – Dreikantschlüssel

triangular cloth – Dreieckstuch (Erste Hilfe)

trigger spark – Zündfunke, auslösender Funke

trigger threshold – Auslöseschwelle (med.)

trillion – Billion (1 000 000 000 000 = 10^{12})

triple point (TP) – Tripelpunkt (TP)

triple-head distributor – Verteiler (Feuerwehr-Gerät), Dreifach-Verteiler

tripod – Dreibeinstativ

tripping device – Auslöseeinrichtung

tritium – Tritium (^{3}H, T)

trolley fire extinguisher – fahrbarer Feuerlöscher

trouble report – Störungsmeldung

trough – Trog, Wanne, Mulde

trowel – Kelle

trunk road – Fernverkehrsstraße

trunked mode operation (TMO) – Netzmodus (Digitalfunk)

trunked radio – Bündelfunk

trunked radio system – Bündelfunksystem

truss – Fachwerk; auch: Bruchband

tube – Rohr, Röhrchen, Röhre; auch: Kanüle, Sonde

tube cross-section – Rohrquerschnitt

tuberculosis (TB) – Tuberkulose (Tb)

tubular hexagon box wrench – Rohrsteckschlüssel

tularaemia – Hasenpest, Tularämie

tunnel fire – Tunnelbrand

turbulence – Turbulenz

turbulent flame – turbulente Flamme

turbulent flow – turbulente Strömung

turn – Krümmung, Bogen, Wende; auch: Auge (bei einem Seil)

turning hook – Wendehaken

turn-out coat – Einsatzanzug

turn-out order – Ausrückeordnung

turn-out reach – Ausrückebereich

turn-out sequence – Ausrückefolge

turnpike – Schlagbaum

turntable ladder (TL), aerial ladder – Drehleiter (DL)

turntable ladder with rescue cage – Drehleiter mit Korb (DLK)

turntable monitor – Wendestrahlrohr

tweezer – Pinzette

twin-stretcher ambulance – Krankentransportwagen, 2 Tragen (2KTW)

two-component adhesive – Zwei-Komponenten-Kleber

two-helper method – Zwei-Helfer-Methode (Erste Hilfe, HLW)

two-man saw – Trummsäge

two-way radio – Funkgerät (FuG)

tying – binden (Knoten)

type approval – Bauartzulassung

type label – Typenschild

type of building – Gebäudeart

type of combustion – Verbrennungsart

type of construction – Bauart

type of damage – Schadensart

type plate – Typenschild

type testing – Typenprüfung

types of decay – Zerfallsarten

typewriter – Schreibmaschine

tyre wrench, wheel brace – Radmutternschlüssel

tyres – Reifen (Kfz)

U

ultrasound – Ultraschall

unambiguous – eindeutig

unbalanced state – Unwucht

unbranched chain reaction – unverzweigte Kettenreaktion

unbreakable – unzerbrechlich

unclean – unrein, unsauber, schmutzig

unconscious behaviour – unbewusstes Verhalten

unconscious person – bewusstlose Person

uncouple – abkoppeln

undercooling – Unterkühlung

undercut – Fällkerbe

understated – unterbewertet

unexploded bomb – Blindgänger

unexploded ordnance (UXO) – Blindgänger

unfenced path/track – Feldweg

unicorn – Einhorn

unified atomic mass unit (*u*) – Atommasseneinheit, vereinheitlichte (*u*)

uniform – einheitlich

uniform store – Kleiderkammer

unit – Einheit

unit leader – Einheitsführer

units under command – unterstellte Einheiten, unterstellte Einsatzkräfte

universal fire-fighting vehicle – Universallöschfahrzeug (ULF)

universal gas constant – universelle Gaskonstante

universal gas equation – universelle Gasgleichung

universal key – Generalschlüssel

universal transversal Mercator projection – universelle transversale Mercator-Projektion (UTM)

unlined delivery hose – ungummierter Druckschlauch

unloaded weight – Leermasse

unpleasant smell – unangenehmer Duft/Geruch

unprotected area – ungeschützter Bereich

unrisky – nicht riskant, unbedenklich

unsafe – unsicher, nicht sicher

unsalaried – ehrenamtlich

unshielded radiation source – unabgeschirmte Strahlenquelle

untying – aufbinden (Knoten, Krawatte)

unused – ungebraucht, unbenutzt

unwritten rule – ungeschriebenes Gesetz

upkeeping – Instandhaltung, Pflege

upper explosion limit (UEL) – obere Explosionsgrenze (OEG)

upper explosion point – oberer Explosionspunkt

upper floor – Obergeschoss (OG)

uranium – Uran

uranium bomb – Uranbombe

uranium enrichment – Urananreicherung

uranium isotope – Uranisotop

uranium radiation – Uranstrahlung

urgency – Dringlichkeit

urgent – dringlich

urticant – Nesselstoff (Kampfstoff, Rotkreuz)

usable – verwendbar

T. Schmiermund, *Fachwörterbuch Feuerwehr und Brandschutz*, https://doi.org/10.1007/978-3-662-64120-0_41

usable power – Nutzleistung, nutzbare Leistung

usage – Nutzung

use of the building – Gebäudenutzung

used air – Abluft

used oil – Altöl, benutztes Öl

useful fire – Nutzfeuer

user – Benutzer, Betreiber, Anwender

user instructions – Gebrauchsanweisung, Gebrauchsanleitung

UV radiation – UV-Strahlung

UV-sensitive – UV-empfindlich

UXO clearance, UXO clearing – Kampfmittelräumung, Räumung eines Blindgängers

UXO-clearing service – Kampfmittelräumdienst

V

vacate (an area) – räumen (einen Bereich)

vaccination – Impfung, Schutzimpfung

vaccination certificate – Impfausweis

vaccination passport – Impfpass

vaccination, first/second – Impfung, erste/zweite

vaccine injection – Impfung

vaccine passport – Impfpass

vacuum – Vakuum

vacuum exhauster – Vakuumpumpe

vacuum-stretcher – Vakuumtrage

valence electron – Außenelektron, Valenzelektron

validity – Gültigkeit

valve – Ventil

valve disk – Ventilscheibe

valve flap – Ventilklappe

van der Waals attraction – Van-der-Waals-Anziehung

vaporization effect – Verdampfungseffekt

vapour barrier – Dampfsperre

vapour density – Dampfdichte

vapour explosion – Dampfexplosion

vapour phase – Dampfphase

vapour pressure – Dampfdruck

vapour, vapours – Dampf, Dämpfe

vapour-air-mixtures – Dampf-Luft-Gemische

variable of state – Zustandsgröße

variedly – abwechslungsreich; auch i. S. v. verschiedenartig

various – verschieden, verschiedene, verschiedenartig

variously – verschieden, verschiedenartig

Vaseline – Vaseline

vault – Gewölbe

V-belt – Keilriemen

vehicle – Fahrzeug

vehicle crew – Fahrzeugbesatzung

vehicle driver – Fahrer, Fahrzeugführer

vehicle equipment – Fahrzeugzubehör

vehicle fire – Fahrzeugbrand, Pkw-Brand

vehicle for transportation of roller containers (ACTS) – Wechselladerfahrzeug (WLF)

vehicle height – Fahrzeughöhe

vehicle length – Fahrzeuglänge

vehicle mass – Fahrzeugmasse

vehicle roof – Fahrzeugdach

vehicle weight – Fahrzeuggewicht

vehicle width – Fahrzeugbreite

veil of smoke – Rauchschleier

vein – Vene

velocity of flow – Strömungsgeschwindigkeit

venetian blinds – Jalousien

vent – entlüften, ablassen (Druck reduzieren)

ventilation – Lüftung, Belüftung, Ventilation; auch: Beatmung

ventilation devices – Be- und Entlüftungseinrichtungen

ventilation duct – Lüftungskanal

ventilation hole – Belüftungsöffnung

ventilation shaft – Lüftungsschacht

ventilation system – Belüftungsanlage, Lüftungsanlage, Ventilationsanlage

T. Schmiermund, *Fachwörterbuch Feuerwehr und Brandschutz*, https://doi.org/10.1007/978-3-662-64120-0_42

ventilator – Ventilator, Lüfter

venting device – Entlüftungseinrichtung

venting pump – Entlüftungspumpe

venting system – Entlüftungseinrichtung

ventricular fibrillation (VF) – Herzkammerflimmern (HKF), Kammerflimmern

verification – Überprüfung, Nachweis, Bestätigung

verify – nachweisen, überprüfen, bestätigen

vertical duct – Installationsschacht

very toxic – sehr giftig

very toxic substance – sehr giftiger Stoff

vesicant agent – Hautkampfstoff

vesicant gas – Hautkampfstoff

vibration dampener – Schwingungsdämpfer

vicinity – Nähe, nahe/nähere Umgebung

victim – Opfer

vigorously reaction – heftige (chemische) Reaktion

violent storm – heftiges Unwetter

viral haemorrhagic fever – virales hämorrhagisches Fieber (VHF)

virgin – neu, ungebraucht, jungfräulich

viricidal – viruzid, virizid

virtual private network (VPN) – Netzwerk, virtuelles, privates

virucidal – viruzid, viruzidal

virulence – Virulenz

viruses – Viren

viscid – zähflüssig, viskos, dickflüssig

viscosity – Viskosität

viscous – zähflüssig, viskos, dickflüssig

visibility – Sichtweite, Sichtverhältnisses

visor – Sichtscheibe, Gesichtsschutzschirm; auch: Sichtscheibe der Atemschutzmaske

visor for helmet – Helmvisier

visual check – Sichtkontrolle, Sichtprüfung

visual inspection – Sichtkontrolle, Sichtprüfung

visual range – Sichtbereich

visual testing (VT) – Sichtkontrolle, Sichtprüfung

vital functions – Vitalfunktionen

vitality – Lebenskraft, Vitalität

void coefficient – Dampfblasenkoeffizient

volatile – flüchtig (Gase, Dämpfe)

volatility – Flüchtigkeit

volatile substances – flüchtige Stoffe

volition – Willenskraft

voltage – elektrische Spannung

voltage crater – Spannungstrichter

voltage drop – Spannungsabfall

voltage funnel – Spannungstrichter

volume burning rate – Volumenabbrandgeschwindigkeit

volume concentration – Volumenkonzentration

volume flow (rate) – Volumenstrom

volume fraction – Volumenanteil

volume yield – Volumenausbeute

volunteer fire brigade – Freiwillige Feuerwehr (FF)

volunteer fire company (VFC) – Freiwillige Feuerwehr (FF)

volunteer fire department – Freiwillige Feuerwehr (FF)

volunteer fire-fighter – Feuerwehrangehörige(r) der Freiwilligen Feuerwehr

volunteers – Freiwillige

vomit – erbrechen

vomit bag – Spuckbeutel, Brechbeutel

vomiting – Erbrechen

vomition – Erbrechen

vomitus – Erbrochenes

W

wages – Lohn, Gehalt

wailing sound – Heulton

walkie-talkie – Handfunkgerät (HFuG), Handsprechfunkgerät

walk-on stability – Begehbarkeit

wall – Wand

wall chart – Wandtafel (Schaubild)

wall effect – Wandeffekt (Löscheffekt)

wall hydrant – Wandhydrant, Wandwasseranschluss

war gas – Kampfgas

War Weapons Control Act – Kriegswaffenkontrollgesetz (KWKG)

warden – Aufseher, Rektor, Direktor

warehouse – Lagerhaus

warfare agent – Kampfstoff, Kampfmittel

warm water heating – Warmwasserheizung

warming – Erwärmen, Erwärmung

warning – Warnung

warning device – Warnsignal

warning lamp – Kontrollleuchte, Warnleuchte, Blinkleuchte

warning light – Warnleuchte

warning sign – Warnzeichen

warning vest – Warnweste

warp – verwerfen (verformen)

wash solution – Waschlauge

washer – Unterlegscheibe

washing room – Waschraum

washland – Retentionsfläche (Hochwasserschutz)

waste – Abfall

waste air – Abluft

waste bag – Abfallsack

waste bin – Abfalleimer

waste container – Abfallbehälter

waste disposal – Abfallbeseitigung

waste heat – Abwärme

waste incineration – Abfallverbtrennung

waste oil – Altöl, Abfallöl

waste paper – Altpapier

waste water – Abwasser

waste-water discharge – Abwasserableitung

waste-water effluent – Abwasserableitung

waste-water pump – Abwasserpumpe

wastewater treatment plant (WWTP) – Abwasseraufbereitungsanlage

water – Wasser

water accident – Wasserunfall

water additive – Löschwasserzusatz

water authority – Wasserbehörde

water barrier – Wassersperre

water carrying fittings – wasserführende Armaturen

water column – Wassersäule

water consumption – Wasserverbrauch

water content – Wassergehalt

water curtain – Wasserschleier

water damage – Wasserschaden

water demand – Wasserbedarf

water distribution – Wasserverteilung

water extinguisher – Wasserlöscher

water finding paste – Wassernachweispaste

T. Schmiermund, *Fachwörterbuch Feuerwehr und Brandschutz*, https://doi.org/10.1007/978-3-662-64120-0_43

water fog – Wasserstrahl, Nebelstrahl

water for fire-fighting – Löschwasser

water gauge glass – Wasserstandsglas

water glass – Wasserglas

water half-life period – Wasserhalbzeit (Löschmittel Schaum)

water hazard class (WHC) – Wassergefährdungsklasse (WGK)

water jet – Wasserstrahl

water knot – Sackstich, Schlaufenknoten, Bandschlingenknoten

water level – Wasserspiegel, Wasserstand; auch: Wasserwaage

water main – Wasserleitung

water meter – Wasserzähler

water miscibility – Wassermischbarkeit

water pollutants – Wasserschadstoffe

water pollution – Wasserverschmutzung

water pressure – Wasserdruck

water pump pliers – Wasserpumpenzange

water removal – Wasserentfernung, Entwässerung

water requirement – Wasserbedarf

water rescue – Wasserrettung

water rescue apparatuses – Wasserrettungsgeräte

water reservoir – Wasserspeicher

water ring primer – Wasserringpumpe

water rising – Anstauen

water separator – Wasserabscheider

water softening – Wasserenthärtung

water solubility – Wasserlöslichkeit

water spray – Sprühwasser

water spray system – Sprühwasser-Löschanlage

water steam – Wasserdampf

water supply – Wasserversorgung, Löschwasserversorgung

water supply for fire-fighting – Löschwasserversorgung, Löschwassereinspeisung

water supply installation – Wasserversorgungsanlage

water supply point – Wasserentnahmestelle, Löschwasserentnahmestelle

water table – Wasserspiegel

water tank – Wasserbehälter

water tanker – Wassertankwagen

water tender – Tanklöschfahrzeug (TLF)

water throughput – Wasserdurchfluss

water tower – Hochbehälter, Wasserturm

water trap – Wasserabscheider

water treatment – Wasseraufbereitung

waterbody – Gewässer

water-crew – Wassertrupp (W-Tr)

water-crew leader – Wassertruppführer, -in (W-TrFü, W-TrFü'in)

water-crew member – Wassertruppmann, -frau (W-TrM, W-TrFr)

water-gate – Schleuse (im Gewässer)

water-jet pump – Wasserstrahlpumpe

water-miscible – wassermischbar

water-shield – Hydroschild

water-squadron – Wassertrupp (W-Tr)

water-squadron leader – Wassertruppführer, -in (W-TrFü, W-TrFü'in)

water-squadron member – Wassertruppmann, -frau (W-TrM, W-TrFr)

water-team – Wassertrupp (W-Tr)

water-team leader – Wassertruppführer, -in (W-TrFü, W-TrFü'in)

water-team member – Wassertruppmann, -frau (W-TrM, W-TrFr)

wave – Welle

wave emission of radiation – Wellenstrahlung

wave length – Wellenlänge

waveband – Wellenbereich (Funk)

wave-corpuscle duality – Welle-Teilchen-Dualismus

wax torch – Wachsfackel

way back – Rückweg

weak – weich

weak interaction – schwache Wechselwirkung

weapons of mass destruction (WMD) – Massenvernichtungswaffen

wearer of breathing apparatus set – Atemschutzgeräteträger (AGT)

weather shelter – Wetterschutz

weaver's hitch – Schotenstich, Schotstek

weaver's knot – Schotenstich, Schotstek

webbing – Gewebeband

webbing brake – Bandfalldämpfer

wedge – Keil

weed killer – Unkrautvernichtungsmittel, Herbizid

weekend shift – Wochenendschicht

weighing – Abwägung

weight – Gewicht, Traglast, Belastung

weight belt (diving) – Bleigurt

weight force – Gewichtskraft

weir – Wehr, Stauwehr

weld – schweißen

welder's goggles – Schweißerbrille

welder's helmet – Schweißerhelm

welding – Schweißen

welding beads – Schweißperlen

welding sparks – Schweißfunken

well – Brunnen

well hole – Treppenauge; auch: Brunnenschacht

wet – nass, feucht, befeuchten, anfeuchten, benetzen

wet extinguishing process – Nasslöschverfahren, mit Wasser löschen

wet rising main – Steigleitung, nasse

wet spot – nasse Stelle, Pfütze

wet water – Netzwasser

wettability – Benetzbarkeit

wetter – Netzmittel

wetting agent – Netzmittel

wheel barrow – Schubkarre

wheel chock – Unterlegkeil

wheel load – Radlast (Kfz)

wheelbarrow – Schubkarre

wheelbase – Radstand

wheeled stretcher – Ambulanzliege, Krankenwagenliege, Fahrtrage, Transportliege

wheeled stretcher carrier – Krankentragen-Fahrgestell

white area – Weißbereich

white spirit – Testbenzin

white-glowing – weißglühend

white-water – Wildwasser

white-water rescue – Wildwasserrettung

whole-body dose – Ganzkörperdosis

whole-body exposure – Ganzkörperbestrahlung

whole-body personal dosimeter – Ganzkörper-Personendosimeter

whole-time fire brigade – Berufsfeuerwehr (BF), „Vollzeit-Feuerwehr"

whole-time firefighter – Berufsfeuerwehrmann/-frau

wick – Docht

wick effect – Dochtwirkung

wide area network (WAN) – Großraumnetzwerk

Wilco!; Will comply! – Wird ausgeführt!

wild fire – Waldbrand

wildfire – Lauffeuer

wildland fire – Waldbrand

winch – Seilwinde, Winde

wind direction – Windrichtung

wind effect – Windeinfluss

wind force – Windstärke

wind gauge – Anemometer, Windmessgerät

wind shelter – Windschutz

wind speed – Windgeschwindigkeit

wind speed meter – Windgeschwindigkeitsmesser

wind strength – Windstärke

wind vane – Windfahne

wind-direction indicator – Windrichtungsanzeiger

windfall – Windwurf

window – Fenster

window centre – Fenstermitte

window ledge – Fenstersims, Fensterleiste

window lintel – Fenstersturz (Baukunde)

window sill – Fensterbrett, Fensterbank, Fenstersims

windthrow – Windwurf

windward – windzugewandt

windward side – windzugewandte Seite

wing – Flügel, Flanke

wipe test – Wischprobe

wire – Draht, Kabel, Leitung

wire brush – Drahtbürste

wire cutters – Seitenschneider, Drahtschneider, Drahtschere

wire loop – Drahtschlinge

wired glass – Drahtglas

wireless access point (WAP) – Zugangspunkt, drahtloser

wireless local area network (WLAN) – Netzwerk, drahtloses, lokales

wireless private area network (WPAN) – Netzwerk, drahtloses, privates

wireless smoke detector – Funkrauchmelder

wireway – Kabelkanal

wiring duct – Kabelkanal

with blue lights switched on – Blaulicht, mit eingeschaltetem …

without risk – ohne Risiko, unbedenklich

wood – Holz

wood gas – Holzgas

wood preservative – Holzschutzmittel

wood protection agent – Holzschutzmittel

wood saw – Holzsäge

wood wool – Holzwolle

wooden beam ceiling – Holzbalkendecke

wooden construction – Holzkonstruktion

work fire brigade (WTF) – Werkfeuerwehr (WF)

work hygiene – Arbeitshygiene

work positioning and restraint belt – Auffang- und Haltegurt

work safety – Arbeitssicherheit

work(ing) gloves – Arbeitshandschuhe

workbench – Werkbank

Worker's Samaritan Foundation – Arbeiter-Samariter-Bund (ASB)

working at heights – Arbeiten in Höhen

working height – Steighöhe

working light – Arbeitsstellenscheinwerfer

working line – Arbeitsleine

working substances – Arbeitsstoffe

working tool – Arbeitsgerät

workplace – Arbeitsplatz

workroom – Arbeitsraum

worst-case accident – größter anzunehmender Unfall (GAU)

wound – Wunde

wound healing ointment – Wundheilsalbe

wound sepsis – Wundsepsis

wreckage – Trümmer, Wrackteile

wrench – Schraubenschlüssel

wrench size – Schlüsselweite (SW) (Werkzeug)

wrench width – Schlüsselweite (SW)

writing tools – Schreibgeräte

written order – schriftlicher Befehl

X, Y, Z

X-ray examination – Röntgenuntersuchung

X-ray radiation – Röntgenstrahlung

X-rays – Röntgenstrahlen

x-way – x-fach

yellow rotating light – Gelblicht (als RKL)

yellow strobe light – Gelblicht (als RKL)

yield – Ausbeute

youth fire brigade – Jugendfeuerwehr

Y-shape – Y-Schlauchanschluss

zero rate – Nullrate

ZIKADE-mnemonic: (approx.: TICDCD-mnemonic: limit *t*ime – prevent *i*ncorporation – avoid *c*ontamination carry-over – shut *d*own – exploit *c*over – keep *d*istance large) – ZIKADE-Merkschema: *Z*eit begrenzen – *I*nkorporation verhindern – *K*ontaminationsverschleppung vermeiden – *A*bschalten – *D*eckung ausnutzen – *E*ntfernung groß halten

zone – Bereich, Gebiet, Abschnitt

zone of combustion – Verbrennungszone

T. Schmiermund, *Fachwörterbuch Feuerwehr und Brandschutz*, https://doi.org/10.1007/978-3-662-64120-0_44

Serviceteil

T. Schmiermund, *Fachwörterbuch Feuerwehr und Brandschutz*,
https://doi.org/10.1007/978-3-662-64120-0

Anhang

In diesem Anhang finden Sie:

- Die Abkürzungen des Lexikon-Teils // The abbrevations oft he encyclopaedia section (◘ Tab. A.1)
- Das deutsche und das internationale Buchstabieralphabet // The German and the international phonetic alphabet (◘ Tab. A.2)
- Die Bedeutung der Nummern der Klassen gefährlicher Güter // The meaning of the numbers of the classes of dangerous goods (◘ Tab. A.3)
- Die Umrechnungen einiger Größen und Einheiten // The conversions of some quantities and units (◘ Tab. A.4)
- Eine Übersicht der in Tabelle verwendeten Abkürzungen // An overview of the abbreviations used in the table (◘ Tab. A.5)

◘ **Tab. A.1** Abkürzungen des Lexikon-Teils // Abbrevations oft he encyclopaedia section

Abkürzung // abbreviation	**Langform // long term**
A&E	accident and emergency
A.O.D.	away on duty
AAO	Alarm- und Ausrücke-ordnung
AB	Abrollbehälter
ABC	atomar, biologisch, chemisch
ac	alternating current
AChEI	acetylcholinesterase-inhibitor
ACTS	abroll container transport system
AED	automated external defibrillator // automatischer externer Defibrillator
AF	Atemfrequenz
AFFF	aqueous film forming foam
AGT	Atemschutzgeräteträger
AKW	Atomkraftwerk
AL	Amtsleiter
AL	Anhängeleiter
ALARA	as low as reasonably achievable
AM	Amplitudenmodulation // amplitude modulation
amu	atomic mass unit
AMV	Atemminutenvolumen
APU	auxiliary power unit
ARD	acute respiratory disease
ARE	akute respiratorische Erkrankung
ARS	acute radiation syndrome
ASB	Arbeiter-Samariter-Bund
ASG	Atemschutzgerät
AS-Gw	Atemschutzgerätewart
ASÜ	Atemschutzüberwachung
ATF	Analytische Task Force/ Analytical Task Force
A-Tr	Angriffstrupp
A-TrFr	Angriffstruppfrau
A-TrFü/'in	Angriffstruppführer/-in
A-TrM	Angriffstruppmann
AZV	Atemzugvolumen

■ Tab. A.1 (Fortsetzung)

Abkürzung // abbreviation	Langform // long term
BA	breathing apparatus
BA-Stelle	Brandausbruchsstelle
BBK	Bundesamt für Bevölkerungshilfe und Katastrophenschutz
Bde	brigade
BE	Bedieneinrichtung
BF	Berufsfeuerwehr
BfS	Bundesamt für Strahlenschutz
BG	Behältergerät
BGB	Bürgerliches Gesetzbuch
BHP	Behandlungsplatz
BHP	British horsepower
BIBA	brought in by ambulance
BioStoffV	Biostoffverordnung
BMS	building management system
BMZ	Brandmeldezentrale
BOS	Behörden und Organisationen mit Sicherheitsaufgaben
bp	boiling point
BS	Brandschutz
BSIng	Brandschutzingenieur
BSL	basic life support
BSO	Brandschutzordnung
BSR	Bereitstellungsraum
BSt	Brandstelle
BtF	Betriebsfeuerwehr
BUA	Brandunterdrückungsanlage
BuK	Brand unter Kontrolle
BVM	bag valve mask
BWA	biological warfare agents

■ Tab. A.1 (Fortsetzung)

Abkürzung // abbreviation	Langform // long term
CA	cardiac arrest
CAF	compressed air foam
CAFS	compressed air foam system
CBRN	chemisch, biologisch, radioaktiv, nuklear // chemical, biological, radioactive, nuclear
CBRNE	chemisch, biologisch, radioaktiv, nuklear, explosiv // chemical, biological, radioactive, nuclear, explosive
CCF	Crimean-Congo fever
CCHF	Crimean-Congo haemorrhagic fever
CCP	cathodic corrosion protection
CDC	Centre for Disease Control {GB}; Center for Disease Control and Prevention {US}
CEST	Central European Summer Time
CET	Central European Time
CFC	chlorofluorocarbons
CHP	combined heat and power station
CIMIC	civil-military co-operation
CIT	crisis intervention team
CIU	chemical incident unit
CKW	Chlorkohlenwasserstoffe
CNS	central nervous system
CO	cardiac output (per minute)
COVID-9	corona virus disease 2019
CPR	cardiopulmonary resuscitation
CRs	cosmic rays

(Fortsetzung)

■ Tab. A.1 (Fortsetzung)

Abkürzung // abbreviation	Langform // long term
CRT	capillary refill time
CSA	Chemikalienschutzanzug
CSI	criticaly safety index
CVA	cerebral vascular accident
CWA	chemical warfare agents
CWÜ	Chemiewaffenübereinkommen
CX	phogen oxim
dc	direct current
DECT	digital enhanced cordless telecommunication
Dekon	Dekontamination
DG	dangerous goods
DG-EA-Code	dangerous goods emergency action code
DL	Drehleiter
DLK	Drehleiter mit Korb
DLM	Dosisleistungsmesser
DLS	Druckluftschaum
DLW	Dosisleistungswarner
DME	digitaler Meldeempfänger
DMO	direct mode operation
DOT	Department of Transportation, United States Department of Transportation
DRK	Deutsches Rotes Kreuz
DT	delay treatment
DTG	Drucklufttauchgerät
DWG	Dosiswarngerät
DWR	Druckwasserreaktor
Dz	disease
E2EE	end-to-end encryption
EA	Einsatzabschnitt
EAC	emergency action code

■ Tab. A.1 (Fortsetzung)

Abkürzung // abbreviation	Langform // long term
EB	electronic beam
ECG	electrocardiogram
ED	effektive Dosis // effective dose
ED	emergency department
EDK	European death knot
EEBD	emergency escape breathing device
EEG	Elektroenzephalogramm // electroencephalogram
EF	Expositionsfaktor/exposure factor
EF	Ebola-Fieber
EFK	Elektrofachkraft
EG	equipment group
EH	Erste Hilfe
EHT	Erste-Hilfe-Training
EIA	environmental impact assessment
EKG	Elektrokardiogramm
EL	Einsatzleitung
ELtr	Einsatzleiter
ELW	Einsatzleitwagen
EMC	electromagnetic compatibility
EMEC	electromagnetic environmental compatibility
EMS	emergency medical service
EMT-B	emergency medical technician-basic
EMT-I	emergency medical technician-intermediate
EMT-P	emergency medical technician-paramedic
EMUV	elektromagnetische Umweltverträglichkeit

■ Tab. A.1 (Fortsetzung)

Abkürzung // abbreviation	Langform // long term
EMV	elektromagnetische Verträglichkeit
EPD	elektronisches Personendosimeter // electronic personal dosimeter
ER	emergency room
ERK	Entrauchungsklappe
ES	Elektronenstrahl
ESD	electrostatic discharge
ESE	elektrostatische Entladung
ESPL	equivalent sound pressure level
ESt	Einsatzstelle
ETW	Einsatztoleranzwert
EUC	end-user certificate
EV	Ermittlungsverfahren
EVD	Ebola virus disease
EVE	Endverbleibserklärung
EVU	Energieversorgungsunternehmen
EW	emergency ward
EX	Explosion …/explosion …
FA	first aider
FACP	fire alarm control panel
FAK	first aid kit
FAT	first aid training
FAT	Feuerwehranzeigetableau
FBC	fluorobromocarbons
FBF	Feuerwehrbedienfeld
FBKW	Fluor-Brom-Kohlenwasserstoffe
FBP	foam branch pipe
FCI	fuel-coolant interaction

■ Tab. A.1 (Fortsetzung)

Abkürzung // abbreviation	Langform // long term
FCKW	Fluorchlorkohlenwasserstoffe
FF	Freiwillige Feuerwehr
Ff	fire-fighter
FFP	filtering facepiece
FFw	Freiwillige Feuerwehr
FG	foam generator
FGH	Feuerwehrgerätehaus
FI	Fehlerstromschutzschalter
FID	Flammenionisationsdetektor // flame ionisation detector
FIUO	for internal use only
FKW	Fluorkohlenwasserstoffe
FLF	Flugfeldlöschfahrzeug
FM	Frequenzmodulation // frequency modulation
FME	Funkmeldeempfänger
FMS	Funkmeldesystem
FOU	fever of undetermined origin
FOUO	for official use only
FP	Feuerlöschkreiselpumpe
Fp	Schmelzpunkt (Festpunkt)
FRTC	Feuerwehr- und Rettungs-Trainingscenter // fire and rescue training center
FRW	Feuer- und Rettungswache
FSD	Feuerwehrschlüsseldepot
FSE	Freischaltelement
FSI	floor-space index
FSK	Feuerschutzkleidung
FSK	Feuerwehrschlüsselkasten

(Fortsetzung)

Tab. A.1 (Fortsetzung)

Abkürzung // abbreviation	Langform // long term
FSS	fire suppression system
FTC	Feuerwehrtrainingscenter // fire training center
FTOH	fluorotelomer alcohols
FTr	foam trailer
FuG	Funkgerät
FüGr	Führungsgruppe
FuSt	Funkstelle
FüSt	Führungsstaffel
FüTr	Führungstrupp
FUU	Fieber unbekannter Ursache
Fw	Feuerwehr
FW	Feuerwache
FwA	Feuerwehrangehörige
FwA	Feuerwehranhänger
FwDV	Feuerwehrdienstvorschrift
FWF	Feuerwiderstandsfähigkeit
FwH	Feuerwehrhaus
GA	Gefahrenanalyse
GA	tabun
GAU	größter anzunehmender Unfall
GB	Großbritannien // Great Britain
GB	sarin
GC	Gaschromatographie // gas chromatography
GD	soman
GDL	gas diffusion layer
GDS	Gasdiffusionsschicht
GDS	gas detection system
GE	genetic engineering
GefStoffV	Gefahrstoffverordnung

Tab. A.1 (Fortsetzung)

Abkürzung // abbreviation	Langform // long term
GenTSV	Gentechniksicherheitsverordnung
GFI	ground fault interrupter
GFK	glasfaserverstärkter Kunststoff
GFZ	Geschossflächenzahl
GGVS	Gefahrgutverordnung Straße
GGVSE	Gefahrgutverordnung Straße Eisenbahn
GLA	Gelenklöscharm
GLT	Gebäudeleittechnik
GM	Gelenkmast
GMLZ	gemeinsames Melde- und Lagezentrum
GMZ	Gefahrenmeldezentrale
GOA	Gone On Arrival
GPS	global positioning system
Gr	Gruppe (group, team-of-nine)
GrFü	Gruppenführer
GRP	glass fiber reinforced plastic
G-RTW	Großraumrettungswagen
GSG	Gefährliche Stoffe und Güter
GTLF	Großtanklöschfahrzeug
GTLF	Großtanklöschfahrzeug
GUP	Gefahrgutumfüllpumpe
GW	Gerätewagen
Gw	Gerätewart
GWA	Gaswarnanlage
GW-AS	Atemschutzgerätewagen
GW-G	Gerätewagen Gefahrgut
H	Hydrant
HA	hazard analysis

Tab. A.1 (Fortsetzung)

Abkürzung // abbreviation	Langform // long term
HAR	Hausanschlussraum
HAT	handheld transceiver
HAV	hepatitis A virus
HAW	high active waste
hazmat	hazardous materials
HBV	hepatitis B virus
HCFC	hydrochlorofluorocarbons
HF	hämorrhagisches Fieber // haemorrhagic fever
HF	Herzfrequenz
HFBC	hydrofluorobromocarbons
HFBKW	Fluor-Brom-Kohlenwasserstoffe, teilhalogeniert
HFC	hydrofluorocarbons
H-FCKW	Fluorchlorkohlenwasserstoffe, teilhalogeniert
H-FKW	Fluorkohlenwasserstoffe, teilhalogenierte
HFuG	Handfunkgerät
HI	hazard information
HiOrg	Hilfsorganisationen
HKF	Herzkammerflimmern
HKW	Heizkraftwerk
HL	hose-layer
HLF	Hilfeleistungslöschfahrzeug
HLKO	Haager Landkriegsordnung
HLW	Herz-Lungen-Wiederbelebung
HMS	Halbmastwurfsicherung (= half clove hitch belay)
HMV	Herzminutenvolumen
HP	horsepower
HR	heart rate
HRR	heat release rate
HSM	Herzschrittmacher
HSR	Hohlstrahlrohr
HTCL	head-tilt and chin-lift manoeuvre
HV	high visibility
HV	high voltage
HVL	half-value layer
HWS	Halswirbelsäule
HWS	Halbwertschicht, Halbwertschichtdicke
HWZ	Halbwertszeit
HX	high expansion foam
i.v.	intravenous // intravenös
IAQ	indoor air quality
IC	ionization chamber
ICH	intensive care helicopter
ICP	incubation period
ICT	information and communication technology
ICVP	International Certificate of Vaccination or Prophylaxis
ID	Infektionsdosis // infectious dose
IdF	Institut der Feuerwehr
IDLH	imminent danger to life and health
IKZ	Inkubationszeit
IR	Infrarot // infrared
IRC	International Red Cross
IRK	Internationales Rotes Kreuz
I-RTW	Infektionsrettungswagen
IS	intrinsic safety

(Fortsetzung)

■ Tab. A.1 (Fortsetzung)

Abkürzung // abbreviation	Langform // long term
IT	immediate treatment
ITH	Intensivtransporthubschrauber
ITW	Intensivtransportwagen
IuK	Information und Kommunikation
JUH	Johanniter-Unfall-Hilfe
Kat	Katalysator
Kat	Katastrophe
KatS	Katastrophenschutz
KatS-DV	Katastrophenschutz-Dienstvorschrift
KdoW	Kommandowagen
KIT	Kriseninterventionsteam
KKS	kathodischer Korrosionsschutz
KKW	Kernkraftwerk
KLF	Kleinlöschfahrzeug
KNG	Kontaminationsnachweisgerät
Kp	Siedepunkt (Kochpunkt)
KRITIS	kritische Infrastrukturen
KTW	Krankentransportwagen
KW	Kranwagen
LA	lightning arrester
LAN	local area network
LAW	low active waste
LE	Löschmitteleinheit
LEL	lower explosion limit
LF	Löschgruppenfahrzeug
LFf	leading fire-fighter
LFG	landfill gas
LG	Leichtschaumgenerator
LH	Luftheber

■ Tab. A.1 (Fortsetzung)

Abkürzung // abbreviation	Langform // long term
LNA	leitender Notarzt
LNG	liquefied natural gas
LNO	liaison officer
LOX	liquid oxygen
LPG	liquified petroleum gas
LSM	lebensrettende Sofortmaßnahmen
LWR	Leichtwasserreaktor/light water reactor
LX	low expansion foam
LZ	Löschzug
MAK	Arbeitsplatzkonzentration, maximale
MANV	Massenanfall von Verletzten
MBS	Mehrbereichsschaummittel
MCI	mass casualty incident, multiple-casualty incident
MDF	middle-density fibreboard
MDL	minimum detection limit
MERV	major emergency response vehicle
MESG	maximum experimental safe gap
MESZ	mitteleuropäische Sommerzeit
MET	Modell für Effekte mit toxischen Gasen
MEZ	mitteleuropäische Zeit
mhp	metric horsepower
MHR	maximum heart rate
MICU	mobile intensive care unit
MMR	Masern-Mumps-Röteln // measles-mumps-rubella
MO	mode of operation
mol	Mol // mole

■ Tab. A.1 (Fortsetzung)

Abkürzung // abbreviation	Langform // long term
mp	melting point
MRV	minute respiratory volume
MSDS	material safety data sheet
MT	minimal treatment
MTBF	meantime between failures
MTF	Mannschaftstransportfahrzeug
MVA	motor vehicle accident
MVV	maximum voluntary ventilation
MX	medium expansion foam
NAW	Notarztwagen
NBC	nuclear, biological, chemical // nuklear, biologisch, chemisch
NBP	Nagelbettprobe
NDL	no decompression limit
NEF	Notarzteinsatzfahrzeug
NfD	Nur für Dienstgebrauch
NFS	Notfallstation
NHE	normal hydrogen electrode
NotSan	Notfallsanitäter
NPP	nuclear power plant
NW	Nennweite
OOB	offset overhand bend
OEG	obere Explosionsgrenze
OG	Obergeschoss
OI	Sauerstoff-Index/oxygen index
OLRD	Organisatorischer Leiter Rettungsdienst
OPA	oropharyngeal airway
OPCW	Organisation for the Prohibition of Chemical Weapons

■ Tab. A.1 (Fortsetzung)

Abkürzung // abbreviation	Langform // long term
OPTA	operativ-taktische Adresse // operative-tactical address
OrgL	Organisatorischer Leiter Rettungsdienst
PA	Pressluftatmer
PAA	peroxyacetic acid
PD	prion disease
PDA	personal distress alarm
PE	Photoeffekt // photoelectrical effect
PECH	Pause, Eis, Compression, Hochlagern
PES	Peroxyessigsäure
PF	Pflichtfeuerwehr
PFC	per-fluorinated/poly-fluorinated chemicals
PFOA	perfluorooctanoic acid
PFOS	Perfluoroctansulfonsäure // perfluorooctanesulfonic acid
PFT	perfluorinated tensids
PI	preliminary investigation
PID	Photoionisationsdetektor // photoionization detector
PLF	Pulverlöschfahrzeug
PLR	passive leg-raising
PM	pacemaker
POL	Polizei
pop.	population
PPE	personal protective equipment
PS	Proteinschaummittel
PS	Pferdestärke
PSA	Persönliche Schutzausrüstung

(Fortsetzung)

Tab. A.1 (Fortsetzung)

Abkürzung // abbreviation	Langform // long term
PSAP	public-safety answering station
PSC	power supply company
PSE	Periodensystem der Elemente
PSNV	psychosoziale Notfallversorgung
PSP	paralytic shellfish poisoning
PTA	patient transport ambulance
PTE	Patienten-Transport-Einheit
PTE	Periodic Table of the Elements
PTO	power take-off
PTU	patient transport unit
PV	photovoltaics
PWR	pressurized water reactor
QF	Qualitätsfaktor // quality factor
RA	Risikoabschätzung
RA	Rettungsassistent
RAM	relative atomic mass
RCCB	residual current operated circuit breaker
RCD	residual current protective device
RD	Rettungsdienst
RDB	Reaktordruckbehälter
RE	risk estimation
RESA	Reaktorschnellabschaltung
RettAss	Rettungsassistent
RettSan	Rettungssanitäter
RetW	Rettungswache
RFNA	red-fuming nitric acid
RFP	request for proposal

Tab. A.1 (Fortsetzung)

Abkürzung // abbreviation	Langform // long term
RG	Regenerationsgerät
RH	Rettungshelfer
RICE	rest, ice, compression, elevation
RIV	rescue intervention vehicle
RKL	Rundumkennleuchte
RL	Richtlinie
RM	radioman
RMM	relative molecular mass
RoE	rules of engagement
rpm	revolutions per minute
RPO	radiation protection officer
RPV	reactor pressure vessel
RR	Restrisiko // residual risk
RS	Rettungssanitäter
RSI	rapid sequence intubation
RTA	road traffic accident
RTB	Rettungsboot
RTH	Rettungshubschrauber
RTW	Rettungswagen
RW	Rüstwagen
RWA	Rauch- und Wärmeabzugsanlage
RW-S	Rüstwagen Schiene
SADT	self-acceleration decomposition temperature
SAR	search and rescue
SARS-CoV-2	severe acute respiratory syndrome coronavirus type 2
SB	spine board
SCBA	self-contained breathing apparatus
SCRAM	safety cut rope axe man

■ Tab. A.1 (Fortsetzung)

Abkürzung // abbreviation	Langform // long term
SCUBA	self-contained underwater breathing apparatus
SD	sea damage
SDB	Sicherheitsdatenblatt
SDS	short data service
SDS	safety data sheet
SEB	Staphylokokken-Enterotoxin B // Staphylococcal Enterotoxin B
SER	Standard-Einsatz-Regel
SHE	smoke and heat exhaust system
SIT	spontaneous-ignition temperature
SiTr	Sicherheitstrupp
SITREP	situation report
SLF	Schaumlöschfahrzeug
SP	Spreizer
SPL	sound pressure level
SPZ	Sprinklerzentrale
SRT	single rope technique
SSB	Strahlenschutzbeauftragter
SSG	Sauerstoffschutzgerät
St	Staffel (team-of-five)
S-Tr	Schlauchtrupp
S-TrFr	Schlauchtruppfrau
S-TrFü/'in	Schlauchtruppführer/-in
StrlSchV	Strahlenschutzverordnung
S-TrM	Schlauchtruppmann
StVO	Straßenverkehrsordnung
STX	saxitoxin
SV	Schlagvolumen // stroke volume
SW	Schlauchwagen

■ Tab. A.1 (Fortsetzung)

Abkürzung // abbreviation	Langform // long term
SZT	seilunterstützte Zugangstechniken
Tb	Tuberkulose
TB	tuberculosis
TCD	thermal conductivity detector
TCE	temperature coefficient of expansion
TETRA	terrestrial trunked radio
TG	tongue and groove joint
TH	Technische Hilfeleistung
TI	transport index
TIC	thermal imaging camera
TKZ	Transportkennzahl
TLD	Thermolumineszensdosimeter // thermoluminescent dosimeter
TLF	Tanklöschfahrzeug
TLV	threshold limit values
TMO	trunked mode operation
TP	Tauchpumpe
TP	Tripelpunkt // triple point
Tr	Trupp (team-of-two/ team-of-three)
TrFr	Truppfrau
TrFü/'in	Truppführer/-in
TrM	Truppmann
TS	Tragkraftspritze
TS	toxic shock
TSA	Tragkraftspritzenanhänger
TSF	Tragkraftspritzenfahrzeug
TUIS	Transport-Unfall-Informations- und Hilfeleistungssystem

(Fortsetzung)

Tab. A.1 (Fortsetzung)

Abkürzung // abbreviation	Langform // long term
u	unified atomic mass unit
UEG	untere Explosionsgrenze
UEL	upper explosion limit
ULF	Universallöschfahrzeug
Upm	Umdrehungen pro Minute
US	Vereinigte Staaten (von Amerika)/United States (of America)
UTM	universelle transversale Mercator-Projektion (Kartenkunde)
UV	Ultraviolett // ultra violet
UVP	Umweltverträglichkeits-prüfung
UXO	unexploded ordnance
VB	vorbeugender Brandschutz
Vb	Verband
VbFü	Verbandsführer
VF	ventricular fibrillation
VFC	volunteer fire company
VHF	very high frequency
VHF	virales hämorrhagisches Fieber // viral haemorrhagic fever
VO	Verbindungsoffizier
Vol.	volunteer
Vol.	Volumen
VPN	virtual private network
VRW	Vorausrüstwagen
VS	Verschlusssache
VT	tidal volume
VT	visual testing
VU	Verkehrsunfall
VwV	Verwaltungsvorschrift

Tab. A.1 (Fortsetzung)

Abkürzung // abbreviation	Langform // long term
VZ	Verschäumungszahl
WAK	Wärmeausdehnungs-koeffizient
WAN	wide area network
WAP	wireless access point
WBK	Wärmebildkamera
WeFü	Wehrführer
WF	Werkfeuerwehr
WFB	works fire brigade
WGK	Wassergefährdungsklasse
WHC	water hazard class
WLAN	wireless local area network
WLD	Wärmeleitfähigkeitsdetektor
WLF	Wechselladerfahrzeug
WMD	weapons of mass destruc-tion
WPAN	wireless private area network
W-Tr	Wassertrupp
W-TrFr	Wassertruppfrau
W-TrFü/'in	Wassertruppführer/-in
W-TrM	Wassertruppmann
WWTP	wastewater treatment plant
Z	Zug
ZFü	Zugführer
ZIKADE	Zeit – Inkorporation – Kon-tamination – Abschalten – Deckung – Entfernung
ZMZ	zivil-militärische Zu-sammenarbeit
ZNS	Zentralnervensystem
ZS	Zivilschutz
ZTr	Zugtrupp

Tab. A.2 Buchstabieralphabet // Phonetic alphabet

	Deutsch (German)	English (Englisch)
a	Anton	Alfa
b	Berta	Bravo
c	Cäsar	Charlie
d	Dora	Delta
e	Emil	Echo
f	Friedrich	Foxtrott
g	Gustav	Golf
h	Heinrich	Hotel
i	Ida	India
j	Julius	Juliett
k	Kaufmann	Kilo
l	Ludwig	Lima
m	Martha	Mike
n	Nordpol	November
o	Otto	Oscar
p	Paula	Papa
q	Quelle	Quebec
r	Richard	Romeo
s	Samuel (Siegfried)	Sierra
t	Theodor	Tango
u	Ulrich	Uniform
v	Viktor	Victor
w	Wilhelm	Whiskey
x	Xanthippe	X-Ray
y	Ypsilon	Yankee
z	Zacharias (Zeppelin)	Zulu
ä	Ärger	-
ö	Ökonom (Österreich)	-
ü	Übermut	-
sch	Schule	-
ch	Charlotte	-
ß	Eszett	-

Tab. A.3 Gefahrgutklassen (Klassen gefährlicher Güter) // hazard classes (classes of dangerous goods)

Klasse // class	Deutsch (German)	English (Englisch)
1	Explosive Stoffe und Gegenstände	explosive substances and articles
1.1	Stoffe und Gegenstände, die massenexplosionsfähig sind	Explosives which have a mass explosion hazard
1.2	Stoffe und Gegenstände, die die Gefahr der Bildung von Splittern, Spreng- und Wurfstücken aufweisen, aber nicht massenexplosionsfähig sind	Explosives which have a projection hazard but not a mass explosion hazard
1.3	Stoffe und Gegenstände, die eine Feuergefahr besitzen und die entweder eine geringe Gefahr durch Luftdruck oder eine geringe Gefahr durch Splitter, Spreng- und Wurfstücke oder beides aufweisen, aber nicht massenexplosionsfähig sind	Explosives which have a fire hazard and either a minor blast hazard or a minor projection hazard or both, but not a mass explosion hazard
1.4	Stoffe und Gegenstände mit geringer Explosionsgefahr	Explosives which present no significant blast hazard
1.5	Sehr unempfindliche massenexplosionsfähige Stoffe	Very insensitive explosives with a mass explosion hazard
1.6	Extrem unempfindliche nicht massenexplosionsfähige Stoffe	Extremely insensitive articles which do not have a mass explosion hazard
2	Gase und gasförmige Stoffe	Gases, including compressed, liquified and dissolved under pressure gases and vapors
2.1	Entzündbare Gase	Flammable gases
2.2	Nicht entzündbare, nicht giftige Gase	Non-flammable and non-toxic gases
2.3	Giftige Gase	Toxic gases
3	Entzündbare flüssige Stoffe	Flammable liquids
4.1	Entzündbare feste Stoffe, selbstzersetzliche Stoffe und desensibilisierte explosive Stoffe	Flammable solids, self-reactive substances and solid desensitized explosives
4.2	Selbstentzündliche Stoffe	Substances liable to spontaneous combustion
4.3	Stoffe, die in Berührung mit Wasser entzündliche Gase bilden	Substances which, in contact with water, emit flammable gases
5.1	Entzünden wirkende Stoffe	Oxidizing substances
5.2	Organische Peroxide	Organic peroxides
6.1	Giftige Stoffe	Toxic substances
6.2	Ansteckungsgefährliche Stoffe	Infectious substances
7	Radioaktive Stoffe	Radioactive material
8	Ätzende Stoffe	Corrosive substances
9	Verschiedene gefährliche Stoffe und Gegenstände	Miscellaneous dangerous substances and articles

Tab. A.4 Umrechnungen von Größen und Einheiten // The conversions of quantities and units

Beziehung		relationship	
deutsche Einheiten	**englisch-amerikanische Einheiten**	**English-American units**	**German units**
Länge // length			
1 km	= 0.621 372 miles	1 mile	= 1,609 342 km
1 m	= 1.0936 yd = 3,28 ft	1 yd = 3 ft	= 0,9144 m
	= 39.37 in	1 ft = 12 in	= 0,3048 m = 30,48 cm
1 mm	= 0.039 37 in	1 in	= 25,4 mm
	= 0.4717 lines {GB}	1 line = 1/12 in {GB}	= 2,117 mm
	= 1.5478 lines {US}	1 line = 1/40 in {US}	= 0,635 mm
Fläche // area			
1 km^2	= 0.3861 mi^2	1 sq. mi = 1 mi^2	= 2,589 988 km^2
1 ha = 100 a = 10^4 m^2	= $3.861 \cdot 10^{-3}$ mi^2		= 258,9988 ha
1 a = 100 m^2	= 0.024 71 A = $3.861 \cdot 10^{-5}$ mi^2	1 A	= 40,468 a
1 m^2	= 1.195 683 yd^2 = 10.7639 ft^2	1 sq. yd = 1 yd^2	= 0,8361 m^2
1 dm^2 = 10^{-2} m^2	= 0.107 639 ft^2 = 15.5 in^2	1 sq. ft = 1 ft^2	= 9,290 34 dm^2
1 cm^2 = 10^{-4} m^2	0.155 in^2	1 sq. in = 1 in^2	= 6,452 cm^2
1 mm^2 = 10^{-6} m^2	= $1.55 \cdot 10^{-3}$ in^2	1 sq. line = $line^2$ {GB}	= 4,4803 mm^2
	= 0.2225 $line^2$ {GB}	1 sq. line = $line^2$ {US}	= 0,403225 mm^2
	= 2.48 $line^2$ {US}		
Volumen // volume			
1 mm^3 = µL	= 0.105449 $line^3$ {GB}	1 cu. line* = $line^3$ {GB}	= 9,4832457 mm^3
	= 3,90552 $line^3$ {US}	1 cu. line = $line^3$ {US}	= 0,256 mm^3
1 cm^3 = mL	0.061 23 in^3	1 cu. inch = in^3	= 16,387 cm^3
	= 33.814 fl.oz.	1 fl.oz. {US}	= 29,573 53 cm^3
1 dm^3 = L	= 61.023 in^3	1 qt {US}	= 0,946 353 dm^3
	= 1.056 qt {US}	1 qt* {GB}	= 1,136 225 dm^3
	= 0.8799 qt* {GB}	1 gal. {US}	= 3,785 dm^3
	= 0.264 20 gal. {US}	1 gal. {GB}	= 4,546 dm^3
	= 0.219 97 gal. {GB}	1 cu. foot = 1 ft^3	= 28,317 dm^3
1 m^3	= 264.172 gal. {US}	1 bbl {US}	= 158,987 dm^3
	= 219.969 gal. {GB}	1 cu. yard = 1 yd^3	= 764,555 dm^3
	= 35.315 ft^3		
	= 6.290 bbl {US}	1 register ton* {GB}	= 2,831 m^3
	= 1.308 cu. yard	1 ocean ton* {US}	= 1,132 m^3

(Fortsetzung)

Tab. A.4 (Fortsetzung)

Beziehung		relationship	
deutsche Einheiten	englisch-amerikanische Einheiten	English-American units	German units
Masse // mass			
1 t = 1000 kg	= 1023 sht	1 sht = 2000 lb	= 907,185 kg
	= 0.9842 ltn	1 ltn = 20 cwt	= 1016,047 kg
1 Ztr.* = 50 kg	= 0.984 21 cwt {GB}	1 cwt {GB} = 112 lb = 1 l.cwt	= 50,802 kg
	= 1.1023 cwt {US}	1 cwt {US} = 100 lb = 1 sh.cwt	= 45,359 kg
1 kg	= 0.1575 st.	1 st. = 14 lb	= 6,3503 kg
	= 2.2046 lb	1 lb	= 0,453 59 kg
1 g = 10^{-3} kg	= 0.03527 oz	1 oz	= 28,3495 g
1 mg = 10^{-6} kg	= 35.27 oz	1 dr = 1/256 lb	= 1,77184 g
	= 0.56439 dr		
Geschwindigkeit // speed			
1 km/h	= 0.62137 mi/h	1 mi/h (mph)	= 1,6093 km/h
1 m/s	= 2.23694 mi/h	1 mi/min	= 95,561 km/h
	= 1.09361 yd/s	1 yd/s	= 0,9144 m/s
	= 3.2808 ft/s	1 ft/s	= 0,3048 m/s
Dichte // density			
1 g/cm³ = 1000 kg/m³	= 62.42977 lb/ft³	1 lb/ft³	= 16,018 kg/m³
	= 10.0225 lb/gal. {GB}	1 lb/gal. {GB}	= 0,099 776 g/cm³
	= 8.34536 lb/gal. {US}	1 lb/gal. {US}	= 0,119 827 g/cm³
	= 0.036127 lb/in³	1 lb/in³	= 27,680 g/cm³
Kraft (Gewicht, Gewichtskraft) // force (weight, weight force)			
1 N = 1 kg·m/s²	= 0.2248 lbf	1 lbf	= 4,448 N
1 kN = 1000 N	= 0.1124 short tf	1 short tf	= 8,9464 kN
	= 0.1004 long tf	1 long tf	= 9,964 kN
Druck // pressure			
1 Pa = 1 N/m²	= $1.450\cdot10^{-4}$ lbf/in²	1 lbf/in² = 1 psi	= $6{,}894\ 75\cdot10^{3}$ Pa
	= $2.953\cdot10^{-4}$ in Hg		= $6{,}894\ 75\cdot10^{-2}$ bar
	= $4.015\cdot10^{-3}$ in H_2O	1 in Hg	= $3{,}386\ 38\cdot10^{3}$ Pa
1 bar = 10^5 Pa	= 14.50 lbf/in²		= $3{,}386\ 38\cdot10^{-2}$ bar
	=		
	= $4.015\cdot10^{2}$ in H_2O		= $2{,}490\ 89\cdot10^{-3}$ bar

Tab. A.4 (Fortsetzung)

Beziehung		relationship	
deutsche Einheiten	englisch-amerikanische Einheiten	English-American units	German units
mechanische Spannung (Festigkeit) // tension, stress			
1 N/mm^2	= $1.450 \cdot 10^2$ lbf/in^2	1 lbf/in^2	= $6{,}894\ 75 \cdot 10^{-3}$ N/mm^2
dynamische Viskosität // dynamic viscosity			
1 Pa·s = 1 $N \cdot s/m^2$	= $2.089 \cdot 10^{-2}$ $lbf \cdot s/ft^2$	1 $lbf \cdot s/ft^2$	= 47,88 02 Pa·s
= 1 kg/(m·s)	= 0.672 lb/(ft·s)	1 lb/(ft·s)	= 1,488 Pa·s
kinematische Viskosität // kinematic viscosity			
1 m^2/s	= 10.76 ft^2/s	1 ft^2/s	= $9{,}29 \cdot 10^{-2}$ m^2/s
Energie, Arbeit, Wärmemenge // energy, quantity of heat			
1 J = 1 W·s = 1 N·m	= 0.7376 ft·lbf	1 ft·lbf	= 1,355 82 J
1 kJ = 1000 J	= 0.9478 Btu	1 Btu	= 1,055 06 kJ
Leistung, Wärmestrom // power, heat flow rate			
1 W = 1 N·m/s = 1 J/s	= $7.367 \cdot 10^{-1}$ ft·lbf/s	1 ft·lbf/s	= 1,355 82 W
	= $4.425 \cdot 10^{1}$ ft·lbf/min	1 ft·lbf/min	= $2{,}259\ 69 \cdot 10^{-2}$ W
	= 3.412 Btu/h	1 Btu/h	= 0,2930 72 W
1 PS* =1 mhp* = 735,999 W	= 0.986 32 BHP	1 BHP*	= 1,08317 PS*
			= 745,70 W
spezifische Wärmekapazität (spezifische Wärme) // specific heat capacity			
1 J/(kg·K)	= $2.388 \cdot 10^{-4}$ Btu/(lb·°F)	1 Btu/(lb·°F)	= $4{,}1868 \cdot 10^{3}$ J/(kg·K)
1 $J/(m^3 \cdot K)$	= $1.491 \cdot 10^{-5}$ $Btu/(ft^3 \cdot °F)$	1 $Btu/(ft^3 \cdot °F)$	= $6{,}71 \cdot 10^{4}$ $J/(m^3 \cdot K)$
Wärmeleitfähigkeit (Wärmeleitzahl) // thermal conductivity			
1 W/(m·K)	= 6.933 $Btu \cdot in/(ft^2 \cdot h \cdot °F)$	1 $Btu \cdot in/(ft^2 \cdot h \cdot °F)$	= 0,1442 W/(m·K)
	= 0.5778 Btu/(ft·h·°F)	1 Btu/(ft·h·°F)	= 1,7307 W/(m·K)
	= $4.815 \cdot 10^{-2}$ Btu/(in·h·°F)	1 Btu/(in·h·°F)	= 20,76 89 W/(m·K)
Wärmestromdichte // heat flux density			
1 W/m^2	= 0.317 $Btu/(ft^2 \cdot h)$	1 $Btu/(ft^2 \cdot h)$	= 3,1546 W/m^2
	= $2.201 \cdot 10^{-3}$ $Btu/(in^2 \cdot h)$	1 $Btu/(in^2 \cdot h)$	= 454,263 W/m^2
Wärmeübergangskoeffizient, Wärmedurchgangskoeffizient // heat transfer coefficient			
1 $W/(m^2 \cdot K)$	= 0.1761 $Btu/(ft^2 \cdot h \cdot °F)$	1 $Btu/(ft^2 \cdot h \cdot °F)$	= 5,678 $W/(m^2 \cdot K)$
Wärmedurchgangswiderstand // thermal resistance			
1 $m^2 \cdot K/W$	= 5.678 $ft^2 \cdot h \cdot °F/Btu$	1 $ft^2 \cdot h \cdot °F/Btu$	= 0,1761 m^2 K/W
Temperaturdifferenz // difference of temperature			
1 K = 1 °C	= 1.8 °F	1 °F	= $0{,}5\bar{5}$ K
			= $0{,}5\bar{5}$ °C

(Fortsetzung)

Tab. A.4 (Fortsetzung)

Beziehung		relationship	
deutsche Einheiten	**englisch-amerikanische Einheiten**	**English-American units**	**German units**
absolute Temperatur // absolute temperature			
Umrechnungsgleichungen		darin bedeuten t_C, t_F, T_K Zahlenwerte für: therein t_C, t_F, T_K mean numerical values for:	
$t_C = \frac{5}{9}(t_F - 32)$	$t_F = \frac{9}{5} \cdot t_C + 32$	t_C	Temperatur/temperature in °C
		t_F	Temperatur/temperature in °F
$T_K = \frac{5}{9} t_F + 255{,}372$	$t_F = \frac{9}{5} (T_K - 255{,}372)$	T_K	Temperatur/temperature in K

* alte, nicht mehr zulässige Einheiten // *old, no longer permissible units

Tab. A.5 In der Tab. A.4 verwendete Einheitenzeichen//Unit symbols used in the Table A.4

Einheitenzeichen // unit symbol	**Einheit // unit**
A	acre
a	Ar
bbl	barrel (= petroleum barrel)
BHP*	british horsepower
Btu	British thermal unit
°C	Grad Celsius // degree Celsius
cm^3	Kubikzentimeter // cubic centimetre
cu.	cubic
cwt	hundredweight
dm^3	Kubikdezimeter // cubic dezimetre
dr.	dram (dram avoirdupois)
°F	Grad Fahrenheit // degree Fahrenheit
fl oz	fluid ounce
ft.	foot // Fuß
g	Gramm/gramm
gal.	gallon // Gallone
h	Stunde // hour

Tab. A5 (Fortsetzung)

Einheitenzeichen // unit symbol	Einheit // unit
ha	Hektar
in	inch // Zoll
in H_2O	inch water column // Höhe der Wassersäule in inch
in Hg	inch mercury column // Höhe der Quecksilbersäule in inch
J	Joule
K	Kelvin
kg	Kilogramm // kilogram
kJ	Kilojoule (= 10^3 J)
kN	Kilonewton (= 10^3 N)
L	Liter // liter
lb	pound (lat. *libra* = pound // Pfund)
lbf	pound-force (‚pound-weight')
l.cwt	long hundredweight
ltn	long ton
m	Meter // metre
m^2	Quadratmeter // square metre
m^3	Kubikmeter // cubic metre
mhp*	metric horsepower (1 mhp = 1 PS)
mi	mile // Meile
min	minute // Minute
mm	Millimeter // millimetre (= 10^{-3} m)
mph	miles per hour = mi/h
N	Newton
Pa	Pascal
PS*	Pferdestärke (1 PS = 1 mhp)
psi	pound-force per square inch (lbf/in^2)
qt*	quart
s	Sekunde // second
sh.cwt.	short hundredweight
sht	short ton // „Schiffstonne"

(Fortsetzung)

Tab. A5 (Fortsetzung)

Einheitenzeichen // unit symbol	Einheit // unit
sq.	square // Quadrat
st.	stone
tf	ton-force (‚ton-weight‘)
W	Watt
yd.	yard
Ztr.*	Zentner

* alte, nicht mehr zulässige Einheiten // *old, no longer permissible units

Printed in the United States
by Baker & Taylor Publisher Services